Lecture Notes on Data Engineering and Communications Technologies

Volume 34

Series Editor

Fatos Xhafa, Technical University of Catalonia, Barcelona, Spain

The aim of the book series is to present cutting edge engineering approaches to data technologies and communications. It will publish latest advances on the engineering task of building and deploying distributed, scalable and reliable data infrastructures and communication systems.

The series will have a prominent applied focus on data technologies and communications with aim to promote the bridging from fundamental research on data science and networking to data engineering and communications that lead to industry products, business knowledge and standardisation.

**** Indexing: The books of this series are submitted to ISI Proceedings, MetaPress, Springerlink and DBLP ****

More information about this series at http://www.springer.com/series/15362

David Baneres · M. Elena Rodríguez ·
Ana Elena Guerrero-Roldán
Editors

Engineering Data-Driven Adaptive Trust-based e-Assessment Systems

Challenges and Infrastructure Solutions

Editors
David Baneres
Faculty of Computer Science, Multimedia and Telecommunications
Universitat Oberta de Catalunya
Barcelona, Spain

M. Elena Rodríguez
Faculty of Computer Science, Multimedia and Telecommunications
Universitat Oberta de Catalunya
Barcelona, Spain

Ana Elena Guerrero-Roldán
Faculty of Computer Science, Multimedia and Telecommunications
Universitat Oberta de Catalunya
Barcelona, Spain

ISSN 2367-4512 ISSN 2367-4520 (electronic)
Lecture Notes on Data Engineering and Communications Technologies
ISBN 978-3-030-29325-3 ISBN 978-3-030-29326-0 (eBook)
https://doi.org/10.1007/978-3-030-29326-0

This Springer imprint is published by the registered company Springer Nature Switzerland AG
The registered company address is: Gewerbestrasse 11, 6330 Cham, Switzerland

Preface

Universities are suffering a deep transformation since the incorporation of the Information and Communication Technologies (ICT) into the teaching-learning process. On the one hand, traditional face-to-face universities have incorporated learning management systems (e.g., Moodle or Blackboard) and learning tools for supporting teachers and learners, giving rise to blended learning. On the other hand, fully online (or virtual) universities have evolved towards more sophisticated ways of e-learning where technology becomes the unique context where the teaching-learning process takes place.

Universities are benefiting from technologies by designing responsive curriculums, incorporating flexible curriculum delivery, incrementing the support to lifelong learners, and improving on assessment, feedback, and accreditation. Thus, the improvement of the teaching-learning process boosted by ICT (also known as technology-enhanced learning) is not referred just to the incorporation of technologies for replicating existing teaching practices, but to the use of ICT for overtaking the existing teaching practices and transforming the learning experience.

In spite of the advances promoted by technology-enhanced learning, the assessment of learners is one of the most crucial ongoing challenges. E-assessment is defined by JISC (Joint Information Systems Committee) as the use of ICT to facilitate the entire assessment process, from designing and delivering assignments to marking (by computers, or humans assisted by digital tools), reporting, storing the results and/or conducting statistical analysis.

Although universities have advanced on e-assessment, on-site final exams continue being the most usual instrument to assess learners in order to ensure their identity and to prove their level of knowledge and competence acquisition. However, this is not aligned with the most common principles of online universities, such as flexibility, mobility, or accessibility. Assessment should not be a limiting factor in e-learning. On the contrary, e-assessment should facilitate the principles cited above. Nevertheless, there are some concerns about the identification and accreditation of learners using current e-assessment tools and mechanisms.

New systems should provide the assessment processes with the necessary features/capabilities to assess learners online reliably, and according to quality principles defined for quality agencies, and society in general; and available technologies developed by industry and research centers, as it would be the case of identification and authorship instruments, as well as other security mechanisms, properly integrated into the teaching-learning process according to pedagogical criteria, and taking into account privacy and legal measures can support their development and integration in the assessment processes.

Therefore, there is a real need to develop trust-based e-assessment systems for Higher Education institutions to conduct all the assessment online. This means for teachers to propose, monitor and assess any type of e-assessment activity being sure that the learners performing such assignments are who claim to be, and they are the legitimate authors of those assignments. Furthermore, this kind of systems should help teachers to prevent and detect different forms of academic dishonest behaviors (e.g., cheating and plagiarism).

Precisely, the TeSLA project (funded by the European Commission's Horizon 2020 ICT program) has leveraged the potential of technology to define and develop an adaptive trust-based e-assessment system that incorporates different mechanisms for e-assessment purposes, thus ensuring the authentication of the learners, as well as the authorship of the assessment activities learners deliver in online learning environments, while avoiding the time and physical space limitations imposed by face-to-face examinations. The TeSLA project has dealt with several aspects regarding the teaching-learning process, supporting formative, summative, and continuous e-assessment models, and taking into account the needs of learners, and especially of those having special educational needs or disabilities, as well as legal and ethical issues, and technological aspects. In order to provide an achievable and realistic solution the consortium was composed of technological companies and research centers (specialized in security and cryptography, biometric recognition, and authorship checking), accrediting quality agencies, and multiple Higher Education institutions across Europe (including online and blended universities), which tested the system into real educational settings by means of pilots.

Mobility of learners and professionals is continuously increasing and, for that reason, a growing percentage of learners are enrolled in fully online, or blended learning programs, in some cases under regulations which specify the conditions to obtain an accredited degree (for example, this is the case of academic programs under the European Higher Education Area). Therefore, engineering data-driven solutions that build a trust relationship between learners and their learning institutions, specifically oriented to improve and enable adaptive e-assessment is an actual need, both from a research and innovation point of view. Such a trust relationship must be constructed based on different and trans-disciplinary perspectives: technological, educational, privacy, and quality assurance. Thus, this book aims to analyze such engineering solutions from those different perspectives.

Also, this book intends to stimulate research and innovation in the e-assessment field, which may allow other academic institutions and organizations to evaluate, apply, and reproduce the book's contributions, thus helping to the advance of the state of the art and providing better understanding of the different problems and challenges faced by the e-learning field concerning to learners' assessment. In this way, industry and academic researchers, as well as practitioners can find interesting the experiences and ideas found in this book.

This book consists of 13 chapters organized into three major areas:

- *Strategies, tools, and technological infrastructure for e-assessment*. The chapters in this area are concerned about the selection, design, implementation, and validation of software solutions seeking to authenticate learner identity and check the authorship of assignments, and assisting the decision-making process of teachers, thus contributing to improve reliability in e-assessment. An engineering data-driven solution is presented by describing the components, features, and challenges across the different chapters.
- *Assessment design and pedagogical aspects in e-assessment*. In this area, the chapters discuss educational approaches to trust-based e-assessment based on practical cases studies in authentic blended and online learning settings, ranging from the management of large pilot studies, the integration of authentication and authorship instruments in assessment activities, accessibility and user experience of learners with special educational needs or disabilities, as well as the evaluation about the acceptance and trust in e-assessment. Note that, an engineering solution without any pedagogical support and analysis will probably be a failure. These chapters contribute to this analysis and they propose guidelines and recommendations for a successful integration of a trust-based system on e-assessment.
- *Privacy, ethics, and quality in e-assessment*. The chapters in this area address the main issues regarding the legal, privacy, and ethical considerations of software developments based on the use of authentication tools that collect biometric data, as well as quality assurance aspects in e-assessment.

The chapters in the first area of *Strategies, tools, and technological infrastructure for e-assessment* are organized as follows:

In chapter "Forensic Analysis Recognition", Pastor López-Monroy, Escalante, Montes-y-Gómez and Baró provide an overview to authorship analysis as a tool for performing forensic analysis, emphasizing the application of this sort of methods in e-learning scenarios with the objective of ensuring authorship verification. Authorship analysis aims to develop computer programs that can tell as much as possible from a person by only looking at his or her written documents through the analysis of the lexical, syntactic, and semantic patterns used. When applied in the e-learning domain, authorship analysis can be used to determine whether written documents have been authored by a specific learner (i.e., authorship verification). Authors analyze different state-to-the-art methods for forensic analysis by using a dataset comparable to authentic submissions performed by learners in an educational setting. The experimental results show the accuracy of such forensic analysis methods of different training-test sets of documents.

Gañan in chapter "Plagiarism Detection" also aims to discuss techniques oriented to ensure authorship verification, in this case by means of anti-plagiarism tools. Plagiarism, i.e., copying some part of other's work and include it as self in his or her own work, is one of the most common ways learners cheat in their school or Higher Education studies. The problem is aggravated due to the easiness to reach and share these contents through the net and computers. Anti-plagiarism tools help to detect similarities between documents that can be considered copy one of another. Although they cannot decide whether the similitudes are really plagiarism or which is the original version, they are a valuable asset in helping teachers to deal with plagiarism cases. The chapter presents the implementation of an anti-plagiarism tool that has been integrated into a more complex e-assessment system, and how the tool can interoperate with other anti-plagiarism tools. In addition, the chapter provides optimizations introduced in the algorithm implemented by the tool which allows to reduce complexity and increase performance.

Chapter "Biometric Tools for Learner Identity in e-Assessment" by Baró, Muñoz-Bernaus, Baneres, and Guerrero-Roldán addresses learners' identity verification by means of biometric recognition tools, which is one of the most challenging topics in online and blended educational institutions. Although there are companies providing services to institutions for secure assessment, those services tend to be intrusive and only focused on the final examination. The approach of the chapter is based on biometrical tools that do not need any special hardware, reducing the requirements on learners' computers, and that they can be used during formative, summative and continuous e-assessment models. Those biometric tools are face recognition, voice recognition, forensic analysis recognition, and keystroke dynamics recognition. The experimental results presented in this chapter are performed on real learner's biometric data obtained from the TeSLA project by analyzing 200 learners from a fully online institution and from a blended institution.

In chapter "Engineering Cloud-Based Technological Infrastructure", Prieto and Gañan discuss architectural challenges posed by cloud-based solutions that must be tackled during the engineering process, in order to achieve the benefits this technology provides (i.e., flexibility, high availability, scalability, costs maintenance reduction, etc.). From an architectural point of view, cloud-based infrastructures require to design carefully the modularization and separation of services, which in turn involve additional work to define communications between components and provide security techniques to guarantee data protection against potential risks like data loss, privacy breaks, and unwanted access, amongst others. This chapter describes as a case study the cloud-based TeSLA system. The development decisions, the architectural design, the advantages, limitations, and learned lessons are fully discussed in this chapter.

Kiennert, Ivanova, Rozeva, and Garcia-Alfaro in chapter "Security and Privacy in the TeSLA Architecture" analyze security and privacy aspects, from a technical standpoint of the TeSLA system. Security is a crucial pillar in such system since privacy data (i.e. biometric data) is transferred from different the Virtual Learning Environments (VLE) of the Higher Education institutions to the TeSLA system, and this data is stored within the cloud-based solution. Thus, an analysis of security requirements as a use case is performed and decisions made to solve them are proposed. Also, a security analysis of the deployment of the TeSLA system on a Higher Education institution is presented based on fuzzy logic on opinion from technical experts who implanted the system in the institution. Potential vulnerabilities and recommendations are presented in this study.

In chapter "Design and Implementation of Dashboards to Support Teachers Decision-Making Process in e-Assessment Systems", Guitart, Rodríguez, and Baró discuss the benefits of analytical systems in the university and review how they have been incorporated in academic institutions for different purposes, thanks to the advances in research fields as learning analytics or educational data mining. The collection and analysis of a large amount of data generated within the university, especially in blended and fully online educational settings, have the potential of generating information that can assist teachers in decision-making processes at different levels. Dashboards are the key visualization component of analytical tools, and they show the information through a friendly, clear, concise and intuitive visual interface. Precisely, the chapter presents the design and implementation of several dashboards, as well as the rationale behind the indicators shown in those dashboards, which integrate data collected by different kinds of authentication and authorship instruments, oriented to assist the decision-making process of teachers, above all in case of suspicion of learners' dishonest academic behavior, thus contributing to improve reliability in e-assessment.

Chapter "Blockchain: Opportunities and Challenges in the Educational Context" by Garcia-Font presents a systematic literature review about blockchain technology, addressing not only the most relevant proposals springing up in recent years related with this technology when applied in the educational and academic domains and classifying them in different categories but also detecting new opportunities and use cases of its application. Moreover, the chapter also discusses the problems associated with blockchain technology and identifies the main difficulties and challenges that future developments embracing this technology will have to address.

The chapters in the second area of *Assessment design and pedagogical aspects in e-assessment* are organized as follows:

In chapter "Design and Execution of Large-Scale Pilots", Peytcheva-Forsyth and Mellar discuss the piloting process across the various stages of the TeSLA e-assessment system development, where more than 20,000 learners of 7 European universities having very different cultures and ways of doing things were involved. The chapter describes the contexts of the pilot institutions and the way these affected the planning, execution, and results of the pilots at project and institutional levels, the development of the communication protocols, the development of contingency plans based on risk assessments, and the reporting procedures. The chapter also presents an outline of the success of the pilots in capturing data for use by the instrument developers, as well as for the development of pedagogic approaches to the integration authentication and authorship instruments in course assessment, and the involvement of learners with special educational needs or disabilities. The experience gained in terms of project management and lessons learned in such complex scenario is undoubtedly valuable contributions for researchers, developers, and institutions interested in piloting engineering solutions under development in e-learning environments involving such huge number of users.

Mellar and Peytcheva-Forsyth in chapter "Integration of TeSLA in Assessment Activities" address the most prominent results, from an educational point, derived from the large scale pilots of the TeSLA e-assessment system carried out in online and blended learning settings. During those pilots, teachers in six European universities constructed case studies of the use of the TeSLA instruments in assessment activities in authentic courses. The chapter provides a thematic analysis of these case studies identified seven common categories of TeSLA instrument use, which varied according to the design of the assessment activity and the teacher's purpose in seeking to authenticate learner identity and/or check the authorship of assignments. The chapter describes these seven categories in detail, and they can serve as a starting point for teachers new to the use of learner authentication and authorship checking instruments to identify appropriate ways to integrate the instruments in assessment activities in their courses. The chapter also identifies some of the issues that arose from the context of the collected data, and implications of the study for the future implementation of TeSLA and similar systems incorporating instruments for learner authentication and authorship checking.

Chapter "Ensuring Diverse User Experiences and Accessibility While Developing the TeSLA e-Assessment System" by Ladonlahti, Laamanen, and Uotinen highlights the importance of recognizing learners' diversity (especially in the case of learners with special educational needs or disabilities), and discuss how to recognize and organize the support for them. Although there are no exact international data on learners with special educational needs or disabilities, it is estimated that between 10% and 15% of learners in Higher Education institutions have some disability or special educational needs. In online universities, the ratio is even higher. Therefore, ensuring accessibility by everyone become an essential part of any technical development and software solution in online education, as it is recognized by the common European legislation, and national legislation which offers the basic guidelines for accessible e-learning practices. This chapter discusses, through a case study, how the accessibility of the e-assessment system developed in the context of TeSLA project has been tested. The wide variety of user experiences, and the participation of learners with special educational needs or disabilities in the pilots conducted along with the project, as well the suitability of the methods followed for testing the system (end-user testing, automated testing tools and test by accessibility specialists) are presented and analyzed.

In chapter "An Evaluation Methodology Applied to Trust-Based Adapted Systems for e-Assessment: Connecting Responsible Research and Innovation with a Human-Centred Design Approach", Okada and Whitelock propose an evaluation methodology underpinned by a Responsible Research and Innovation approach combined with human-centered design to evaluate the opinion of the different TeSLA stakeholders. The system was used by learners, teachers, technical staff, and institutional leaders. Thus, their opinions are crucial for the further development of the TeSLA trust-based e-assessment system to fully meet the educational expectations and requirements. The chapter proposes the methodology to gather the data obtained from the different Higher Education institutions and performs a broad analysis of the research question proposed in the chapter.

The chapters in the third and last area of *Privacy, ethics, and quality in e-assessment* are organized as follows:

Knockaert and De Vos in chapter "Ethical, Legal and Privacy Considerations for Adaptive Systems" address a very relevant topic for any research and innovation project or software development that deals with personal data. The use of any information related to an identified or identifiable person by a software necessarily implies the compliance with some kind of regulation that, in the case of the European Union, the legal framework in place is the General Data Protection Regulation (GDPR). Authors argue the importance that a legal framework should be necessarily complemented with ethical considerations. When developing an e-assessment system, those aspects become particularly important, especially whether it is intended to use biometric tools for learners' identification. Data captured and processed in such a case belong to a specific category of personal data known as sensitive data. The GDPR also reinforces requirements for security measures to ensure the integrity and confidentiality of these personal data. The chapter addresses main privacy aspects both from a legal and ethical point of view in the e-assessment domain. To begin with, the possibility to obtain a valid consent from learners to process their personal data, by analyzing the legal conditions, and how the ethical considerations could reinforce the requirements to have a free, specific, informed and unambiguous consent. Following this, the transparency obligation principle is discussed. Such principle aims to strengthen the learners' control over the processing of their personal data. Finally, the chapter analyzes the implications of openness, and the ethical framework to implement in order to teachers can properly use the information provided by the e-assessment system concerning learners' identification. The understanding of such information and the delivered indicators, as well as how to interpret and integrate them into internal rules of each academic institution is discussed.

Finally, chapter "Underpinning Quality Assurance in Trust-Based e-Assessment Procedures" by Ranne, Huertas-Hidalgo, Roca, Gourdin, and Foerster discuss quality assurance aspects in online and blended Higher Education, including accreditation, audit and evaluation of academic programs. The chapter highlights the importance of assessment as a key aspect of curriculum design, and how assessment methods must be planned so that they align well with intended learning outcomes. Specifically, the chapter analyzes ways to best evaluate and assure quality in e-assessment procedures, in particular in order to ensure reliable and secure learner authentication and authorship in online and blended learning environments. A framework for quality assurance in trust-based e-assessment processes in accordance with these objectives is presented, based on the standards and guidelines for quality assurance available in the European Higher Education Area, thus assuring the applicability of the e-assessment system to all online and blended learning environments, while respecting the shared European framework for quality assurance.

The book covers technical and educational research and innovation perspectives that contribute to the advance of the state of the art and provide a better understanding of different problems and challenges in the field of e-assessment, and in consequence, in e-learning. We expect that industry and academic researchers, as well as practitioners, can benefit from the experiences and ideas found in this book.

We would like to thank the authors of the chapters and also the reviewers for their invaluable collaboration and prompt responses to our inquiries, which enabled the completion of this book. We also gratefully acknowledge the feedback, assistance, and encouragement received from the editorial staff of Springer, Thomas Ditzinger, Holger Schaepe, Anja Seibold, and Jennifer Sweety Johnson as well as the book Series Editor Dr. Fatos Xhafa.

Barcelona, Spain

David Baneres
M. Elena Rodríguez
Ana Elena Guerrero-Roldán

Acknowledgements This work was supported by H2020-ICT-2015/H2020-ICT-2015 TeSLA project "An Adaptive Trust-based e-assessment System for Learning", Number 688520.

Contents

Contributors

Okada Alexandra The Open University, Milton Keynes, UK

David Baneres EIMT, Universitat Oberta de Catalunya, Barcelona, Spain

Xavier Baró EIMT, Universitat Oberta de Catalunya, Barcelona, Spain

Nathan De Vos CRIDS/NADI Faculty of Law, University of Namur, Namur, Belgium

Whitelock Denise The Open University, Milton Keynes, UK

Hugo Jair Escalante Instituto Nacional de Astrofísica, Óptica, y Electrónica, Tonantzintla, Puebla, Mexico;
Centro de Investigación y de Estudios Avanzados del IPN, Mexico City, Mexico

Martin Foerster European Quality Assurance Network for Information Education, Düsseldorf, Germany

David Gañan EIMT-UOC Rambla Poblenou, Barcelona, Spain

Joaquin Garcia-Alfaro Institut Mines-Telecom, Institut Polytechnique de Paris, Paris, France

Victor Garcia-Font Internet Interdisciplinary Institute (IN3), Universitat Oberta de Catalunya (UOC), CYBERCAT-Center for Cybersecurity Research of Catalonia, Barcelona, Spain

Anaïs Gourdin European Association for Quality Assurance in Higher Education, Brussels, Belgium

Ana Elena Guerrero-Roldán EIMT, Universitat Oberta de Catalunya, Barcelona, Spain

Isabel Guitart Hormigo Universitat Oberta de Catalunya, Barcelona, Spain

Esther Huertas Hidalgo Agència per a la Qualitat del Sistema Universitari de Catalunya, Barcelona, Spain

Malinka Ivanova College of Energy and Electronics, Technical University of Sofia, Sofia, Bulgaria

Christophe Kiennert Institut Mines-Telecom, Institut Polytechnique de Paris, Paris, France

Manon Knockaert CRIDS/NADI Faculty of Law, University of Namur, Namur, Belgium

Merja Laamanen University of Jyväskylä, Jyväskylä, Finland

Tarja Ladonlahti University of Jyväskylä, Jyväskylä, Finland

A. Pastor López-Monroy Centro de Investigación en Matemáticas (CIMAT), Guanajuato, Gto, Mexico

Harvey Mellar UCL Institute of Education, University College London, London, UK

Manuel Montes-y-Gómez Instituto Nacional de Astrofísica, Óptica, y Electrónica, Tonantzintla, Puebla, Mexico

Roger Muñoz Bernaus EIMT, Universitat Oberta de Catalunya, Barcelona, Spain

Roumiana Peytcheva-Forsyth Sofia University, Sofia, Bulgaria

Josep Prieto-Blazquez EIMT-UOC Rambla Poblenou, Barcelona, Spain

Paula Ranne European Association for Quality Assurance in Higher Education, Brussels, Belgium

Roger Roca Manlleu, Barcelona, Spain

M. Elena Rodríguez Universitat Oberta de Catalunya, Barcelona, Spain

Anna Rozeva Department of Informatics, Technical University of Sofia, Sofia, Bulgaria

Sanna Uotinen University of Jyväskylä, Jyväskylä, Finland

List of Figures

Forensic Analysis Recognition

Plagiarism Detection

Biometric Tools for Learner Identity in e-Assessment

Engineering Cloud-Based Technological Infrastructure

Security and Privacy in the TeSLA Architecture

Design and Implementation of Dashboards to Support Teachers Decision-Making Process in e-Assessment Systems

Blockchain: Opportunities and Challenges in the Educational Context

Design and Execution of Large-Scale Pilots

Ensuring Diverse User Experiences and Accessibility While Developing the TeSLA e-Assessment System

An Evaluation Methodology Applied to Trust-Based Adapted Systems for e-Assessment: Connecting Responsible Research and Innovation with a Human-Centred Design Approach

Ethical, Legal and Privacy Considerations for Adaptive Systems

List of Tables

An Evaluation Methodology Applied to Trust-Based Adapted Systems for e-Assessment: Connecting Responsible Research and Innovation with a Human-Centred Design Approach

Ethical, Legal and Privacy Considerations for Adaptive Systems

Forensic Analysis Recognition

A. Pastor López-Monroy, Hugo Jair Escalante, Manuel Montes-y-Gómez and Xavier Baró

Abstract Automatic systems for the analysis of textual information are highly relevant for the sake of developing completely autonomous assessment and authentication mechanisms in online learning scenarios. Since written texts within e-learning systems are mostly associated with exams and evaluations, it is critical to authenticate the identity of participants. Although there are some efforts from the natural language processing community to develop automated tools for assessing education in a very high semantic level (e.g., to provide feedback in argumentative tasks, to evaluate coherence, etc.), solutions are still far from being implemented because of the complexity of the associated problems and the unstructured information contained in written language. Efforts on authentication mechanisms, on the other hand, are more promising to be implemented in e-learning platforms because of the bulk of work devoted to the analysis of authorship. Despite this fact, it is only recently that authorship analysis is being started to be used in e-learning scenarios. This chapter provides an introduction to authorship analysis as a tool for performing forensic analysis, emphasizing the application of this sort of methods in e-learning scenarios. State of the art representations for text analysis are evaluated and compared in the authorship verification task. The goal of these experiments is to determine the best way to represent texts in the context of authorship verification. Finally, an overview of open challenges and research opportunities in this direction are outlined and briefly discussed.

A. P. López-Monroy
Centro de Investigación en Matemáticas (CIMAT), 36023 Guanajuato, Gto, Mexico
e-mail: pastor.lopez@cimat.mx

H. J. Escalante (✉) · M. Montes-y-Gómez
Instituto Nacional de Astrofísica, Óptica, y Electrónica, 72840 Tonantzintla, Puebla, Mexico
e-mail: hugo.jair@gmail.com

M. Montes-y-Gómez
e-mail: mmontesg@inaoep.mx

H. J. Escalante
Centro de Investigación y de Estudios Avanzados del IPN, 07360 Mexico City, Mexico

X. Baró
Universitat Oberta de Catalunya, Barcelona, Spain
e-mail: xbaro@uoc.edu

D. Baneres et al. (eds.), *Engineering Data-Driven Adaptive Trust-based e-Assessment Systems*, Lecture Notes on Data Engineering and Communications Technologies 34,
https://doi.org/10.1007/978-3-030-29326-0_1

Keywords Authorship verification · Authorship attribution · Textual forensic analysis · Online learning

Acronyms

BoT	Bag of Terms
DOR	Document Occurrence Representation
FT	Fast-Text
MOOCs	Massive Open Online Courses
POS	Part-Of-Speech
TCOR	Term Co-Occurrence Representation
W2V	Word to Vector Representation

1 Introduction

Nowadays, written information is still among the main communication channels in everyday life, massive amounts of textual data is being generated every day, making it impossible to manually analyze them. Online education is not the exception, with the ubiquity of e-courses in universities and the establishment of Massive Open Online Courses (MOOCs), e-learning platforms are able to reach millions of potential students. This vast amount of people involved, together with the associated activities in such platforms makes text the dominant information source for e-learning systems.

Written text in e-learning systems is mostly associated with exams and assessment activities. Where these documents can be generated by students, course designers or professors. In this regard, automatic systems for the analysis of textual information are highly relevant for the sake of developing completely autonomous assessment and authentication mechanisms. In fact, there are efforts from the natural language processing community to develop automated tools for assessing education at a very high semantic level. For instance to assess or provide feedback in argumentative tasks [6] and to evaluate coherence in academic texts [14]. However, efforts on the development of authentication mechanisms from the analysis of textual information in the context of e-learning have not been studied in depth. All this, despite the maturity of research in tasks such as Authorship Verification [15], Authorship Attribution [23], and Author Profiling [13].

The goal of authorship analysis tasks is to develop computer programs that can tell as much as possible from a person by only looking at their written documents. Specifically, verification and attribution tools aim at modeling the writing style of authors, where writing style can be exposed by looking at the lexical, syntactic and semantic patterns that people use when writing [23]. More specifically, authorship verification is the task of determining whether written documents have been authored by a specific person or not [15]. This is a very complex task that, in order to be solved properly, requires of a careful analysis of the writing style of authors. Automatic methods for authorship verification use a sample of documents and, by means of machine learning techniques, generate a coarse model of authors' writing style, see e.g., [4]. While promising results have been obtained with these techniques, in general scenarios online education poses additional complications that require of tailored authorship verification techniques.

Although there are several challenges inherent to the online education scheme, authorship verification is critical to guarantee a trustable assessment of learning activities. In this direction, the present chapter deals with the authorship verification problem in the context of e-learning. We review related work on authorship verification and elaborate on the specific complications that e-learning has for authorship analysis methodologies. In order to determine the effectiveness of state of the art methodologies for text representation in the context of authorship analysis, a comparison among several techniques is presented. Experimental results suggest it is feasible to implement authorship verification mechanisms into e-learning platforms with representations that are not computationally expensive.

The remainder of the chapter is organized as follows. Next section provides background information, including a review of related work on authorship verification and a discussion on the inherent difficulty of developing authorship analysis methods in the context of e-learning. Section 3 describes in detail the methodologies considered in the comparative study, together with the experimental results of an empirical evaluation. Finally, Sect. 4 outlines conclusions and highlights research opportunities in the field.

2 Background

We consider the problem of determining the authorship of written documents in the context of online learning. Specifically, for a given a document under analysis we have to make a yes/no decision indicating whether the document has been written by a particular author. Since we want this process to be automated, an authorship model has to be built from sample (training) texts written by the author of interest. The generated model then can classify the authorship of new documents. The considered scenario is described in Fig. 1.

In the remainder of this section, we review related work on authorship attribution and elaborate on the specific challenges that e-learning poses for authorship analysis methodologies.

2.1 Author Verification

As previously described, the main idea behind authorship analysis is that by measuring some features or style markers from texts, we can distinguish between documents

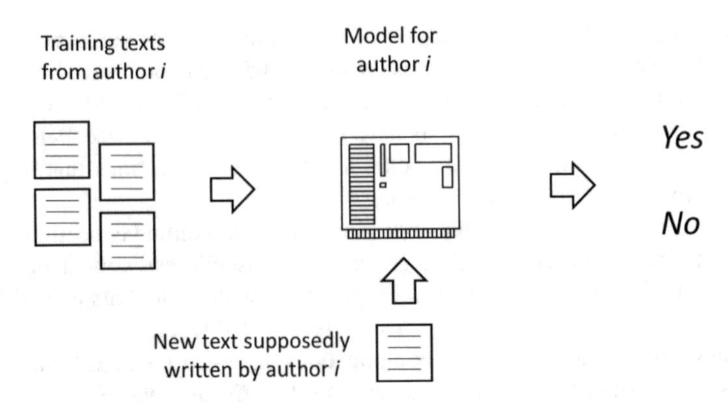

Fig. 1 Authorship verification scenario considered in this work. A set of documents are available for building an authorship verification model. The generated model can provide yes/no answers on new documents

written by different authors [23]. Unlike the authorship attribution task, where the goal is to determine the author of an anonymous document from a given set of candidate authors, the author verification task focuses on the problem of deciding whether or not a sample text was written by a *specific author*, given a few other documents authored by him or her [18]. In other words, it regards to the confirmation (or rejection) of the hypothesis that a given person is the author of the text [27].

In general, the author verification problem is a special case of the authorship attribution task, where there is exactly one candidate author with undisputed text samples. Therefore, as a classification problem, it is more difficult than attribution. It mainly corresponds to a *one-class classification* problem, where the samples of known authorship by the author in question form the target class. All texts written by other authors are viewed as the outlier class, a huge and heterogeneous class [25]. Additionally, this task is further complicated by the fact that the writing style of authors may change by theme differences or because of a stylistically drift over time [8].

There are two main approaches for author verification: the *intrinsic* and *extrinsic* models. In the intrinsic model, methods only use the known documents by the given author and the unknown document to arrive at their decision. These methods approach the verification task as a one-class classification problem. On the other hand, in the extrinsic model, methods also consider texts by other authors. They mainly transform the problem from a one-class task to a binary or multiclass classification problem, where the documents from other authors are considered as negative examples [25]. It is important to highlight that the most successful methods to date follow the extrinsic verification model, and consider a negative class formed by documents of other authors found in external resources such as the Web. For this reason we have adopted an extrinsic approach for the evaluation of text representations for authorship verification in Sect. 3.2. In the following, we describe the three main paradigms for authorship verification proposed in the literature.

The unmasking method: it is one of the most popular intrinsic methods [9]. As described in [26], its main idea is to build a classifier to distinguish the questioned document from the set of known documents, then to remove the most important features and repeat this process. In case the questioned and known documents are by the same author, the accuracy of the classifier significantly drops after a small number of repetitions while it remains relatively high when they are not by the same author. This method is supported on the observation that the differences between two documents from the same author may be reflected only in a relatively small number of features, despite possible differences in theme and genre [8]. Experimental results have shown that this method is very effective when dealing with long documents but fails with short documents.

Author verification by multiple binary classifiers: This is the most simple and popular approach regarding the application of an extrinsic verification model. In an initial work, Luyckx and Daelemans [15] approximated the author verification problem as a binary classification task by considering all available texts by other authors as negative examples. Then, extending this method, Escalante et al. [4] applied particle swarm model selection to select a suitable classifier for each author. The underlying idea of this method is that the individual problems that such classifiers face are different for distinct authors and, therefore, that using the same feature set and learning algorithm for all of the authors does not guarantee that the individual models are the best ones for each author. Experimental results give evidence that classifiers selected with this approach are advantageous over selecting the same classification model for all of the authors involved.

The impostors method is perhaps the best well-known and successful extrinsic method for author verification [10]. As described in [26], it handles the verification task as a multiclass classification problem by introducing additional authors, the so-called impostors, using documents found in external sources such as the Web. Basically, it generates a very large set of texts by authors that did not write the questioned document, to transform the problem into an open-set author attribution problem with many candidates [18]. Modifications to this method obtained the best performing results at the PAN forum 2014 and PAN 2015 editions [25, 26]. From a different point of view, author verification methods also differ from each other in the kind of features used for text representation. According to the overview papers of the Author Identification Tasks from PAN 2013 to 2015 [7, 25, 26], the majority of the participant methods consider combinations of different types of low-level features. In particular, the most common features were: *character-level* features such as punctuation mark counts, prefix/suffix counts as well as character n-grams (i.e., character sequences of length n); *lexical-level* features such as vocabulary richness measures, sentence/word length counts, stopword frequency, and word n-grams. Syntactic and semantic features require deeper linguistic analysis, consequently only few methods incorporate these kinds of features. Particularly, n-grams of Part-of-Speech (POS) Tags are the most popular *syntactic-level* features.

2.2 *Authorship Verification in Online Learning Scenarios*

As previously mentioned, to the best of our knowledge the authorship verification problem has not been studied in the context of e-learning. Note that the impact of these methods can have in enhancing the current authentication mechanisms (e.g., via login-password or in the best cases with the use of additional information as described in other chapters of this book). The successful development and implementation of authorship analysis methodologies in this setting requires the disentangling of the considered scenario and of the assumptions and relaxations that have to be made so that state of the art techniques can be used. In the following, we elaborate on the most important aspects that make authorship analysis complex in the context of e-learning.

Perhaps the biggest challenge for automatic authorship verification techniques in the context of online learning is the scarcity of samples. In general, the performance of automatic methods, such as those described in Sect. 2.1, is directly related with the number of samples used to build the authorship model. In online learning environments, however, collecting samples through Enrollment activities is not easy. For instance, it is unrealistic to ask students to provide a few dozens of sample documents to build an accurate authorship verification instrument. Instead, instruments have to be adapted to work with a few sample documents at the beginning and to implement incremental learning and adaptive mechanisms that allow them to improve the model with new samples that authors can provide on the fly.

Another major challenge has to do with the lack of agreement between sample documents used to build the model and the documents that the instrument has to analyze when it is in operation. Enrollment activities usually require students to write about a generic topic to obtain sample text that can be used for modeling their writing style. Whereas writing style is not dependent on the thematic content, students may use a different language structure when writing informal (e.g., sample documents required from enrollment activities) versus formal documents (e.g., like those that the system will be exposed to when in operation). Therefore, author verification methods should be robust to this mismatch in the writing style of documents.

Other associated challenges have to do with the type of documents that will be under analysis, for instance compare the contents, format and style that is used in math texts to that used in essay like documents. The length of documents is another challenge when trying to automatically analyze authorship in written documents, the less one writes the less information is available to undercover the writing style of authors. Finally, the language itself comprises another challenge, this mostly attaining authorship verification in general.

2.3 Discussion

In general, standard methods for authorship verification are applicable to e-learning, especially if the documents under analysis convey information about the writing style of authors and enough sample documents are available for building the verification model. From the review of related work, intrinsic and extrinsic solutions are available. Whereas extrinsic methods are dominant as they have reported among the best results in authorship verification in standard collections, they are limited in the sense that when developing a real authorship verification system, one rarely has at hand sample documents from alternative authors. Techniques gathering documents from the Web overcome this limitation to some extend. However, in the context of e-learning, it is much complicated to obtain relevant publicly available documents from the Web (at least realistic enough as to effectively contrast those available in a real online learning platform). For this reason, in some cases, intrinsic methods would be advantageous over extrinsic ones. One should note, however, that this is subject to the specific scenario and availability of the right information/data. For example, if the e-learning scenario could provide sample documents of other authors (e.g., other students in the same class). Hence, we think there is no general rule on the implementation of authorship verification methods in e-learning scenarios. Still, a sound advice in this direction is to develop/implement the method according to the very specific scenario.

In the next section, we empirically compare the authorship verification performance of different text representations that are known to capture authorship information. We evaluate these variants in an extrinsic manner, although all of the considered representations can be used with both, extrinsic and intrinsic variants. The goal of the empirical evaluation is to give the reader some insight into the usefulness of competitive text representations that can be used with most authorship verification solutions, including those reviewed in Sect. 2.1.

3 Text Representations in Authorship Verification (AV)

As previously mentioned, there are several aspects to consider into authorship verification (AV) systems for dynamic online platforms. One of these aspects is that the amount of text evidence can be small at the beginning, but it is expected to grow as the user interacts with the platform, which could make it possible to improve the classification model. The way to take advantage of that property will be aligned to the application domain, the degree interaction of the users as well as the expected time to give accurate classifications.

In this work, we aim to study and identify feasible strategies for modeling text evidence of users interacting with online platforms. Specifically, we focus in the evaluation of different methodologies that have proved to be useful for representing textual information in authorship analysis. In fact, the considered techniques allow the incremental improvement of the representation, which is particularly appealing to the considered scenario.

All of the considered representations are transverse to the authorship verification method under analysis. Meaning that they could be implemented with most existing authorship verification methods. We consider text representation methodologies that have shown remarkable performance for similar classification problems, but also with acceptable computing cost in terms of space and time. The idea is that considered strategies can work with short texts, but also efficiently aggregate new textual evidence to improve the classification. Having in mind that the final goal is to enhance the user experience by continuously improving the learning and classification stages.

The evaluated representations of documents are intuitively described and motivated in Sect. 3.1. Since we adopted the extrinsic evaluation approach, we built authorship verification models based on classifiers. The considered classification algorithms we evaluated were the Support Vector Machines and k-Nearest Neighbors methods, which have shown stable and acceptable performance for AV in different domains [1].

3.1 Text Representation

A critical aspect of any text mining application is the way the documents/texts are represented. Since most classification models require of vector inputs to work, most of the text representation methods transform documents into numerical vectors. There are a wide variety of methodologies to derive vector representations from texts, each of which captures content at different levels (e.g., lexical, syntactic, enven semantic). In the following, we describe the text representations that were considered in this study.

3.1.1 Bag of Terms

The Bag of Terms is a vector representation for documents where each term in the vocabulary is a dimension of the vector space. That is, the representation of a document doc_k is given by a vector $\mathbf{d}_k \in \mathbb{R}^m$. Where the vocabulary $\mathcal{V}$ is formed by all the different terms that appear in the documents available to build the AV model, that is $\mathcal{V} = \{v_1, \ldots, v_m\}$. The definition of what a term is depends on the domain and application, most commonly terms are associated to words. The values $d_{k,j}$ associated with each element of the vector $\mathbf{d}_k$ indicate the relevance of the corresponding word for a document. Under a Boolean weighting scheme, the elements of the BoT representation take values $d_{k,j} \in \{0, 1\}$ to indicate the absence and presence of terms in the document. Other weighting schemes include term frequency and inverse-document term frequency schemes. Where the former can be seen as an histogram of terms occurrence in each document, and the later a normalized version of the former that penalizes common terms.

BoT is one of the most effective and efficient methodologies for representing texts in classification problems. In fact, the BoT using the appropriated textual features (terms) for each domain, is often competitive or better than much more elaborated methodologies. In this work, we evaluate the performance of the BoT using words as textual features, but also using other stylistic features such as character n-grams.

3.1.2 Document Representations Based on Term Vectors

There are several alternatives for text representation that are as effective and efficient as the BoT. Among the most relevant ones are those that built document representations from term vectors. These representations have proved to be very effective in associated authorship analysis tasks [13]. More formally, let $\mathcal{D} = \{(doc_1, y_1), \ldots, (doc_n, y_n)\}$ be the set of training instances. In other words, $\mathcal{D}$ is a dataset of $n-$pairs of documents (doc_k) and a binary label (y_k) indicating their authorship. Document representations based on term vectors usually have two steps. In the first step, each term (for example, words) $v_i \in V$ is mapped into a vector space in $\mathbb{R}^r$, that is, $\mathbf{t_i} \in \mathbb{R}^r$. In that new space, each r dimension is a feature that contributes to the characterization of the term v_i according to a specific methodology (see Table 1). Intuitively, the idea is to map terms into a space capturing term semantics to some degree. In a second step, the representation for a document doc_k is obtained by aggregating the term vectors associated to every term occurring in doc_k. More formally, the representation for document doc_k is computed as $\mathbf{d_k} = \sum_{v_i \in doc_k} \alpha_i \cdot \mathbf{t_i}$, where α_i is the normalized frequency of the term $v_i \in doc_k$.

Table 1 Word vector representations for experimentation

Rep.	Ref.	Description
DOR	[11]	The aim of DOR is to model the semantics of the terms by measuring their occurrence in documents. The intuitive idea is to observe how each term was used in other documents of the corpus. For this, each term v_i is represented by a vector $\mathbf{t_i} = \langle t_{i,1}, \ldots, t_{i,\lvert\mathcal{D}\rvert} \rangle$. Under this representation, each $t_{i,k}$ value captures the association degree between each term t_i and each training document doc_k. This is usually by computing the frequency of t_i in doc_k
TCOR	[11]	In TCOR distributional semantics of terms is captured by computing the co-occurrences of terms in the training documents. Each term v_i is mapped to a vector $\mathbf{t_i} = \langle t_{i,1}, \ldots, t_{i,\lvert\mathcal{V}\rvert} \rangle$. Where each $t_{i,k}$ value indicates the contribution of the term v_k to characterize term v_i. This is usually by computing the number of documents in which v_k and v_i co-occur
W2V	[17]	Word2Vec (W2V) learns a vector space $\mathbb{R}^r$ in which words lie, this space is leaned by analyzing the sequential occurrence of words in large corpora. The model is an efficient two-layer neural networks trained to reconstruct the linguistic contexts of words. The main goal is that semantically related words should have similar term vectors in $\mathbb{R}^r$
FT	[2]	FastText (FT) is similar to W2V with the difference that the term vectors are subwords (e.g., character 2-grams or 3-grams) used to built new word vectors by means of linear operations. The latter procedure makes it possible to form word vectors for out of vocabulary words by averaging the vectors of the contained subwords (n-grams) in the target term

What makes different to each representation under this formulation is the way in which the terms vector space is estimated. In the following, we briefly describe each of the variants considered in this study, for further information we refer the reader to the corresponding references. The term vector representations considered in this study are summarized in Table 1.

Document occurrence representation (DOR). The aim of Document Occurrence Representation is to model the semantics of the terms by modeling their occurrence in the training instances [11]. The intuitive idea is to observe how each term was used in other documents. For this, each term v_i is represented by using a vector $\mathbf{t_i} = \langle t_{i,1}, \ldots, t_{i,|\mathcal{D}|} \rangle$. Under this representation, $t_{i,k}$ captures the association degree between each term t_i and each training document doc_k. Equation 1 describes DOR.

$$t_{i,k} = df(v_i, doc_k) \cdot \log \frac{|\mathcal{V}|}{|\mathcal{N}_k|} \tag{1}$$

In that formula, $\mathcal{N}_k \subseteq \mathcal{V}$ is the set of different terms in doc_k, and $df(v_i, doc_k)$ is defined as follows:

$$df(v_i, doc_k) = \begin{cases} 1 + \log(\#(v_i, doc_k)) & \text{if } \#(v_i, doc_k) > 0 \\ 0 & \text{otherwise} \end{cases} \tag{2}$$

where $\#(v_i, doc_k)$ is the frequency of v_i in doc_k. The intuition is that the frequency of the term v_i in doc_k is important, but inversely weighted by the number of different terms ($\mathcal{N}_k \subseteq \mathcal{V}$) contained in doc_k. These term vectors are normalized by using $l2$ [19].

Term Co-Occurrence Representation (TCOR). In the Term Co-Occurrence Representation the semantics of terms is captured by computing the co-occurrences between terms in the training documents [11]. In this strategy, each term v_i is mapped to a vector $\mathbf{t_i} = \langle t_{i,1}, \ldots, t_{i,|\mathcal{V}|} \rangle$. In TCOR, each $t_{i,k}$ indicates the contribution of the term v_k to characterize v_i. Equation 3 describes the TCOR formulation.

$$t_{i,k} = tff(v_i, v_k) \cdot log \frac{|\mathcal{V}|}{|\mathcal{T}_i|} \tag{3}$$

in the above formula $v_k \in \mathcal{T}_i$ and $\mathcal{T}_i \subseteq \mathcal{V}$ is the set of different terms co-occurring with v_i in at least one document and *tff* is as follows:

$$tff(v_i, v_k) = \begin{cases} 1 + \log(\#(v_i, v_k)) & \text{if } \#(v_i, v_k) > 0 \\ 0 & \text{otherwise} \end{cases} \tag{4}$$

where $\#(v_i, v_k)$ is the number of documents in which words v_i and v_k co-occur. The resultant $\mathbf{t_i}$ vector is normalized by using $l2$ [19].

Embedded representations. The purpose of embedded or distributed representations is to learn a vector space in which terms can lie, and in such a way that the vector space captures semantics of terms as available in large corpora of documents. The idea consists in having similar word vectors in $\mathbb{R}^r$ for semantically related words. For example, *dog* vector should be relatively close to the *cat* vector, since both are domestic animals [17]. These models are shallow, two-layer neural networks trained to reconstruct the linguistic contexts of words. In the following we describe two popular embedding methodologies evaluated in this work; Word2Vec (W2V) [17] and FastText (FT) [2].

In W2V word vectors are trained to predict the contextual words. This strategy is called *skipgram* and the objective is to maximize the log likelyhood of Expression 5. In this strategy T represents the sequence of words in the corpus, and C_t is a set of contextual words defined according to a sliding fixed size window centered in w_t. The probability of $p(w_c|w_t)$ is estimated by using hierarchical softmax [17] under a binary classification framework, where the goal is to predict if a word w_c is in the context of w_t. The words in the context window C_t are positive instances, whereas negative samples are randomly selected from the vocabulary.

$$\sum_{t=1}^{T}\sum_{c\in C_t} logp(w_c|w_t) \tag{5}$$

FT take advantage of the internal structure of the words by means of the n-grams (subsequences of characters of size n) contained in words. For example, for the word "*played*" and subwords of size 3, there will be one vector for each of the following terms: "*<pl*", "*pla*", "*lay*", "*aye*", "*yed*", "*ed>*", and also "*played*". The advantage of this procedure is that more information is captured for each word, including content and style (e.g., the "*ed>*" capture the usage of past tense). Furthermore, at test phase it will be possible to build unseen term vectors during the training data (e.g., by averaging subword vectors).

3.2 Experimental Methodology

Traditionally, the evaluation of AV methods has been using very small data sets to build models that determine if a questioned document was written by an author. However, the amount of textual information for users in online platforms varies according to each subject. In some cases the textual evidence could be small but in others can be large. In this research, our interest is in evaluating models that could be effective in most of the cases. Thus, we favor models that according to the average performance represent the best alternative. In the following subsections we describe in detail each of the relevant aspects in our experimental methodology.

3.2.1 Dataset

In our evaluation, we adopted a dataset for authorship analysis that contains more documents per author than most of the AV collections (having no more than 10 authors, possibly as few as one). The idea of having more documents is to artificially simulate different realistic scenarios (e.g., unbalanced information). The dataset we used is the so called c10 corpus that was originally used by Plakias et al. [20]. The c10 dataset consists of documents from the Reuters Corpus Volume 1 (RCV 1) [12]. The collection has 10 authors with documents in a category with texts about corporate and industrial news. Note that having only documents of the same category reduces the topic factor (all of the documents are associated to the same theme) and focus the evaluation in aspects related to authorship analysis. In this corpus, each author has 50 documents for training, and 50 other documents designated for testing.

For evaluating the different alternatives for AV, we built six different versions of the dataset following the methodology adopted in [5, 20]. Three versions of the dataset are balanced, meaning that each author has the same number of documents. The first, second and third versions have 50, 20, and 10 documents for training per author, respectively. We also built three imbalanced versions, having 2:10, 5:10, 10:20 training documents per author , where a:b means, minimum a and maximum b documents per author. For example, in the case 2:10, there would be two authors with two documents, and 2 authors with ten documents, the rest of the authors will have a number of documents between (a, b) according to a Gaussian distribution used as in [20]. Please note that for the evaluation all of the 50 test documents available for all of the authors were considered.

3.2.2 Classification and Evaluation

For classification, we considered the k-Nearest Neighbors (kNN), and the Support Vector Machines (SVM) classifiers, both of them have shown outstanding performance in text classification problems. The kNN is a very simple yet effective classifier that belongs to the family of instance-based algorithms, or lazy learners. kNN gives the classification by comparing test instances with samples in the training data, usually the prediction is consensus from the class of the most similar k objects. kNN can be used with a wide range of similarity functions and voting schemes to predict the class. In the other hand, SVM is a classifier that focus the learning stage in estimating an optimal separating hyperplane between positive and negative documents (e.g., positive documents are those belonging to author under analysis). More formally, let $(\mathbf{d}_i, y_i)$ be pairs of training instances, where $\mathbf{d}_i \in \mathbb{R}^r$ is the document representation and $y \in \{-1, +1\}$ are class labels. The SVM determines a mapping from training documents to target classes by means of Eq. 6.

$$f(\mathbf{d}) = sign\left(\sum_i \alpha_i y_i k(\mathbf{d}_i, \mathbf{d}) - b\right) \tag{6}$$

where α_i is a learned weight for the label y_i of the training document doc_i. And $k(\mathbf{d}_i, \mathbf{d})$ is a kernel function used to map the input document-vectors $(\mathbf{d}_i, \mathbf{d_j})$ into a different feature space. The latter step is called the kernel trick and allows to measure the similarity of instances doc_i and doc_j in different spaces. The parameters α and b, usually are learned using optimization techniques. For text classification problems the linear kernel is the most used function.

For the evaluation we used the f_1 measure, defined as follows:

$$f_1 = \frac{2 \times precision \times recall}{precision + recall} \tag{7}$$

In f_1 measure, $precision = \frac{TP}{TP+FP}$ and $recall = \frac{TP}{TP+FN}$, with TP, FP and FN being the true-positive, false-positive and false-negative rates, respectively. The macro average of f_1 measure is computed over the positive classes for each version of the dataset. This is the average of the f_1 measure performance for each author in test set when training with the different versions of the training data. The f_1 measure is desirable when the target classes have variable number of instances or the classes are imbalanced. For representing documents, we evaluated the representations described in Table 1. All of the previous strategies were evaluated using the word and character based textual features. The relevance of these features have been shown in different authorship analysis works [24]. For example, the usage of stopwords have proven to be helpful. Also, some specific character based textual features have shown to be effective to capture the style. For example, character n-grams are sequences of characters of size n that can capture stylistic preferences in the tense of verbs (e.g., *ed* for past, *ing* for gerund), or specific usage of prefixes and suffixes.

3.3 Experiments and Results

This section presents the most relevant experiments in our evaluation. In these experiments we evaluate the classification performance by using the macro average f_1 measure of the positive class for each version of the collection. In the first experiment the aim is to use words as the base term to build each representation. It is worth to mention that, stopwords and punctuation marks were not removed. The latter is with the idea of capturing stylistic textual patterns [24] of the authors. Similarly, in experiments using character n-grams we did not remove any character. The decision of keeping all text to extract words or character n grams are motivated by previous findings in authorship analysis [24]. In those works, researchers have found that contrary to conventional thematic classification, where frequent terms are removed or penalized by weighting schemes (e.g., tf-idf), in authorship analysis the stylistic patterns are precisely in very frequent terms used by authors.

Table 2 Evaluation of SVM and 1NN using different document representations built on words. The f-measure performance for each author is computed in each version of the dataset, and the macro average is showed on each column

Model		Instances per author						
		Balanced			Imbalanced			
		50	10	5	10:20	5:10	2:10	Avg
SVM	BoT	**75.67**	**60.23**	36.17	**58.92**	48.95	37.79	**52.95**
	TCOR	65.91	58.76	38.44	56.63	**49.84**	**40.92**	51.75
	DOR	65.06	60.17	**39.62**	55.61	48.61	39.28	51.39
	W2V	35.91	47.70	31.56	37.30	27.67	21.29	33.57
1NN	BoT	65.79	**59.09**	**49.14**	**60.08**	**50.54**	**41.97**	**54.43**
	TCOR	**67.99**	53.54	33.10	55.20	45.46	28.55	47.30
	DOR	66.38	48.16	37.76	54.50	41.15	25.85	45.63
	W2V	37.83	38.42	33.31	40.25	35.37	30.64	35.96

The best result obtained by each classifier is shown in bold for each column

In Table 2, we show the classification performance of the proposed alternatives using words as base features. From these results, it can be seen that the best average performance was 1-NN and BoT.[1] In fact it is somewhat surprising that 1NN-BoT was only outperformed by SVM-BoT in the balanced version of the dataset with 50 and 10 documents. Also note that, the representations based on term vectors only obtained better results when few documents were available for training. For example, note that SVM-TCOR and SVM-DOR outperformed SVM-BoT only in some versions of the dataset were 5 or fewer documents were available for training, which confirms their effectiveness when little text is available [11].

In a second experiment we used character 3-grams as base textual features to built the document representations. Character 3-grams have previously been used in other works, showing that they could outperform other syntactical and lexical features including: Part-of-Speech-Tags (POS), word n-grams, character n-grams with other values of n, etc. [24].

Table 3 presents the most relevant experimental results. In those experiments the SVM-BoT outperformed the other methodologies including the 1NN-BoT. We infer that character 3-grams features added a significant number of dimensions with more stylistic information. These results confirms what other authors have found about the robustness and suitability of character sequences for this task [24]. It is worth noting that with the inclusion of character 3-grams, virtually all the representations showed improvements. However, it seems that the added dimensions to the BoT had much more impact in the representation than in other methodologies.

[1]The k value for k-NN was selected from preliminary experiments by varying it from 1 to 3.

Table 3 Evaluation of SVM and 1NN using different document representations built on character 3-grams. The f-measure performance for each author is computed in each version of the dataset, and the macro average is showed in each column

Model		Instances per author						
		Balanced			Imbalanced			
		50	10	5	10:20	5:10	2:10	Avg
SVM	BoT	**73.47**	**64.60**	44.08	**63.67**	**56.09**	41.07	**57.16**
	TCOR	51.30	54.15	36.96	52.87	51.95	41.52	48.12
	DOR	65.15	63.21	**45.79**	57.14	53.06	**43.04**	54.56
	FT	28.86	9.53	17.84	31.26	19.78	5.54	18.80
1NN	BoT	**66.19**	**58.42**	**53.81**	**56.40**	**53.34**	**44.69**	**55.47**
	TCOR	60.96	52.10	48.79	51.54	47.24	38.75	49.89
	DOR	62.82	51.83	36.02	46.45	46.17	36.57	46.64
	FT	39.70	37.82	26.65	42.11	39.34	32.86	36.41

The best result obtained by each classifier is shown in bold for each column

3.4 *Discussion*

In spite of term-vectors representations have shown to be effective alternatives in other text classification problems, the stylistic factor involved in authorship analysis is hard to model in a framework where each term has only one vector. In other words, the intuition is that the style is tight to each author (class), therefore there should be an strategy to encode different styles for each term. The latter might be the reason that embedding representations, such as W2V and FT (the version of W2V based on *n*-grams), are not obtaining better results. In fact, we performed preliminary experiments using pretrained term vectors (e.g., word vectors learned from wikipedia), but the results were in essentially the same or lower.[2] Results in that direction confirmed the relevance of the stylistic factor for authorship, which is totally disregarded if pretrained vectors are used.

The textual representation methodologies evaluated in previous sections can be easily adapted to work with other authorship verification methods/scenarios. In fact, the only component that needs to be modified is the classification model. For example, for closed authorship analysis where the task is to discriminate among a set of predefined authors, the classifier should be modified to work into a single label multiclass classification framework. For example, change the binary SVM by multiple SVMs under a One-Versus-One (OVO) voting scheme or a One-Versus-All (OVA) formulation [21]. In the case where only documents of the author are available and the the task is to verify the authorship of one questioned document, then a One-Class SVM [16] should be adapted, or a threshold should be defined for the distance function of kNN.

[2]The idea of using pretrained term vectors was to alleviate the small amount of data for training.

4 Conclusions and Future Work

Authorship analysis methods deal with problems that require modeling the writing style of author for different purposes, most notably for forensic analysis. This problem has been the subject of study by a sector of the natural language processing community. Thanks to such efforts nowadays one can count on effective methodologies for the representation texts and the modeling of authorship. Despite the progress, it is somewhat surprising that the community has not paid enough attention to certain domains in which authorship analysis techniques are more than needed. This is the case of e-learning, where vast amounts of texts are available and where automated techniques for assessment and authentication are required.

Among the contributions of this chapter, we can find the following: we have formulated the authorship verification scenario associated to e-learning scenarios. The difficulties inherent to this particular domain have been identified, and pointers that aim at serving as guidelines for the implementation of this sort of systems have been outlined. We reviewed the state of the art in authorship verification modeling, with emphasis on methods. Besides, we performed an empirical evaluation of methodologies for text representations that can be incorporated in authorship verification systems.

The main outcome of this study is that of highlighting the relevance that authorship analysis has into the e-learning domain. The associated scenario poses several challenges to the natural language community that are summarized as follows:

- **Ad-hoc textual representations**. Studies as those presented in this paper evidence the need of powerful text representations that can capture discriminative and descriptive information that can be exploited by authorship analysis tasks. Hence a promising research venue is that of devising and/or learning effective methodologies for representing documents with the characteristics found in e-learning scenarios.
- **Authorship modeling methods capable of working with few samples**. A particular difficulty that e-learning represents for authorship verification is the lack of enough samples. Hence, zero or few-shot learning methodologies can have an important impact into the field.
- **Incremental and lifelong learning models for authorship analysis**. e-learning scenarios require of adaptive methods for modeling authorship, this in terms of both aspects: incremental availability of samples as the user interacts with the platform and dynamism and *drift* of the writing style of authors. Clearly, lifelong [22] or continuous learning [3] methodologies will be decisive to deploy robust authorship analysis models in e-learning.
- **Large scale authorship verification models in the wild**. Undoubtedly, e-learning platforms are associated with massive numbers of users, this requires that authorship analysis techniques should be scalable in the number of potential authors they can deal with. Likewise, methods for efficiently processing big data in this particular context will be important in the forthcoming years.

We foresee further research will be conducted in the next few years addressing this concerns, that will eventually translate into mechanisms that will strengthen authorship verification methods in the context of e-learning.

References

1. Aggarwal C, Zhai C (2012) A survey of text classification algorithms. In: Aggarwal CC, Zhai C (eds) Mining text data. Springer, New York, pp 163–222
2. Bojanowski P, Grave E, Joulin A, Mikolov T (2017) Enriching word vectors with subword information. Trans Assoc Comput Linguist 5:135–146
3. Chen Z, Liu B (2016) Lifelong machine learning. Morgan & Claypool Publishers, San Rafael
4. Escalante HJ, Montes-y-Gómez M, Pineda LV (2009) Particle swarm model selection for authorship verification. In: 14th Iberoamerican conference on pattern recognition progress in pattern recognition, image analysis, computer vision, and applications, CIARP 2009, Guadalajara, Jalisco, Mexico, November 15–18, 2009. Proceedings', pp 563–570
5. Escalante HJ, Solorio T, Montes-y-Gómez M (2011) Local histograms of character n-grams for authorship attribution. In: The 49th annual meeting of the association for computational linguistics: human language technologies, proceedings of the conference, 19–24 June, 2011, Portland, Oregon, USA, pp 288–298. http://www.aclweb.org/anthology/P11-1030
6. Gorrostieta JMG, López-López A, López SG (2018) Automatic argument assessment of final project reports of computer engineering students. Comput. Appl. Eng. Educ. 26(5):1217–1226. https://doi.org/10.1002/cae.21996
7. Juola P, Stamatatos E (2013) Overview of the author identification task at PAN 2013. In: Working notes for CLEF 2013 conference, Valencia, Spain, September 23–26, 2013. http://ceur-ws.org/Vol-1179/CLEF2013wn-PAN-JuolaEt2013.pdf
8. Koppel M, Schler J (2004) Authorship verification as a one-class classification problem. In: Machine learning, proceedings of the twenty-first international conference (ICML 2004), Banff, Alberta, Canada, July 4–8, 2004. https://doi.org/10.1145/1015330.1015448
9. Koppel M, Schler J, Bonchek-Dokow E (2007) Measuring differentiability: unmasking pseudonymous authors. J Mach Learn Res 8:1261–1276. http://dl.acm.org/citation.cfm?id=1314541
10. Koppel M, Winter Y (2014) Determining if two documents are written by the same author. JASIST 65(1):178–187. https://doi.org/10.1002/asi.22954
11. Lavelli A, Sebastiani F, Zanoli R (2004) Distributional term representations: an experimental comparison. In: CIKM. ACM, New York, pp 615–624
12. Lewis DD, Yang Y, Rose TG, Li F (2004) RCV1: A new benchmark collection for text categorization research. J Mach Learn Res 5:361–397
13. López-Monroy AP, Montes-y-Gómez M, Escalante HJ, Pineda LV, Stamatatos E (2015) Discriminative subprofile-specific representations for author profiling in social media. Knowl-Based Syst 89:134–147. https://doi.org/10.1016/j.knosys.2015.06.024
14. López SG, López-López A (2017) Computational methods for analysis of language in graduate and undergraduate student texts. In: Higher education for all. From challenges to novel technology-enhanced solutions - first international workshop on social, semantic, adaptive and gamification techniques and technologies for distance learning, HEFA 2017, Maceió, Brazil, March 20–24, 2017, Revised Selected Papers', pp 32–51
15. Luyckx K, Daelemans W (2008) Authorship attribution and verification with many authors and limited data. In: COLING 2008, 22nd international conference on computational linguistics, proceedings of the conference, 18–22 August 2008, Manchester, UK, pp 513–520. http://www.aclweb.org/anthology/C08-1065
16. Manevitz LM, Yousef M (2001) One-class SVMs for document classification. J Mach Learn Res 2:139–154

17. Mikolov T, Sutskever I, Chen K, Corrado GS, Dean J (2013) Distributed representations of words and phrases and their compositionality. In: Advances in neural information processing systems, pp 3111–3119
18. Milios MJE, Keselj V (2014) Author verification using common n-gram profiles of text documents. In: Proceedings of COLING 2014
19. Pedregosa F, Varoquaux G, Gramfort A, Michel V, Thirion B, Grisel O, Blondel M, Prettenhofer P, Weiss R, Dubourg V, Vanderplas J, Passos A, Cournapeau D, Brucher M, Perrot M, Duchesnay E (2011) Scikit-learn: machine learning in Python. J Mach Learn Res 12:2825–2830
20. Plakias S, Stamatatos E (2008) Tensor space models for authorship identification. In: Hellenic conference on artificial intelligence, Springer, Berlin, pp 239–249
21. Rifkin R, Klautau A (2004) In defense of one-vs-all classification. J Mach Learn Res 5:101–141. http://dl.acm.org/citation.cfm?id=1005332.1005336
22. Silver DL, Yang Q, Li L (2013) Lifelong machine learning systems: Beyond learning algorithms. In: Lifelong machine learning: papers from the 2013 AAAI Spring Symposium, pp 49–55
23. Stamatatos E (2009a) A survey of modern authorship attribution methods. JASIST 60(3):538–556. https://doi.org/10.1002/asi.21001
24. Stamatatos E (2009) A survey of modern authorship attribution methods. J Am Soc Inf Sci Technol 60(3):538–556
25. Stamatatos E, Daelemans W, Verhoeven B, Juola P, López-López A, Potthast M, Stein B (2015) Overview of the author identification task at PAN 2015. In: Working notes of CLEF 2015 - conference and labs of the evaluation forum, Toulouse, France, September 8–11, 2015. http://ceur-ws.org/Vol-1391/inv-pap3-CR.pdf
26. Stamatatos E, Daelemans W, Verhoeven B, Stein B, Potthast M, Juola P, Sánchez-Pérez MA, Barrón-Cedeño A (2014) Overview of the author identification task at PAN 2014. In: Working notes for CLEF 2014 conference, Sheffield, UK, September 15–18, 2014, pp 877–897. http://ceur-ws.org/Vol-1180/CLEF2014wn-Pan-StamatosEt2014.pdf
27. Stamatatos E, Kokkinakis G, Fakotakis N (2000) Automatic text categorization in terms of genre and author. Comput Linguist 26(4):471–495. https://doi.org/10.1162/089120100750105920

Plagiarism Detection

David Gañan

Abstract There have always been students who cheat and copy content from other students or external sources, but nowadays with the available technologies the students can share their works easier, and they have access to a vast amount of information through the Internet, so the capability and easiness to plagiarize have increased considerably. In consequence, detecting plagiarism has become even harder work for teachers. It is very difficult to manually detect a copy between students (unless the cheat is evident), especially in online environments where the number of students tends to be high. And if we consider that the student can plagiarize from students of previous courses or even from the Internet, the manual detection becomes nearly impossible. In order to aid with this detection, there are different anti-plagiarism techniques and tools that assist the teacher providing some hints or evidences of plagiarism in the students' work. This chapter discusses about the different approaches of anti-plagiarism tools available in the literature, and explains in more detail the solution applied in the TeSLA project, and the experiences obtained from its usage.

Keywords Anti-plagiarism tools · Authorship · e-learning · Student copy detection · SIM algorithm

Acronyms

API	Application Program Interface
ASCII	American Standard Code for Information Interchange
JSON	JavaScript Object Notation
OCR	Optical Character Recognition
REST	REpresentational State Transfer
SIM	Software (SIM)ilarity Tester

D. Gañan (✉)
EIMT-UOC Rambla Poblenou, 156. 08018 Barcelona, Spain
e-mail: dganan@uoc.edu

D. Baneres et al. (eds.), *Engineering Data-Driven Adaptive Trust-based e-Assessment Systems*, Lecture Notes on Data Engineering and Communications Technologies 34,
https://doi.org/10.1007/978-3-030-29326-0_2

TPT TeSLA Plagiarism Tool
UOC Universitat Oberta de Catalunya (Open University of Catalonia)
VLE Virtual Learning Environment

1 Introduction

Plagiarism (or copying some part of other's work and include it as self in his or her own work) is one of the most common ways students cheat in their school or high education studies. This is even more frequent in online environments, where there is no direct supervision of the teacher when the students are doing their activities.

Nowadays is quite easy to plagiarize internet sources or activities from other students, due to the easiness to reach these contents through the net and mobile devices. The high number of students, the great number of available information sources, and the online factor of e-learning environments, reduces the chance of the instructor to detect plagiarism on student's submissions. This is why some plagiarism detection tools (also known as anti-plagiarism tools) are required to help teachers detecting it.

Anti-plagiarism tools help detecting similarities between documents that can be considered copy one of another. However such tools cannot decide whether the similitudes are really plagiarism or which is the original version, they just detect and point the facts, but the final decision must be taken by the instructor. There are different types of plagiarism detection tools including those matching text, programming code or even images, and they can be applied to many different contexts, especially on education or research.

This chapter discusses about plagiarism detection tools and presents a concrete case of one of these tools and its integration into a system for student identity and authorship verification called TeSLA.[1] The rest of the chapter is organized as follows: Sect. 2 discusses about different available plagiarism detection techniques and anti-plagiarism tools. Sections 3 and 4 explain the case tool and the integration into the TeSLA system respectively. Finally, Sect. 5 presents conclusions and some further work identified to improve the tool.

2 Related Work

Plagiarism can be found in almost all human contexts and works. There is much research work in academic context about plagiarism [10, 18, 27], because students who plagiarize, do not acquire the expected knowledge from the activity, nor learn to think and develop their writing skills by themselves, and because the number of

[1] https://tesla-project.eu/.

plagiarism cases in this field increases each time more, in part because of the availability of the new technologies [24, 32]. Other important fields in which plagiarism is common are: (i) research publishing or publishing in general, not only from other authors [6], but also from oneself [9] or (ii) programming, either in professional or academic contexts [4, 31].

The authors in Alzahrani et al. [3], made a great study on plagiarism status consolidating many other works in this field. They classify plagiarism basically in two types: literal and intelligent plagiarism, which in turn are subclassified into different types of manipulations like exact copy, near copy or modified copy (with restructuration) for literal plagiarism, and text manipulation, translation or idea adoption for intelligent plagiarism. Some techniques for intelligent plagiarism are paraphrasing (expressing the same meaning with different words) or summarizing the original content or changing its structure.

Humans can suspect about plagiarism between two documents when it is quite evident, but detecting plagiarism between many documents it is not possible because it is almost impossible to memorize all the text to match similitudes among them. In order to help humans to detect plagiarism it is required to automate the process using plagiarism detection tools [3, 30]. Such tools can be seen as a black box receiving a document to be analyzed and a collection of reference documents to compare with. The result of the analysis typically consists in a value that represents a plagiarism percentage and a list of fragments or sections found suspicious to be copied, and the sources where similitudes were found.

Plagiarism detection can be divided into three different approaches: extrinsic, intrinsic and cross-language detection [3, 22, 29]. On the one hand, extrinsic detection, focused on detecting literal plagiarism, analyzes the input document in comparison with the set of documents provided as reference. This approach divides both the input documents and the reference documents into smaller pieces: n-grams (contiguous sequences of n characters from a given text), words or sentences. Then compares them to find different types of plagiarism previously discussed like literal copy, paraphrasing or idea adoption. There are many works tackling this approach like Gupta [16], Barrón-Cedeño and Rosso [5], Yerra and Ng [34] or Kasprzak et al. [21]. On the other hand, intrinsic detection looks inside the same document to find suspicious plagiarism content analyzing the writing style of the author [36]. This kind of plagiarism detection is also referenced frequently as forensic analysis (for a detailed account see Chap. 1 in this volume). Finally, cross-language detection tackles the problem of finding plagiarism between documents written in different languages [30]. This type of analysis uses a combination of machine translation algorithms with some of the extrinsic analysis techniques.

In Kraus [22] the author introduces new approaches to plagiarism detection based on the results of the International Competition on Plagiarism Detection, a yearly competition that compares several plagiarism detection tools for accuracy, performance and other factors [29]. Among these approaches is worth to mention the citation-based plagiarism detection [11, 12], which is an specific technique for research publications that uses the order of citations instead of the text content to find similarities between papers, and the semantic similarity, which tries to reduce the number of comparisons between documents required to find plagiarism by first classifying documents into some categories, and only comparing the input document with the source documents that have the same category [19].

Plagiarism detection tools use different textual features to characterize a text in order to analyze it. For extrinsic analysis Alzahrani et al. [3] and Kraus [22] distinguish between (i) lexical features which can be n-grams, word-grams or even sentences, (ii) syntactic features which consist on tagging words depending on their role in the sentence (verbs, nouns, adjectives, etc.) (iii) semantic features which takes into account the meaning of the words searching for synonyms, antonyms, hypernyms and hyponyms and (iv) structural or tree features which consist on finding the structure of the document (headers, sections, subsections, paragraphs, etc.). Stylometric features are used for intrinsic analysis (also known as forensic analysis), and consist on analyzing the writing style of the author with some metrics like frequency of words, average sentence length, synonyms, average paragraph length, etc. If there are changes of the writing style within the document they can be suspicious of plagiarism.

Some of the most referenced detection tools in the academic context are CodeMatch [13], JPlag [25], MOSS [7], Plaggie [1], Sherlock,[2] SIM [14, 15] and YAP [33]. There are also well-known commercial tools, some specific for publishing plagiarism like CrossCheck [35] used by Elsevier and Springer among other publishers, Docoloc[3] used by the EDAS conference management system and iThenticate,[4] others for education like Turnitin[5] and others specific for internet plagiarism detection like URKUND.[6] Is obvious that no tool is good for each purpose; the selection of the right tool depends on the requirements. There are several works comparing different plagiarism detection tools [2, 17, 23, 26] from different points of views and purposes.

3 A Plagiarism Tool for TeSLA Project

The TeSLA system (An Adaptive Trust-based e-assessment System for Learning) is a framework for student identity and authorship verification. The aim of the TeSLA project is to construct a system capable of identifying students during the realization of e-learning activities, in order to check the authorship of the work they deliver to the teacher. In order to attain this objective, the TeSLA system uses some components called instruments, which monitorize some data of the students during or after the activity execution in order to identify the students and verify they are the authors of their own activities. Such instruments include face and voice recognition, keyboard pattern recognition (or keystroke dynamics), forensic analysis (analysis of the writing

[2] https://warwick.ac.uk/fac/sci/dcs/research/ias/software/sherlock/.

[3] https://www.docoloc.com/.

[4] http://www.ithenticate.com/.

[5] https://www.turnitin.com/.

[6] https://www.urkund.com/.

style) and anti-plagiarism. In regard of the anti-plagiarism tool, the initial aim was to adapt the anti-plagiarism tool used in the UOC[7] (Universitat Oberta de Catalunya) virtual campus.

The UOC campus has an integrated anti-plagiarism tool called PacPlagi [28] which allows teachers to compare students' works in order to find similitudes. The PacPlagi tool is highly coupled to the UOC campus and it would be very difficult to integrate it into another system, so the decision was not to integrate the whole tool but the same anti-plagiarism algorithm that it uses inside, the SIM algorithm [14, 15], which will be introduced in the next section. The name used for the new software component is TPT (TeSLA Plagiarism Tool) and will be used in the rest of the document for reference.

3.1 The SIM Algorithm

The SIM algorithm (SIM stands for software SIMilarity tester) was introduced in the related work section as one of the most referenced anti-plagiarism tools in the academic context. Using the classification made in Alzahrani et al. [3] mentioned in the related work section, the SIM algorithm would be defined as a literal or extrinsic plagiarism detection tool, provided that it only checks the structure of the documents in order to find similitudes between them.

The algorithm takes two plain text files A and B as input, and return two results P_{AB} and P_{BA} corresponding to the percentage of similitude of A compared to B and vice versa. Optionally, is also possible to get a summary of the similitudes between the two documents. It is important to note that P_{AB} and P_{BA} can be different values, since although the common part between both documents is the same, the size of the document can be different and therefore the percentage of each document considered as plagiarism is different. Principally two types of input files are supported: text and code (C, C++, Java, Pascal, Modula-2, Lisp and Miranda languages). Internally, the SIM algorithm uses a grammar (one different for each programming language or plain text) to tokenize the input documents, then builds a forward reference table that is used later in order to detect the similitudes between documents.

The main reason for choosing this algorithm as the one that will be used in the TeSLA project is because it is the one that uses the PacPlagi tool, which was the tool that was initially expected to be integrated. The other principal type of plagiarism analysis, which is intelligent or intrinsic analysis, is covered by the forensic analysis tool adopted by the TeSLA project. In any case, the SIM algorithm is generally well referenced in the academic context and has good results in terms of accuracy and performance compared to other similar tools. Some references including those comparisons were already introduced in the related work section.

[7] https://www.uoc.edu/.

3.2 The TPT Tool

In the context of a course students may have to perform some writing activities. Some of these activities can be performed directly on the VLE (Virtual Learning Environment) or by the means of an editor tool and then delivered through the VLE. In any case the work delivered by a student for the activity is considered as a submission. A submission can be a unique digital document, or more than one document usually submitted inside a compressed file. As new submissions are sent by students, they are also sent to the TPT tool for processing them.

The TPT tool processes submissions sequentially as they are received. The result of processing a submission is a number ranged from 0 to 1 indicating the percentage of plagiarism of the submission (0 means no plagiarism). In order to calculate this value, the submission has to be compared with other submissions previously processed by the tool, so after a submission is processed it is stored in the database for this purpose.

There are some parts of a submission that can be detected as plagiarism but they are not (is what is known as false positives). For example, the submissions of the students may contain totally or partially the statement of the activity. This will increase the percentage of similitude between submissions, but it cannot be considered plagiarism. To prevent this, the instructor can provide the statement of the activity so it is removed from submissions of that activity before to perform the comparison (as explained in the following subsection about the configuration of the tool). In contrast, the tool is not able to distinguish between plagiarism and acknowledged content copy, nor determining which document was the original version and which one the copy. It can only detect similarities and provide enough information to the instructors so they can make decisions about the plagiarism.

For the sake of modularization, the processing of a submission is split into several chained steps or phases which are shown in Fig. 1, and explained in more detail in the following subsections. Each step takes the output of the previous one, processes it, and returns the corresponding result which will be the input for the next step.

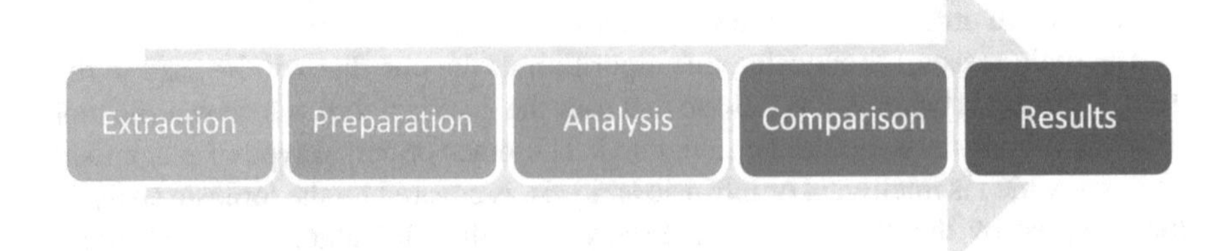

Fig. 1 Submission processing steps

3.2.1 Configuration

The configuration of the TPT tool is not part of the submission processing, but a previous step that the instructor should take in order to setup how the tool will behave. The configuration of TPT is defined per activity, so each activity (independently of the subject or course itbelongs) can have a different configuration. These settings can be set through the configuration pages of the activity in the VLE. On the one hand, TPT can be configured to perform five types of analysis (non-exclusive):

1. Text: treats all content as plain text.
2. C: only analyzes C language related content.
3. CPP: only analyzes CPP language related content.
4. Java: only analyzes java language related content.
5. Text only: only analyzes text documents.

On the other hand, instructors can select which is the criterion to compare one submission with other submissions, depending on the course, subject and activity of such submissions. Specifically one submission can be compared:

1. With other submissions of the same course, subject and activity of the current semester (default option).
2. With other submissions of the same course, subject and activity of any semester (including the current one).
3. With other submissions of a given list of courses, subjects and activities.

Finally, instructors can provide a document containing the statement of the activity. The text in this document will be searched and removed from all submissions in order to prevent finding plagiarism if the students copy some parts of the statement in the submission.

3.2.2 Extraction

In this step the TPT tool checks if the student submission is a compressed file and if so, it extracts its content. If a compressed file is found, its content is also extracted recursively. The TPT supports handling different types of compressed files like zip, gz, tar.gz. Other types of compressed files like .rar could be supported in the future using new extraction libraries. The result of this step is a folder structure containing all the documents contained in the submission (see Fig. 2).

3.2.3 Preparation

The preparation of a submission involves different manipulation phases. First the text content of each file in the submission has to be extracted because the SIM algorithm only works with plain text files (or files codified using ASCII set of characters). Some files in the submission content can be already plain text files so they are

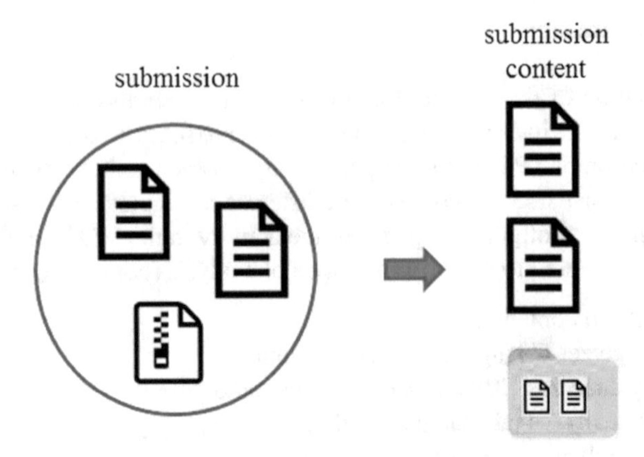

Fig. 2 Submission content extraction

already prepared, but some others (binary files) may also contain text in a format that is not understandable for the SIM algorithm directly. The TPT tool is able to extract text from file documents like pdf or docx (see Fig. 3). The pdf extraction also supports OCR (Optical Character Recognition) for pdf documents based on images. As a future work, new types of documents can be supported by including

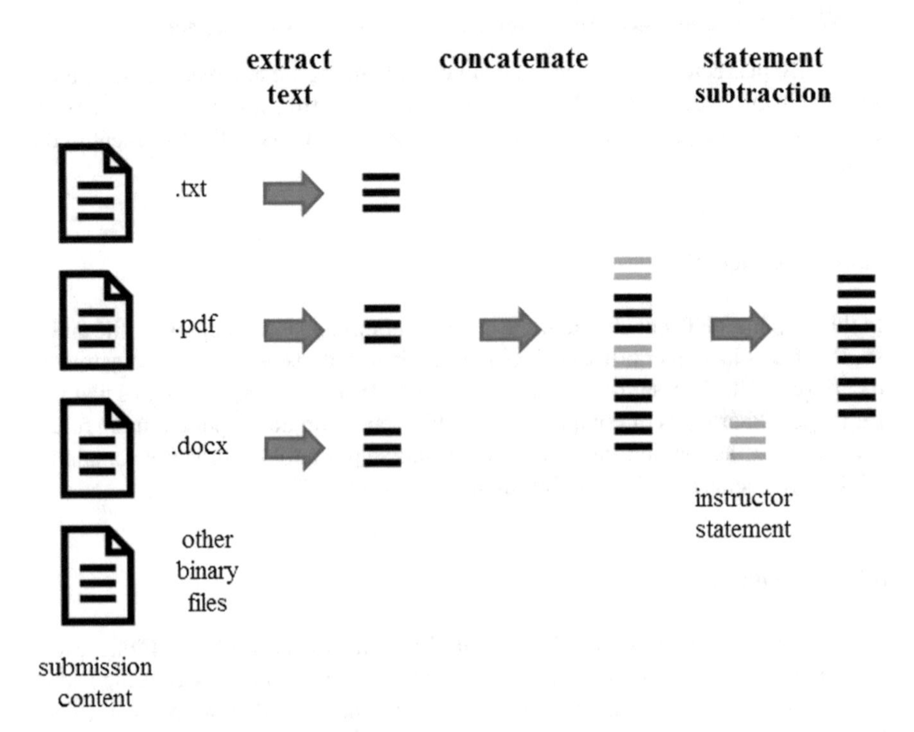

Fig. 3 Text extraction, concatenation and statement subtraction

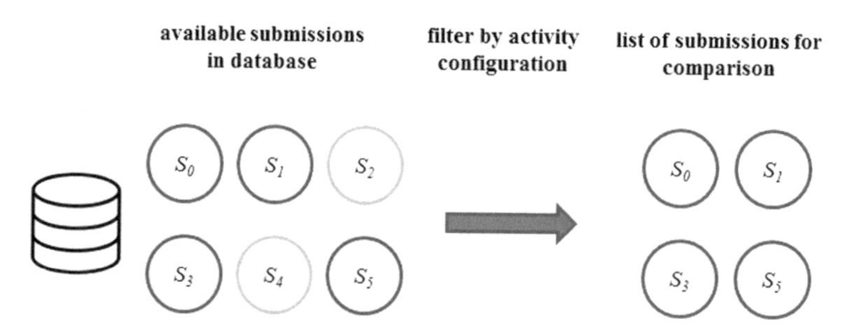

Fig. 4 Analysis of required comparisons

new text extraction libraries. The complete list of supported file types of the TPT is the following: txt, csv, docx, doc, pdf, html, htm, c, xml, cpp, hpp, java, jsp, and in general any text document. The rest of binary files are discarded. It is worth to mention that there are techniques to compare other kinds of binary files that are not currently supported by the TPT tool (for example images). This is another of the improvements to be considered in the future.

Once all the text is extracted from the supported files of the submission, it is concatenated into a single file, which is easier to handle than multiple files, and gives better performance (see Sect. 3.3 about optimizations for more details). Finally, if the instructor attached some statement files to the configuration of the activity (see Sect. 3.2.1), the corresponding text is removed from the content of the submission in order to avoid detecting plagiarism if the student copies some parts of this statement in the submission.

3.2.4 Analysis

The objective of the analysis step is to determine with which submissions in the database the submission that is being processed must be compared to (see Fig. 4). Ideally, when a submission is processed, it should be compared with all the already processed submissions stored in the database. However, as the number of submissions in the database increases, the number of comparisons required for each new submission would increase exponentially (Sect. 3.3 approaches this problem in more detail). Besides that, not always is strictly necessary to compare a submission with all the available submissions in the database, but just with other submissions of the same activity, subject or course. This is one of the settings of the TPT tool which the instructor can configure, as described in Sect. 3.2.1.

3.2.5 Comparison

This step performs all the comparisons from the list returned by the analysis step. For each comparison, the SIM algorithm is executed with the text of the two correspond-

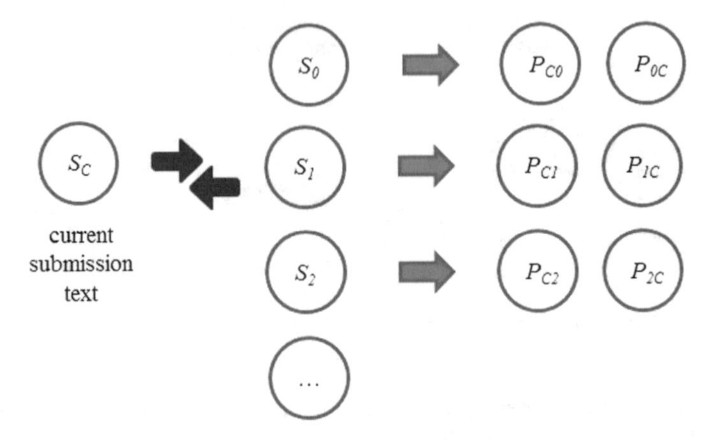

Fig. 5 Comparison step

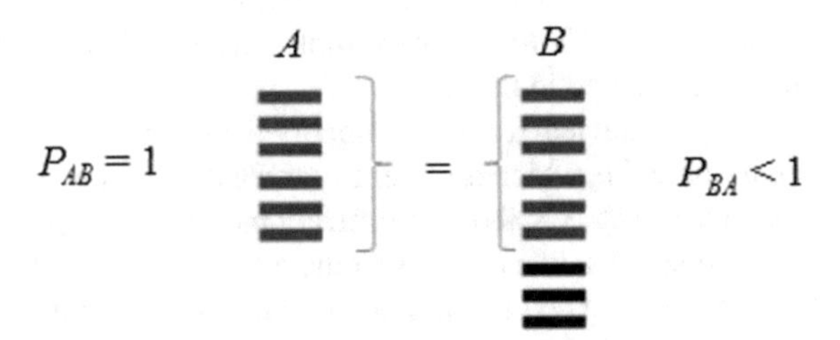

Fig. 6 The amount of similarity in *A* is greater than in *B*

ing submissions as parameters and returns the corresponding result (see Fig. 5). The result of comparing two submissions *A* and *B* consist of two numerical values P_{AB} and P_{BA} corresponding to the percentage of *A* in *B,* and the percentage of *B* in *A* respectively (in a scale between 0 and 1, where 0 means no plagiarism and 1 means the entire submission is a copy of the other). It is important to understand that P_{AB} and P_{BA} are, in general, different values; for example, consider *B* contains a copy of *A* plus other content, the value for the comparison between *A* and *B* will return a value of 1 because all the content in *A* is also contained in *B*, while the comparison between *B* and *A* will return a value less than 1 because it contains some content not included in *A* (see Fig. 6).

3.2.6 Results

Once the comparison step finishes, the final plagiarism percentage for the submission is calculated as the maximum value of all partial comparisons results (see Fig. 7). Other types of aggregation formulas to calculate the submission result from the partial comparisons results can be considered, e.g. pondering each partial result with the length of the corresponding documents to not penalize so much the final result if

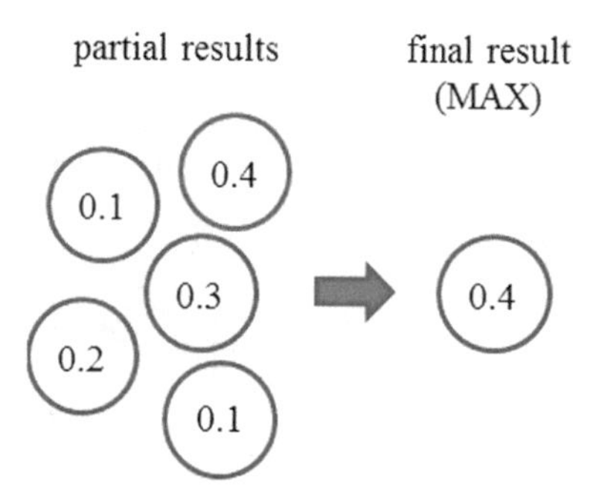

Fig. 7 Calculation of final result as the maximum value of partial results

there are short documents with a high number of similitudes. This can be considered as an improvement to be addressed as further work.

It is important to notice that the percentage of plagiarism of a submission can change along the time, because when new incoming submissions are processed, new partial comparison results will be available so the submission result can be affected. Consider the example depicted in Fig. 8, where submission *A* is processed at time t_1, with a plagiarism percentage of 0.1. When submission *B* is processed at time t_2, it is compared with *A*, and the percentage found of *B* in *A* is 0.4, so the global plagiarism of *A* (considering all the submissions, including *B*), now is of 0.4. So each time a new submission is processed, all the submission results (of submissions involved in the comparison step) should be reviewed. As the formula used for calculating the submission result is the maximum of the partial results, it will only change if the new partial result is greater than the actual submission result.

All the results calculated by the TPT tool will be available for instructors through the VLE, in the results page of each activity. The main results page shows a list of submissions for the activity, with the result of the TPT tool process (see Fig. 9). The evaluation result column shows the percentage of plagiarism detected for the corresponding submission (expressed by means of a value between 0 and 100), and each submission row is colored according to that percentage (red, orange and green for high, mid and low level of plagiarism respectively).

The buttons with the label 'view' in the last column of the previous list, allow the instructor to see the audit data for one specific submission. The audit data consists

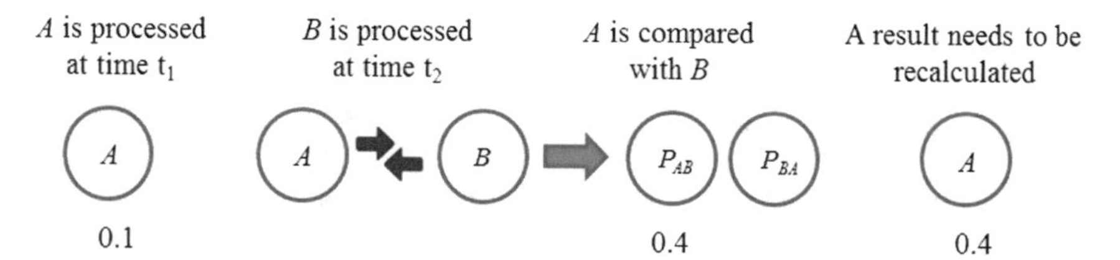

Fig. 8 Plagiarism percentage results can change as new submissions are processed

evaluation result	start date	end date	audit data
85	February 16th 2017, 12:12:19 pm	February 16th 2017, 12:12:21 pm	view
83	February 16th 2017, 12:16:43 pm	February 16th 2017, 12:17:01 pm	view
80	February 16th 2017, 13:23:56 pm	February 16th 2017, 13:24:13 pm	view
75	February 16th 2017, 11:45:57 pm	February 16th 2017, 11:46:09 pm	view
60	February 16th 2017, 10:34:25 pm	February 16th 2017, 10:35:03 pm	view
53	February 16th 2017, 11:45:35 pm	February 16th 2017, 11:45:50 pm	view
42	February 16th 2017, 12:23:14 pm	February 16th 2017, 12:23:54 pm	view
31	February 16th 2017, 12:07:03 pm	February 16th 2017, 12:07:56 pm	view
25	February 16th 2017, 13:11:04 pm	February 16th 2017, 13:12:03 pm	view
9	February 16th 2017, 11:04:53 pm	February 16th 2017, 11:05:14 pm	view
0	February 16th 2017, 10:02:34 pm	February 16th 2017, 10:03:12 pm	view
0	February 16th 2017, 12:43:42 pm	February 16th 2017, 12:44:14 pm	view

Fig. 9 The list of submissions for an activity and results from the TPT tool processing. The color helps the instructor to focus on suspicious submissions

of some information or evidence about how the given result was calculated. In a first place the instructor will see a list of all the comparisons performed with other submissions and the corresponding partial values retrieved on each comparison (see Fig. 10). In this view, the student information appears anonymized for privacy reasons. The instructor can navigate deeper on a particular comparison by clicking the corresponding row. Then the comparison view appears, showing both submissions, compared side-by-side (see Fig. 11). Lines of each text are numbered, and the part

Comparisons:

Student	Type	Result	Details
5b7f712c8ef161daad27409508d29d7cb3bd090157ff78b67deaba80184371d2	text_only	0	More details
5b7f712c8ef161daad27409508d29d7cb3bd090157ff78b67deaba80184371d2	text_only	0	More details
5b7f712c8ef161daad27409508d29d7cb3bd090157ff78b67deaba80184371d2	text_only	0	More details

Fig. 10 List of partial comparisons and partial results

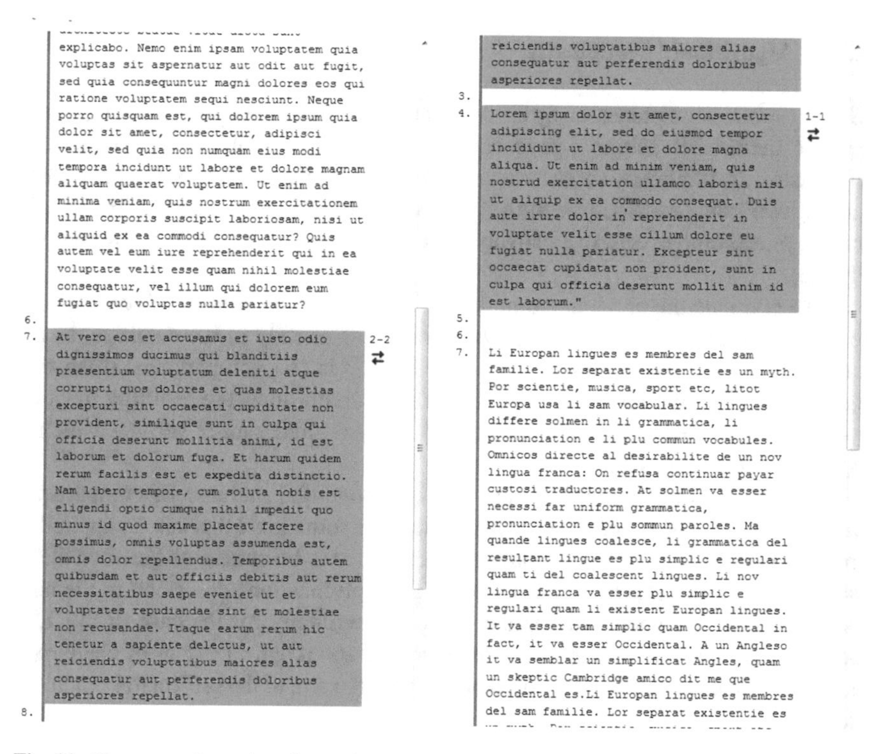

Fig. 11 The comparison view shows the contents of the two submissions side by side

of the text where similarities where found is highlighted with a red background. At the right side of each highlighted block there is a synchronize icon, which will scroll both submissions so the corresponding text blocks are aligned in both sides. If the same text block on one submission appears multiple times in the other one, multiple synchronization icons will be available.

3.3 *Performance Problems and Optimizations*

In order to determine the similitudes of one document with the already processed documents, many comparisons must be done and as already mentioned, this number of comparisons increases with each new submitted document. This section describes this problem with more detail, and explains some optimizations applied to TPT in order to reduce the number of required comparisons.

3.3.1 Algorithmic Complexity

The first submission processed by the tool, name it *A*, does not require any comparison because there is no submission to compare with in the database yet. The second submission *B*, requires just one comparison with *A*; the third submission *C*, will require two comparisons, one with *A* and one with *B*; fourth submission *D*, will require three comparisons respectively with submissions *A*, *B* and *C*, and so on (see Fig. 12).

If we continue the previous example, we can see that the number of comparisons for a given submission is equal to the submissions already in the database, so the number of comparisons required for submission *D* is *D*-1 comparisons. If we sum the number of total comparisons required for *N* submissions, the resulting function is equivalent to the number of edges of a complete graph with *N* vertices which is exponential (see Fig. 13).

Although it is not possible to avoid this exponential increase of comparisons when the number of submissions is high, there are some hints that can be applied in order to reduce the number of submissions that have to be compared between them, as explained in the following subsections. The optimizations described in the next sections where designed and applied ad hoc, specifically for the case of the TPT tool, although there are references in the literature about optimizations of plagiarism techniques on large datasets [8, 20], there was not the aim of the project to implement them.

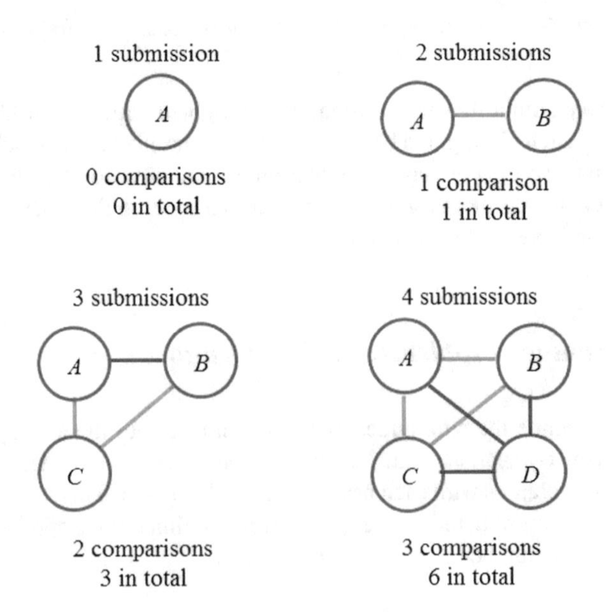

Fig. 12 Number of comparisons of first processed submissions

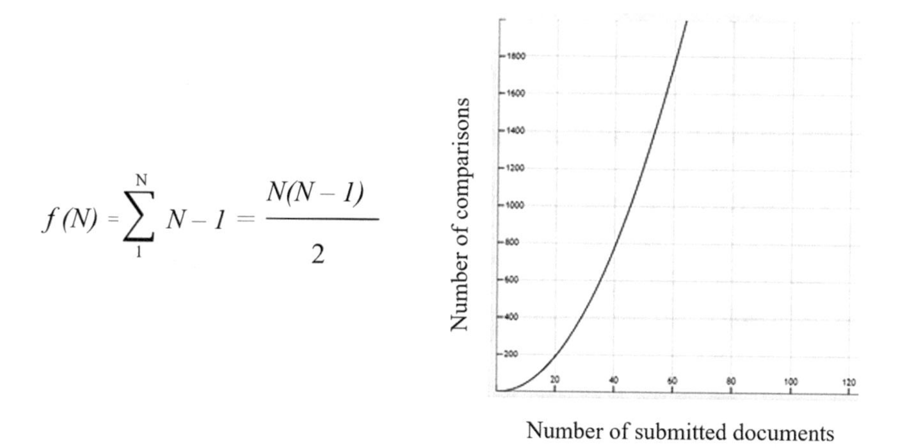

Fig. 13 The number of comparisons required for analyzing N submissions is exponential

3.3.2 Reduction of Comparisons

The first approach to tackle the exponential problem is to reduce the number of comparisons to be done. As already mentioned before, the comparison of two documents A and B is not a symmetric operation because the P_{AB} result (percentage of plagiarism of A regarding to B) can be different from P_{BA} result (percentage of plagiarism of B regarding to A). The SIM algorithm is able to return both results from the same call so it is not necessary to make one call to compare A with B and another to compare B with A. This reduces the number of comparisons to the half.

Comparing two submissions can involve many document comparisons as long as each submission can include more than one document. One of the initial decisions in order to reduce comparisons between submissions was to concatenate the text content from all the documents in the same submission. In this way the final result is more accurate, and the number of comparisons required between two submissions is reduced to one. The graphic in Fig. 14 shows the difference in performance between comparing submissions file by file or by concatenating them. As can be seen, the concatenation approach is able to process more submissions in the same time than comparing each file separately.

Submissions are related with assessment activities inside a course, so another hint to reduce the number of comparisons is not to compare submissions from different activities or courses if not necessary (e.g. a math assessment activity and a literature review). In any case, the instructor can specify which documents should be included in the comparison by means of one of the options already mentioned in Sect. 3.2.1 about TPT tool configuration options. Applying this filter, the number of comparisons can be divided, in the best case, by the number of activities in the system.

Students commonly have the chance of sending several submissions for the same assessment activity along the time (in case, these submissions are prior to the deadline

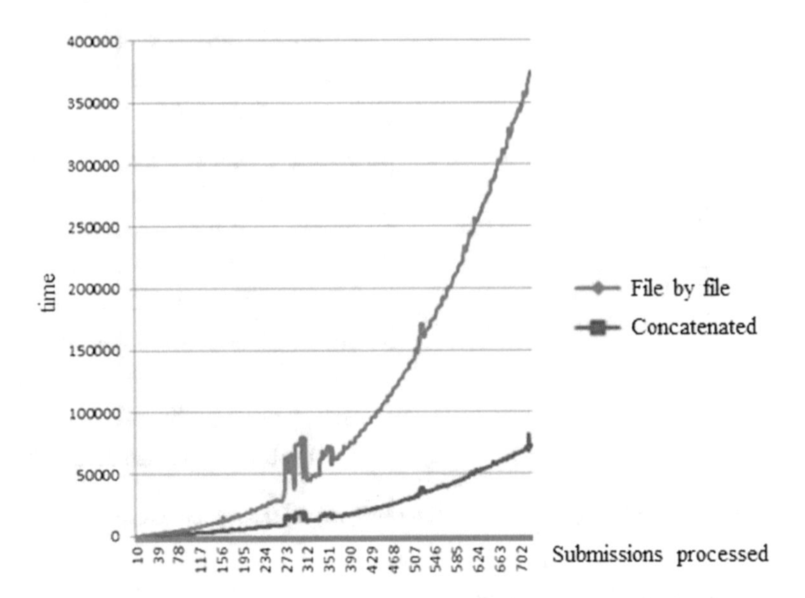

Fig. 14 Comparison between concatenated versus file by file processing

of the assessment activity). Another hint to reduce comparisons is to only consider the last submission of a student as the good one, and ignore the older ones. This is a debatable decision, provided that intermediate deliveries could contain parts of plagiarism eliminated in the last submission, but in fact the last submission is the one which instructor will consider, so it can be assumable for the sake of performance.

Finally, extracting the statement of the activity from the submissions also can reduce the size of the documents to be compared so it also reduces the time required by the SIM algorithm to return a result.

3.3.3 Parallelization of the Process

With the application of the improvements explained in the previous subsections, the number of total required comparisons were reduced considerably, but as already mentioned, the problem is still exponential. Another approach to tackle it and improve performance is to parallelize some parts of the process.

The separation of the process described in Section 3.2 was defined with the idea of parallelization in mind. On the one hand, each phase or step can be encapsulated and executed independently from the other ones in parallel; this allows to process until five submissions at the same time, one on each step. For example one submission S_1 can be in the preparation step, while S_2 is in the extraction step. When finished, S_1 will enter the analysis step, S_2 will be in preparation step and the extraction step will start processing a new submission S_3. And so on, like as in an assembly line (see Fig. 15). However the order of the steps must be applied to a submission in the correct order.

On the other hand, most of the steps are scalable; meaning that more than one instance of the step can be run in parallel (at the same time) on different submissions. For example, multiple instances of the comparisons step can handle multiple comparisons at the same time, so the whole time of processing all comparisons is reduced by the number of running instances. As can be seen in Fig. 16, three instances of the

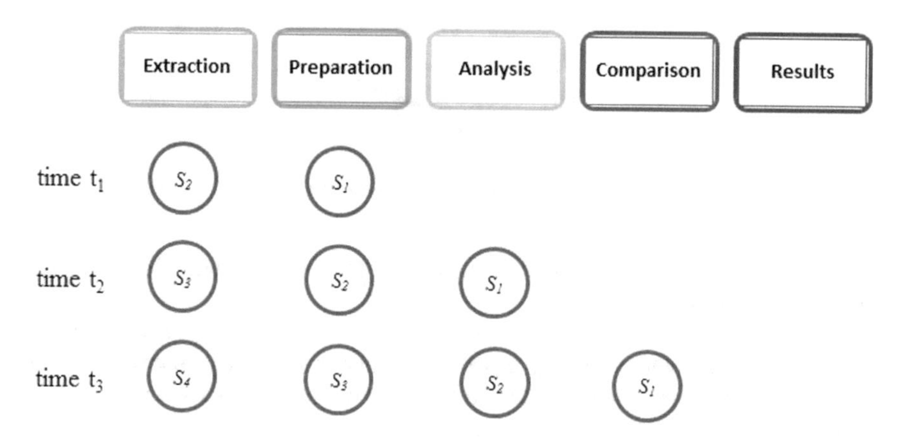

Fig. 15 Different steps of the process are executed in parallel

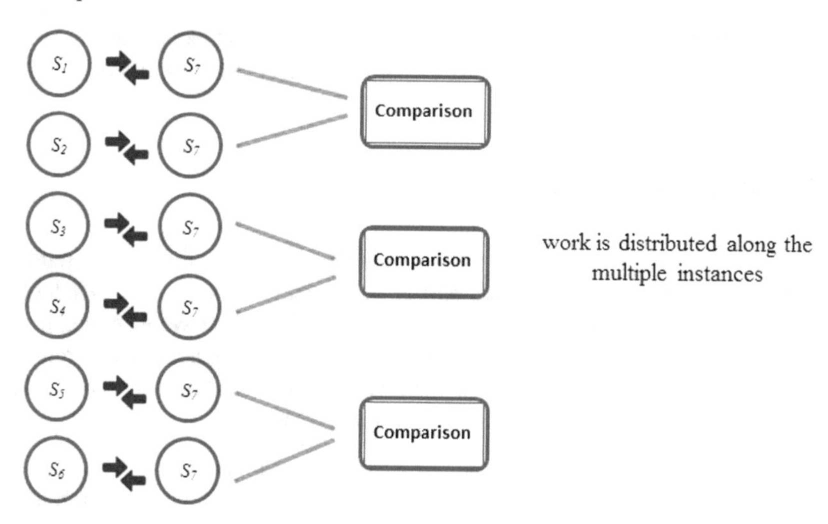

Fig. 16 Different instances of the comparisons step executed in parallel

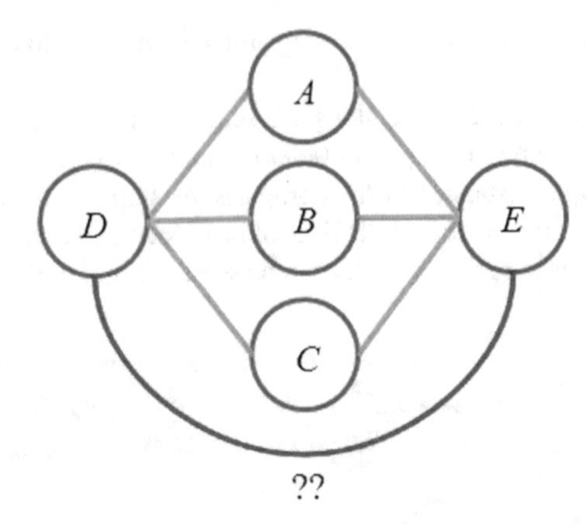

Fig. 17 *D* and *E* submissions have also to be compared between them

comparison step can process three times quicker the same amount of comparisons than only one instance.

Nevertheless, not all the steps in the process are parallelizable; the exception is the analysis step. When this step processes a submission, it calculates the set of submissions the new submission has to be compared with. Let's suppose that the analysis step is executed in parallel for submissions *D* and *E*. As shown in Fig. 17, in both cases the analysis step determines the same comparisons (*A*, *B* and *C*), but what happens between *D* and *E*? As both submissions where not already processed when the analysis step started handling them, they are not considered and the comparison between *D* and *E* is not included (when it should). If the analysis step was executed first on submission *D* and next on submission *E*, this comparison would be included, so this step cannot be run simultaneously for more than one submission.

4 Integration and Results

The architecture of the TeSLA system was designed with the aim of reusability and extensibility in mind, so each part of the system (including the TPT tool) was defined as an isolated component, and some public interfaces were defined for communication between components. Furthermore, each component of the architecture is encapsulated into a Docker[8] container, which is a technology that provides many benefits in a distributed architecture like (among others):

1. Scalability. Each component can be executed simultaneously.
2. Integrability. Each component can use its own environment and communicates with others through a well defined interface.

[8]https://www.docker.com/.

3. Ease of deployment. The components of the system can be prepared and started easily without hard configurations.

The implementation of the TPT tool was divided into two assets: a library, which implements all the features of the tool itself, and a wrapper component that handles all the integration stuff with the rest of the TeSLA system. Both the wrapper and the tool library were developed in Python programming language. This was a general decision for the overall TeSLA project because it was a well common language for all the participants. Furthermore, Python provides many available libraries that simplified the implementation of the different requirements.

The communications between the TPT tool and the rest of the system were implemented through a REST APIs (Representational State Transfer) using JSON format (JavaScript Object Notation) to exchange data. All messages between components of the TeSLA system required a secure channel so all of them used server and client certificates to encrypt communications and to enforce trust; certificates allow client/server component to know if the server/client component is the right one (it is not an attacker or unauthorized one).

During the integration of the TPT tool into the TeSLA system, it was also integrated with an internet anti-plagiarism tool called URKUND, already mentioned in Sect. 2. URKUND is a commercial online service which offers solutions both for schools, universities and corporations. Once the TPT tool finishes the preparation step of a submission, the text is sent to the URKUND API and a reference identifier for the request is retrieved, which can be used to query the status of the request. Once the result for the request is ready, the TPT tool has to update the internal result (calculated with the results of comparisons) with the result of the internet comparison.

In the first phase of development, the TPT tool was tested internally using unitary tests for checking the correct operation of the tool. As a result of these first tests, the performance problem discussed in Sect. 3.3 was identified and further performance tests were prepared. Such tests consisted in executing the tool multiple times with a set of test submissions and measuring the time taken to complete each of the steps in the process. Figure 14 shows the result of one of the most effective optimizations applied, consisting in concatenating the text of all the documents inside a submission.

The TPT tool was also tested as part of the TeSLA system, during four programmed pilots, with the participation of instructors and students from 7 high education institutions. From the technical point of view, the aim of the first pilot was to make a first integration test in a production environment. The second and third pilots helped to test new added features and to check the system scalability capacities by increasing the number of participants in the pilot. Finally the aim of fourth pilot was to provide a final version of the system and make stress tests with the participation of thousands of students.

One of the purposes of the data gathered during TeSLA pilots is to help improve the identification and authorship instruments, including the TPT tool. There are still no results from these tests, but the approach will be two-faced. On the one hand more performance tests will be executed in parallel with a specific research on the field about other strategies or methods to reduce or tackle the algorithmic complexity. On the other hand, the results calculated by the tool during pilots will be analyzed in order to increase the accuracy of the tool and reduce false positives. This will be complemented with further research of other plagiarism tools and techniques that could improve the quality of the results.

5 Conclusions and Further Work

The aim of this chapter was to discuss about plagiarism tools and present one of these tools and its integration with the TeSLA system. As seen, the tool has some performance problems due to the number of comparisons between documents required to determine the percentage of plagiarism for that document, and how this number of comparisons increase exponentially as the number of already processed documents grows.

Some improvements have been explained to reduce the number of comparisons and to increase the performance of such comparisons. Although this is a problem that affects other plagiarism tools in general, the improvements explained in this chapter were applied ad hoc; they are not fruit of a specific investigation because it was not the aim of the project. However, although the improvements reduced the number of comparisons considerably, there are still performance problems with the number of students is quite big, like in online environments, so one of the lines of future research should be to study how to tackle this problem.

Other improvements that can be considered as further work in order to add new functionalities to the tool where described, like increasing the number of compatible documents supported or applying other anti-plagiarism techniques already discussed at the beginning of the chapter to enable image or cross-language plagiarism detection, among others.

References

1. Ahtiainen A, Surakka S, Rahikainen M (2006) Plaggie: GNU-licensed source code plagiarism detection engine for Java exercises. In: Proceedings of the 6th Baltic Sea conference on computing education research: Koli Calling 2006. ACM, pp 141–142
2. Ali AMET, Abdulla HMD, Snasel V (2011) Overview and comparison of plagiarism detection tools. In: DATESO, pp 161–172
3. Alzahrani SM, Salim N, Abraham A (2012) Understanding plagiarism linguistic patterns, textual features, and detection methods. IEEE Trans Syst Man Cybern Part C Appl Rev 42(2):133–149
4. Arwin C, Tahaghoghi SM (2006) Plagiarism detection across programming languages. In: Proceedings of the 29th Australasian computer science conference, vol 48. Australian Computer Society, Inc, pp 277–286
5. Barrón-Cedeño A, Rosso P (2009) On automatic plagiarism detection based on n-grams comparison. In: European conference on information retrieval. Springer, Berlin, Heidelberg, pp 696–700

6. Baždarić K (2012) Plagiarism detection–quality management tool for all scientific journals. Croatian Med J 53(1):1–3
7. Bowyer KW, Hall LO (1999) Experience using "MOSS" to detect cheating on programming assignments. In: FIE'99 29th Annual frontiers in education conference, vol 3. IEEE, pp 13B3–18
8. Burrows S, Tahaghoghi SM, Zobel J (2007) Efficient plagiarism detection for large code repositories. Softw Pract Experience 37(2):151–175
9. Collberg CS, Kobourov SG, Louie J, Slattery T (2003) SPLAT: a system for self-plagiarism detection. In: ICWI, pp 508–514
10. Ercegovac Z, Richardson JV (2004) Academic dishonesty, plagiarism included, in the digital age: a literature review. Coll Res Libr 65(4):301–318
11. Gipp B, Meuschke N, Beel J (2011) Comparative evaluation of text-and citation-based plagiarism detection approaches using guttenplag. In: Proceedings of the 11th annual international ACM/IEEE joint conference on digital libraries. ACM, pp 255–258
12. Gipp B (2014) Citation-based plagiarism detection. In: Citation-based plagiarism detection. Springer Vieweg, Wiesbaden, pp 57–88
13. Glanz L, Amann S, Eichberg M, Reif M, Mezini M (2018) CodeMatch. In: Software engineering und software management
14. Grune D, Huntjens M (1989) Detecting copied submissions in computer science workshops. Vrije Universiteit, Informatica Faculteit Wiskunde & Informatica, p 9
15. Grune D (2006) The software and text similarity tester SIM
16. Gupta D (2016) Study on extrinsic text plagiarism detection techniques and tools. J Eng Sci Technol Rev 9(5)
17. Hage J, Rademaker P, van Vugt N (2010) A comparison of plagiarism detection tools. Utrecht University, Utrecht, The Netherlands, p 28
18. Hawley CS (1984) The thieves of academe: plagiarism in the university system. Improving Coll Univ Teach 32(1):35–39
19. Hussain SF, Suryani A (2015) On retrieving intelligently plagiarized documents using semantic similarity. Eng Appl Artif Intell 45:246–258
20. Jadalla A, Elnagar A (2008) PDE4Java: plagiarism detection engine for java, source code: a clustering approach. IJBIDM 3(2):121–135
21. Kasprzak J, Brandejs M, Kripac M (2009) Finding plagiarism by evaluating document similarities. In: Proc. SEPLN, vol 9, no 4, pp 24–28
22. Kraus C (2016) Plagiarism detection-state-of-the-art systems (2016) and evaluation methods. arXiv preprint arXiv:1603.03014
23. Lancaster T, Culwin F (2004) A comparison of source code plagiarism detection engines. Comput Sci Educ 14(2):101–112
24. Ma HJ, Wan G, Lu EY (2008) Digital cheating and plagiarism in schools. Theory Into Pract 47(3):197–203
25. Malpohl G (2005) JPlag: detecting software plagiarism. URL http://www.ipd.ukade, 2222
26. Martins VT, Fonte D, Henriques PR, da Cruz D (2014) Plagiarism detection: a tool survey and comparison. In: OASIcs-OpenAccess series in informatics, vol 38. Schloss Dagstuhl-Leibniz-Zentrum fuer Informatik
27. Meuschke N, Gipp B (2013) State-of-the-art in detecting academic plagiarism. Int J Educ Integrity 9(1)
28. Núñez MF, Duran i Cals J (2011) El pec-plagio. Definición, prevención y detección. http://hdl.handle.net/10609/8575
29. Potthast M, Barrón-Cedeño A, Eiselt A, Stein B, Rosso P (2010) Overview of the 1st international competition on plagiarism detection. In: SEPLN 2009 workshop on uncovering plagiarism, authorship, and social software misuse (PAN09), CEUR-WS.org
30. Potthast M, Barrón-Cedeño A, Stein B, Rosso P (2011) Cross-language plagiarism detection. Lang Resour Eval 45(1):45–62
31. Rosales F, García A, Rodríguez S, Pedraza JL, Méndez R, Nieto MM (2008) Detection of plagiarism in programming assignments. IEEE Trans Educ 51(2):174–183

32. Sutherland-Smith W (2005) The tangled web: internet plagiarism and international students' academic writing. J Asian Pac Commun 15(1):15–29
33. Wise MJ (1996) YAP3: improved detection of similarities in computer program and other texts. In: ACM SIGCSE bulletin, vol 28, no 1. ACM, pp 130–134
34. Yerra R, Ng YK (2005) A sentence-based copy detection approach for web documents. In: International conference on fuzzy systems and knowledge discovery. Springer, Berlin, Heidelberg, pp 557–570
35. Zhang HY (2010) CrossCheck: an effective tool for detecting plagiarism. Learn Publish 23(1):9–14
36. Zu Eissen SM, Stein B (2006) Intrinsic plagiarism detection. In: European conference on information retrieval. Springer, Berlin, Heidelberg, pp 565–569

Biometric Tools for Learner Identity in e-Assessment

Xavier Baró, Roger Muñoz Bernaus, David Baneres and Ana Elena Guerrero-Roldán

Abstract Biometric tools try model a person by means of its intrinsic properties or behaviours. Every person has a set of unique physical traits derived from genetics and vital experience. Although there are many traits that can be used to verify the identity of a learner, such as the voice, appearance, fingerprints, iris, or gait among others, most of them require the use of special sensors. This chapter presents an analysis of four biometric tools based on standard sensors and used during the TeSLA project pilots. Those tools are designed to verify the identity of the learner during an assessment activity. The data for on-site and on-line institutions is used in order to compare the performance of such tools in both scenarios.

Keywords Biometrics · Learner identification · Identity verification · Face recognition · Voice recognition · Keystroke dynamics recognition · Forensic analysis

Acronyms

AUC	Area Under the ROC Curve
CMVN	Cepstral Mean and Variance Normalization
FA	Forensic Analysis
FN	False Negative
FP	False Positive

X. Baró (✉) · R. Muñoz Bernaus · D. Baneres · A. E. Guerrero-Roldán
EIMT, Universitat Oberta de Catalunya, Barcelona, Spain
e-mail: xbaro@uoc.edu

R. Muñoz Bernaus
e-mail: rmunozber@uoc.edu

D. Baneres
e-mail: dbaneres@uoc.edu

A. E. Guerrero-Roldán
e-mail: aguerreror@uoc.edu

D. Baneres et al. (eds.), *Engineering Data-Driven Adaptive Trust-based e-Assessment Systems*, Lecture Notes on Data Engineering and Communications Technologies 34,
https://doi.org/10.1007/978-3-030-29326-0_3

FPR	False Positive Rate
FR	Face Recognition
GMM	Gaussian Mixture Model
KS	Keystroke Dynamic Recognition
MFCC	Mel Frequency Cepstral Coefficients
ROC	Receiver Operating Characteristic
SVM	Support Vector Machine
TN	True Negative
TP	True Positive
TPR	True Positive Rate
VAD	Voice Activity Detector
VR	Voice Recognition

1 Introduction

Learner identity during assessment activities is one of the most challenging topics in online and blended educational institutions. Although the use of plagiarism systems has been widely adopted by many institutions, the use of tools to verify the identity continues being an exception. The simplest and usual method for learner identity verification is using a login authentication. This mechanism does not guarantee the user identity, since password is an information that can be shared.

More secure approaches are those based on biometric information of the learner, since this information cannot be shared. Authors in [1, 8, 20] propose using other identification methods based on biometric recognition. Biometric recognition involves fingerprint detection [15], keystroke dynamics [14, 28], voice recognition [21, 23] and face recognition [25, 26]. The use of more sophisticated sensors such as depth cameras [27] or wearable devices (e.g., smart watches), which are becoming very popular nowadays, could also be considered. They can provide complementary information about individuals (e.g., in a soft-biometric way) which could be used, for instance, for the recognition of advanced patterns such as hand motions [11] or heart rate variability [17].

In the last years, many companies started providing services to institutions for secure assessment. Some of those services are related to the creation of secure work places where the activity of the learner is monitored and restricted. Others provide online service to monitor learners using the same process, as Kryterion,[1] ProctorU,[2] or Pearson VUE.[3] Those services are intrusive and are only focusing on the final examination.

We can found some tools implementing biometric tools, as Kryterion, which also offers learner's authentication by means of biometric control (only facial recognition

[1]Kryterion website: https://www.kryteriononline.com.

[2]ProctorU website: http://www.proctoru.com.

[3]Pearson VUE website: http://home.pearsonvue.com.

and writing patterns) that allows to detect the identity of the learner. Some MOOC platforms, like Coursera,[4] already offers a simple and not correlated service of authentication based on only one biometric control to provide verified certificates for authenticated learners. In the concrete case of Coursera, it was using keystroke dynamics and face recognition, however, recently published news state that keystroke has been removed.

One of the technological challenges of the TeSLA Project[5] was to adapt state of the art biometric tools to the educational context. The selected tools were those that did not required any special hardware, reducing the requirements on learners' computers. Assuming that most of current computers have keyboard, webcam and microphone, learner identity in the TeSLA system is checked analyzing the face, voice, keystroke dynamics patterns and writing style. In [7] an analysis of perception of such systems in a real educational scenario is introduced.

After introducing the basics of the biometric tools used for learner identification, in this chapter we analyze the accuracy of those tools using the data captured during the TeSLA project pilots. First, a brief introduction to the biometric tools is provided. In the following, the experimental design and data sanitation are explained to finally present the results and conclusions.

2 Learner Identity Verification

Biometric tools try to model a person by means of its intrinsic properties or behaviours. Every person has a set of unique physical traits derived from genetics and vital experience. Many of those traits are the base of biometric tools that try to identify a person using the voice, appearance, fingerprints, iris, gait, etc.

Despite which are the learner data used for the identification, the architecture of a biometric tool is standard. All biometric tools require a special process called enrolment, where learners are registered to the system and a model for the learner is created. Once a learner is registered, tools can compare new incoming data with the previously learnt model in order to verify the identity of the learner.

Figure 1 shows the architecture of a biometric tool. Input data (picture, audio, document, keystroke patterns, ...) is first converted to a numeric representation by a feature extraction process. The complexity of this process will depend on each tool, and have the goal of creating a set of quantifiable properties (features) of the input data. When a new learner is registered to the system, those features are used to create the biometric model of the learner. In the case of identity verification, the features extracted from the input data are compared with the model to generate the final result. Depending on the biometric tool, this final result can be a binary classification output or a similarity value.

---

[4]Coursera website: https://www.coursera.com.

[5]TeSLA Project website: http://tesla-project.eu.

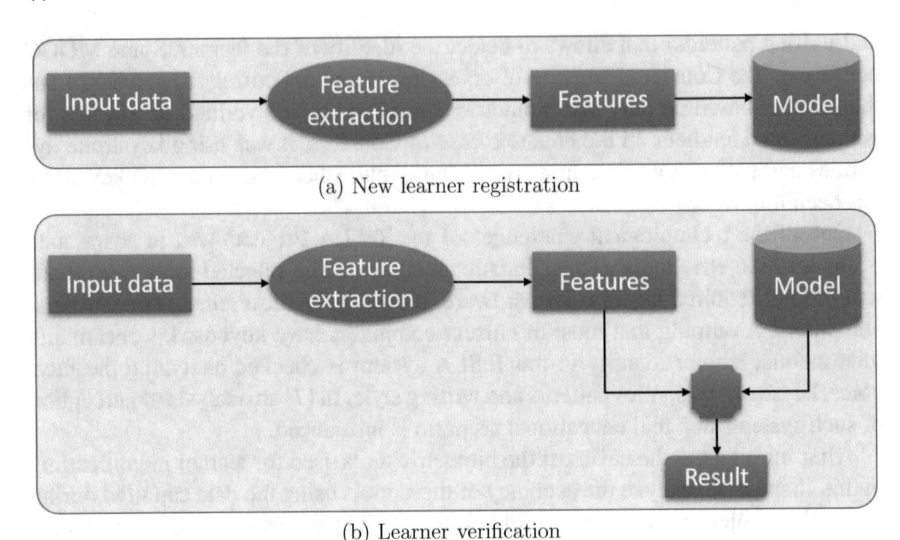

Fig. 1 Biometric tool architecture

The registration process for learners is a critical part of the use of a biometric tool. The system may ensure the legitimacy of provided data and its quality, otherwise the results obtained during the verification process will be useless. On-site institutions can afford this process on their facilities, ensuring the identity of the learner and providing a controlled environment to ensure data quality. In the case of fully virtual institution, this process cannot be performed on the institution facilities, and additional measures are required. TeSLA system assumes the second scenario, and therefore quality checks on the enrolment data has been deployed, and is recommended the use of such tools for continuous assessment activities during consecutive academic periods, mitigating the problem of legitimacy of provided data.

Although the list of available biometric tools is large, in the case of the TeSLA project, we considered only those tools without special sensor requirements. Selected tools for identity verification were face recognition, voice recognition, keystroke dynamics recognition and forensic analysis. All these tools work with a basic computer configuration with a keyboard, webcam and microphone.

The rest of this section introduce the biometric tools used during the TeSLA pilots. Although the results can slightly change using other implementations for the same biometric tools, most of the problematic and findings applying those concrete implementations are mainly extensible to other available implementations. The goal of the chapter is to analyse the considerations of porting such biometric tools to the learning context. Therefore, to compare different implementations of the same biometric tools is not addressed.

2.1 Face Recognition

Face Recognition tool identify a learner by means of the face. Input images are analysed in order to detect the region where the face is. This process is known as face detection, and we use Histogram of Oriented Gradients or HOG [2] for this task. Basically this method compares how light is a pixel with respect to its neighbours, which is interpreted as a gradient vector. The face is divided into regions, and each region is described by means of an histogram of the gradient vectors in the region, obtaining a final description of the face. Using a huge dataset of faces for training, we obtain a model of the gradients of a face, and we use this model in order to classify each region of the input image as containing a face or not. Those regions containing a face are the ones selected for the recognition process.

In order to reduce the variability of the faces due to scale or rotations, all images are aligned using affine transformations, so that the eyes and lips are always in the same place in the image. To be able to align faces, we detect the face landmarks using approach presented in [9]. This method extracts 68 landmarks corresponding to specific parts of the face: the top of the chin, the outside edge of each eye, the inner edge of each eyebrow, etc.

Finally, a Deep Convolutional Neural Network is used to generate a 128 features vector for each face, following the approach presented in [22]. During the learner registration, the generated feature vectors are stored as part of the learner model. During the verification process, the feature vector extracted from the input image is compared with all the vectors stored in the model, and the distance of the most similar vector is used to compute the similarity value returned as the result of the verification.

2.2 Voice Recognition

Voice Recognition biometric tool verifies learners based on their voice-characteristics. Given a voice-sample, an energy-based Voice Activity Detector (VAD) is first used to select the segments of the voice-sample that contain speech information. The detected voiced segments are then represented using standard 19 Mel Frequency Cepstral Coefficients (MFCC) (log energy features) (see [19]) computed over 20 ms Hamming-windowed frames (with 10 ms shifts). This descriptor is augmented with the first and second order derivatives.

The resulting 60 cepstral coefficients are normalized using Cepstral Mean and Variance Normalization (CMVN) (see [18]). The normalized coefficients are finally transformed to a compact version of GMM super-vectors, so-called I-vectors or identity-vector (see [3]), based on total-variability modeling. Thus, each speech-sample of a learner is represented by a single I-vector. For details of the speech I-vector extraction process, the reader is referred to [10]. This feature-representation has been adopted in the TeSLA VR biometric tool because I-vectors have been shown to perform well for speaker-recognition tasks based on short speech-samples. The current version of the VR instrument uses I-vectors of length 400.

For each enrollment voice-sample, the corresponding *enrollment* I-vector is extracted and stored in a database, along with a unique identifier for the corresponding learner.

During learner-verification, random voice-samples of length 15 seconds are extracted from the speech-stream of the learner. In biometrics jargon these samples are referred to as *probe* samples. Since this is an identity-verification task, each probe sample has an associated *claimed identity* (of the learner in question). For each probe-sample, the corresponding probe I-vector is compared to each enrollment I-vector of the claimed identity, using the Cosine-similarity measure (see [24]).

The *match-score* assigned to the probe-sample is the highest similarity value among the similarity-values of the 15 enrollment I-vectors for the claimed identity. If the match-score is higher than a preset threshold, the probe-sample is assumed to come from the claimed speaker, otherwise the match is rejected.

2.3 Keystroke Dynamic Recognition

Keystroke dynamics recognition biometric tool verifies learners identity using timing information from pressed and released keys during an assessment activity, that is, using the typing rhythm patterns of the learner [5]. In order to describe the typing patterns, the amount of time each key is held down (dwell time), and the elapsed time between the release of the first key and the depression of the second (flight time) are computed. Dwell and flight times are considered the atomic features, and are merged to form n-graphs that represent consecutive keystrokes for $n \in \{2, 3, 4\}$. Feature vectors containing the n-graphs are build for each 125 characters.

During the registration process, feature vectors containing the n-graphs are stored as the learner model. During the identity verification process, provided feature vectors are compared with the ones stored in the model and a response is generated. When analysed patters are considered genuine, the learner model is updated in order to capture changes on its typing pattern. A full description of this biometric tool is provided in [16].

2.4 Forensic Analysis

The aim of forensic analysis tool is to verify the authorship of documents written by a specific learner. For this, the input texts are analyzed in order to capture style information instead of content. This author verification procedure is performed by finding patterns among very frequent stylistic terms, for example; stopwords used by each author. Despite the stylistic terms are particular to each language, most of them can be automatically captured in several languages by using subsequences of characters [6]. In text mining this can be achieved by using char 3-grams (sequences of characters of size 3), which can capture stylistic preferences in the tense of verbs (e.g., ed for past, ing for gerund), specific usage of prefixes and suffixes, or patterns in punctuation marks.

The use of 3-grams has shown to be very effective in several authorship analysis tasks, and the evaluation presented in Chapter 1 combines them with different distributional document representations [13] that could be used under different scenarios. The Forensic Analysis tool for TeSLA was implemented considering that there are few documents and words to train the model. Therefore, one of the strategies of the tool before to encode sample documents is to generate subsamples of text by sliding a window of fixed size (e.g., 100 characters) with slipping (e.g., 10 characters) over each train document. Then, each of those subsamples is represented with a vector under a specific strategy and fed into a classifier. The developed tool can be easily switched to work with different distributional term representations (e.g., Term Co-occurrence matrices, histograms, Word2Vec, etc.).,[6] which are helpful according the document type and context (e.g., short/large documents, social media texts).

The implemented tool assumes few train documents and generate texts subsamples by extracting 3-grams at character level. Then each instance is represented by using an histogram of normalized frequency of the 3-grams in each subsample. Finally, the classifier consists in a One-Class Support Vector Machine (SVM) that predicts whether the query document under analysis belongs to the author. As explained in Chapter 1 depending of the context of application and nature of documents, the tool could be switched to exploit a different document representation that could be more convenient in other scenarios. For example, if the available data is very class unbalanced, or much more documents become available.

3 Experimental Design

During the TeSLA pilots, where more than 20,000 learners were involved, we collected images, audios, keystroke patterns and documents from real assessment activities. All this data was processed and results delivered to instructors. As a live project, TeSLA system evolved during the pilots, with the addition of data filters and the change of some data formats to better adapt to the assessment requirements. Moreover, during the first pilots, some technical issues impacted on data quality and in some cases prevented to capture data on the initial pilots.

[6]Open source framework developed by authors for [12]: https://github.com/lopez-monroy/FeatureSpaceTree.

For privacy and legal reasons, not all the institutions participating in the pilots were able to share the data from their learners. From the institutions that shared the data, we have selected for the experiments one fully virtual university and on on-site university, which are representative in terms of activities and learner profiles. In total, data from 810 learners (71,839 samples) is used for the experiments. All the data is considered genuine data since learners participated in pilots in a voluntary manner. To evaluate the capability of the biometric tools detecting cheating attempts, an experiment crossing the data between different learners is designed. In this section, we present the data sanitation process performed to remove invalid data, which simulates the filters added to the TeSLA system, the experimental design and the evaluation metrics that will be used to analyze the results on next section.

3.1 Data Sanitation

Before a sample captured by the TeSLA system is processed by a biometric tool, there are multiple checks and filters:

Legal aspects: TeSLA system cannot store any data from a learner unless an informed consent of this learner granting the capture and use of the data is signed.

Enrolment status: Before a learner can perform an assessment activity using any biometric tool, a registration process is required. If the biometric model is not ready, an error is generated for any received data.

Data quality: Biometric tool developers have provided the TeSLA system with a set of quality check tools for the sample. If any data sample do not pass such quality checks, an error is generated and the sample is not evaluated.

When legal aspects are not satisfied, the data is not captured, therefore we only need to deal with enrolment status of learners and data quality checks. As we want to analyse the biometric tool responses, we will remove all the data from learners with an incomplete biometric model and those individual samples not passing quality checks. The goal of this sanitation process is to simulate as close as possible to the TeSLA scenario, and therefore, to get indicators about the confidence on each biometric tool responses.

Let's define as $\mathcal{X} = \{X_1, X_2, \ldots, X_N\}$ the set of all samples collected during assessment activities for a certain biometric tool, where X_i correspond to the set of all samples belonging to learner i, and N is the total number of learners. In a similar way, let's define $\mathrm{E} = \{E_1, E_2, \ldots, E_N\}$ as the set of all the samples collected during the registration process, being E_i the set of enrolment samples for learner i (Table 1).

Table 1 Quality filters applied for each biometric tool

<table>
<tr><th>Biometric tool</th><th>Enrolment ($\mathcal{Q}_E$)</th><th>Verification ($\mathcal{Q}_V$)</th></tr>
<tr><td rowspan="3">FR</td><td colspan="2">Format is $mp4$[a], png or $jpeg$</td></tr>
<tr><td colspan="2">Image contains pixel variability, black images are discarded</td></tr>
<tr><td colspan="2">Image has one and only one face in it</td></tr>
<tr><td>KS</td><td colspan="2">No quality filters</td></tr>
<tr><td rowspan="5">VR</td><td colspan="2">Format is wav, $weba$, oga or $mp3$</td></tr>
<tr><td colspan="2">The length of the sample is longer than 10 ms</td></tr>
<tr><td colspan="2">Sampling frequency is over 16 MHz</td></tr>
<tr><td>No silence fragments larger than 1.5 s at the start or end of the sample</td><td rowspan="2"></td></tr>
<tr><td>Total silence in the sample is smaller than 1.5 s</td></tr>
<tr><td rowspan="2">FA</td><td colspan="2">Format is txt, zip[b], gz[b], tar[b], $tar.gz$[b], rar[b], pdf, doc, $docx$, odt, csv, $html$, htm, xml or $java$</td></tr>
<tr><td colspan="2">The length of the document is longer than 10 words</td></tr>
</table>

[a]Each frame is considered a sample for quality checks, enrolment and identity verification
[b]Compressed files are extracted and all valid documents are concatenated as a single document

Table 2 Enrolment percentage estimation

Biometric tool	Enrolment estimation ($\mathcal{E}$)
FR	$\mathcal{E}_i^{FR} = min\left(1, \frac{\lvert\tilde{E}_i\rvert}{15}\right)$ (2) where $\lvert\tilde{E}_i\rvert$ is the number of samples in $\tilde{E}_i$
KS	$\mathcal{E}_i^{KS} = min\left(1, \frac{\lvert\tilde{E}_i\rvert}{15}\right)$ (3) where $\lvert\tilde{E}_i\rvert$ is the number of samples in $\tilde{E}_i$
VR	$\mathcal{E}_i^{VR} = min\left(1, \frac{\lvert\tilde{E}_i\rvert}{12}\right)$ (4) where $\lvert\tilde{E}_i\rvert$ is the number of samples in $\tilde{E}_i$
FA	$\mathcal{E}_i^{FA} = min\left(1, \frac{\sum_{x \in \mathcal{E}_i} max(300, \lvert\lvert x \rvert\rvert)}{1000}\right)$ (5) where $\lvert\lvert x \rvert\rvert$ is the number of words in the sample x

Let's define the quality check functions $\mathcal{Q}_E(x) \rightarrow \{0, 1\}$ and $\mathcal{Q}_V(x) \rightarrow \{0, 1\}$ for enrolment and identity verification respectively as:

$$\mathcal{Q}_{E/V}(x) = \begin{cases} 1 \text{ if quality filters in Table 1 are passed.} \\ 0 \text{ otherwise.} \end{cases} \quad (1)$$

As biometric tools require a model of the learners in order to generate a prediction, an estimation of the learner model completeness is performed. Let's define $\widetilde{\mathrm{E}} = \{\tilde{E_1}, \tilde{E_2}, \ldots, \tilde{E_N}\}$ where $\widetilde{\mathrm{E}}_i \subseteq E_i$ as the set of all the samples $e \in E_i$ such as $\mathcal{Q}_E(e) = 1$. The enrolment estimator function for a learner i is defined as $\mathcal{E}_i \rightarrow [0, 1]$, computed using the valid enrolment samples $\tilde{E_i}$ (see Table 2).

Finally, let's define as $\tilde{\mathcal{X}} = \{\tilde{X_1}, \tilde{X_2}, \ldots, \tilde{X_N}\}$, where $\tilde{X_i} \subseteq X_i$ is the set of all samples $x \in X_j$ such as $\mathcal{Q}_V(x) = 1$.

3.2 Experimental Setting

We assume that all the data collected during the pilots is genuine data, both for enrolment and assessment activities. That is, we assume that all data belongs to the learner that performed the registration process and the assessment activities. When enrolment data provided during the registration is used to train a model for one of the learners, we expect that this model will provide a positive verification of all the data provided in the assessment activities. In the same way, we assume that every learner is unique on the system, and therefore, if we provide the data captured for one learner on assessment activities to a model trained for other learner, we expect that the verification of this data will fail. A graphical representation is shown in Fig. 2, where we have a biometric tool trained for three different learners (1, 2 and 3). The small

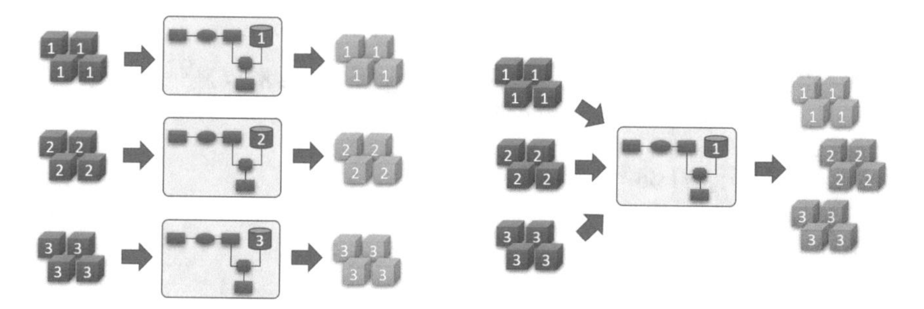

Fig. 2 Expected behaviour for a biometric tool. At left, data for learners 1, 2 and 3 are provided to instrument trained for such learners, therefore, those data samples are classified as correct (green). At right, data from learner 1, 2 and 3 is sent to instrument trained for learner 1, therefore, we expect learner 1 samples accepted (green) and samples from learners 2 and 3 rejected (orange)

boxes representing the data collected for those learners in assessment activities are verified by the tool and the expected result is shown in colours (green for successful verifications and orange for failed verifications). Sending the data from one learner to a biometric tool with the model of another learners simulates a cheating attempt.

In a more formal way, a biometric tool can be defined as a function:

$$\mathcal{F}(x, \mathcal{W}_i) \rightarrow \mathbb{R} \in [0, 1] \tag{6}$$

where $x \in X_k$ is a data sample for learner k and $\mathcal{W}_i$ are the parameters learnt for learner i during the registration. That is, $\mathcal{W}_i$ is the model of the learner i for the biometric tool $\mathcal{F}$.

Given the datasets $\widetilde{\mathrm{E}}$ and $\widetilde{\mathcal{X}}$ with all the valid samples captured during enrolment and assessment activities respectively, a new dataset $\mathcal{K}$ is computed with the responses of the biometric tool $\mathcal{F}$ for genuine sample and cheating sample. As is shown in Algorithm 1, for each learner i with an estimated enrolment percentage $\mathcal{E}_i = 1$, we train the model $\mathcal{W}_i$ for this learner and apply the biometric tool to all the genuine samples of the learner. For each response of the biometric tool, a row with the response and value 1 is added to a matrix $\mathcal{K}$. Finally, we apply the biometric tool with the model learnt for this learner to the genuine samples of other different learner. In this case, for each response, a row with the response and value -1 is added to a matrix $\mathcal{K}$. As a result we obtain a matrix $\mathcal{K}$ with 2 columns and as many rows as computed responses. Note that the first column contains the responses of the biometric tool and the second column a value indicating if the response is for a genuine sample 1 or a cheating attempt -1.

```
Data: Datasets Ẽ and X̃
Result: Dataset 𝒦 = {K_1, ..., K_L} with L ≤ N where K_l ∈ ℝ^{M,2} and M is
        the number of evaluations for learner l;
l := 0;
for i ∈ {1, ..., N} do
    if ℰ_i = 1 then
        m := 0;
        l := l + 1;
        𝒲_i := trainModel(Ẽ_i);
        for x ∈ X_i do
            m := m + 1;
            K_i(m, 1) := ℱ(x, 𝒲_i);
            K_i(m, 2) := +1;
        end
        if i < N then
            for x ∈ X̃_{i+1} do
                m := m + 1;
                K_i(m, 1) := ℱ(x, 𝒲_i);
                K_i(m, 2) := −1;
            end
        else
            for x ∈ X̃_1 do
                m := m + 1;
                K_i(m, 1) := ℱ(x, 𝒲_i);
                K_i(m, 2) := −1;
            end
        end
        𝒦(l) := K_i;
    end
end
```

Algorithm 1: Learner sample cross-evaluation

3.3 Metrics

Dataset $\mathcal{K}$ contains the responses of a biometric tool when genuine and cheating samples are provided as input. In addition, for each response we have the expected verification value, encoded as $+1$ for genuine inputs and -1 for cheating attempts. Using this information we can compute the performance of the such biometric tool and analyze its behavior.

Given an input sample x, a learner model $\mathcal{W}_i$ and a biometric tool $\mathcal{F}$, the decision function $\mathcal{V}$ can be defined as:

$$\mathcal{V}(x, \mathcal{W}_i, Thr) = \begin{cases} +1 \text{ if } \mathcal{F}(x, \mathcal{W}_i) \geq Thr \\ -1 \text{ otherwise} \end{cases} \quad (7)$$

where $Thr \in [0, 1]$ is the threshold value that is applied to the response of the biometric tool $\mathcal{F}$. Note that this function converts the response value of a biometric tool into a final binary prediction, being $+1$ for genuine input samples and -1 for cheating samples. Considering the dataset $\mathcal{K}$, this function can be redefined as:

$$\hat{\mathcal{V}}(k, Thr) = \begin{cases} +1 \text{ if } k(1) \geq Thr \\ -1 \text{ otherwise} \end{cases} \quad (8)$$

where k corresponds to a row of $K_i \in \mathcal{K}$, and $k(1)$ is the first element of the row, containing the biometric tool response. Given a threshold value Thr, we can classify all the samples according its real label and the prediction as:

True Positive (TP): Genuine samples correctly predicted as genuine, that is, $\hat{\mathcal{V}}(k, Thr) = \mathrm{k}(2) = +1$.
True Negative (TN): Cheating attempts correctly predicted as cheating attempts, that is, $\hat{\mathcal{V}}(k, Thr) = \mathrm{k}(2) = -1$.
False Positive (FP): Cheating attempts wrongly predicted as genuine samples, that is, $\hat{\mathcal{V}}(k, Thr) = +1$ and $\mathrm{k}(2) = -1$.
False Negative (FN): Genuine samples wrongly predicted as cheating attempts, that is, $\hat{\mathcal{V}}(k, Thr) = -1$ and $\mathrm{k}(2) = +1$.

If all the responses in the dataset $\mathcal{K}$ are classified using those descriptions, the result is a confusion matrix (see Fig. 3). Note that the number of samples in each of the four cells of the matrix will depend on the selected threshold value.

Given the confusion matrix, let's define the *True Positive Rate (TPR)* as:

$$TPR = \frac{TP}{TP + FN} \quad (9)$$

the *False Positive Rate (FPR)* as:

$$FPR = \frac{FP}{FP + TN} \quad (10)$$

and the accuracy as:

$$Acc = \frac{TP + TN}{TP + TN + FP + FN} \quad (11)$$

Considering all the possible threshold values, we can build the Receiver Operating Characteristic curve or ROC curve [4] (see Fig. 4), computing the TPR and FPR for

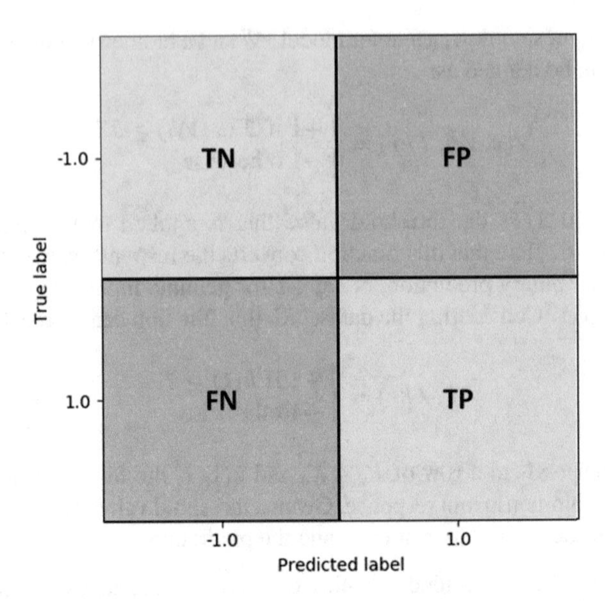

Fig. 3 Confusion Matrix. Green cells correspond to the correctly classified samples, while red ones are classification errors

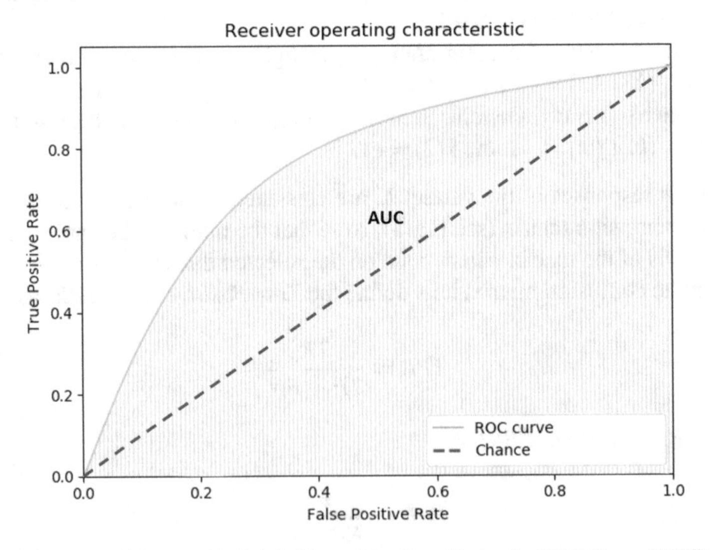

Fig. 4 ROC curve and the graphical definition of the Area Under the ROC Curve (AUC)

each threshold value. The ROC curve is a useful tool to compare the discrimination capabilities of a prediction system.

The Area Under the ROC Curve (AUC) can be interpreted as the probability that a biometric tool will rank a randomly chosen genuine sample with a higher value than a randomly chosen cheating attempt.

4 Results

In this section we present the results obtained for each biometric tool following the previously defined experimental setting and computing the provided metrics. Results are analyzed tool by tool, considering the particularities of the data and tools. As biometric tools responses are in the range [0, 1], the evaluation of all tools is done selecting a threshold value of 0.5 (see Eq. 7). Finally, using the ROC curve an optimized threshold is computed for each tool.

4.1 *Face Recognition*

The data of 267 learners is analyzed, with a total of 47, 524 images containing genuine and cheating attempts. As can be seen in Fig. 5a, where the responses of the instrument for genuine and cheating attempts are displayed for each learner, there is a clear difference between the response for genuine and cheating attempts. Looking at the histogram of responses in Fig. 5b, most of the responses for genuine samples are over 0.5, while for cheating attempts most are under 0.4.

As we can see in the ROC curve (Fig. 6), the performance of the biometric tool is near perfect, having an estimated AUC of 1.

Finally, if we analyze the decision capabilities of the face recognition tool, we see that using a threshold of 0.5 (see Fig. 7a), we obtain a an accuracy of 97.7%. Most of the failed predictions are concentrated on the false negative cell, meaning that instrument tend to reject genuine pictures, while only in few cases cheating attempts are considered genuine. We found that the optimum threshold to maximize the accuracy is found at 0.47 (see Fig. 7b), which reduce the number of false negatives, but increases the number of false positives. The optimized accuracy raises up to a value of 98.2%.

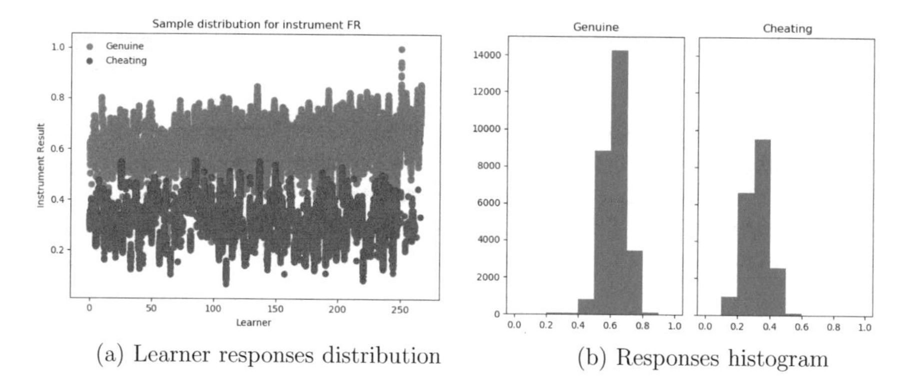

(a) Learner responses distribution (b) Responses histogram

Fig. 5 Visualization of the face recognition tool responses

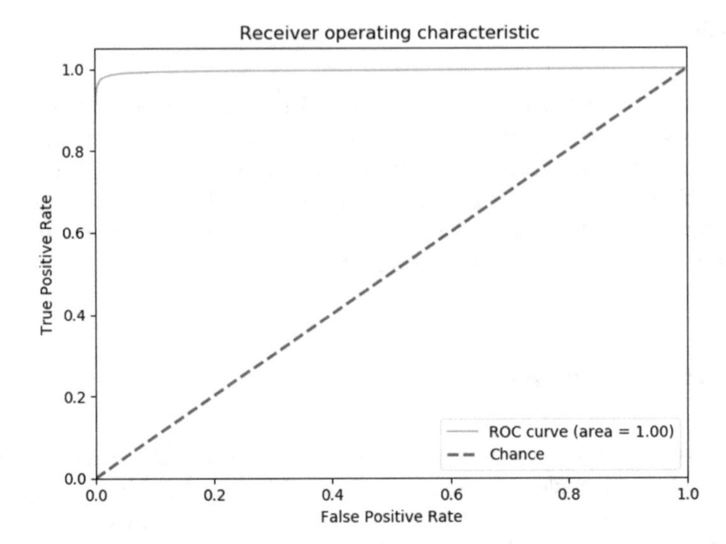

Fig. 6 ROC curve for face recognition tool

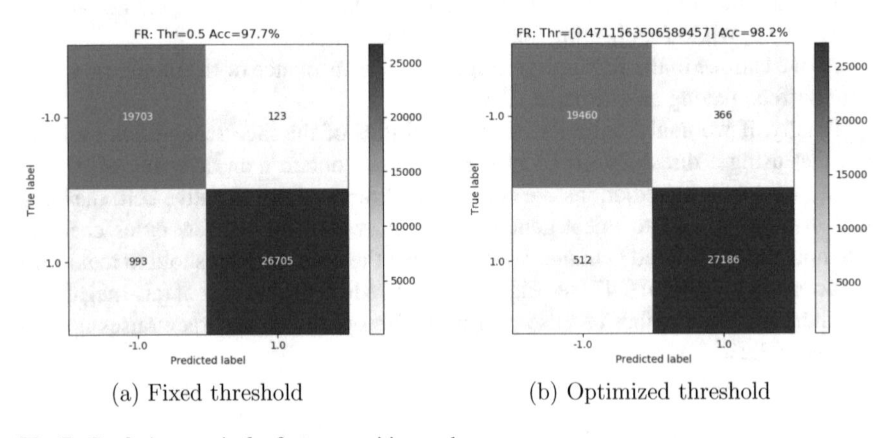

(a) Fixed threshold (b) Optimized threshold

Fig. 7 Confusion matrix for face recognition tool

The good results obtained by face recognition tool are explained in part by the use of filters previous to the use of this tool. As has been explained in this chapter, images with a bad quality or with an invalid number of faces are discarded by the filters implemented in the TeSLA system. From an instructor point of view, the results show that when a recognition value is provided, the confidence on this result is really high.

4.2 Keystroke Dynamics Recognition

For keystroke dynamic recognition we analyzes 23, 709 samples from 476 learners. If we look to the response of this biometric tool by learner (see Fig. 8a) we can see that responses for genuine samples receive higher values than the cheating attempts. However, there is a big overlapping between between both. This fact is confirmed by the histogram of responses (see Fig. 8b), where we can see that in the interval [0.4, 0.6] we have many responses from both, genuine and cheating attempts.

Looking at the ROC curve for this biometric tool (see Fig. 9), is clear that even the overlap between genuine and cheating attempts, the discrimination capabilities of the tool is quite good, obtaining a AUC value of 0.86.

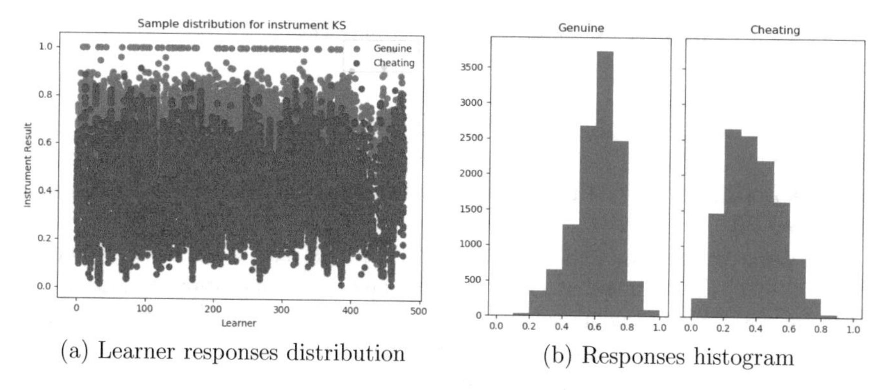

(a) Learner responses distribution (b) Responses histogram

Fig. 8 Visualization of the keystroke dynamics recognition tool responses

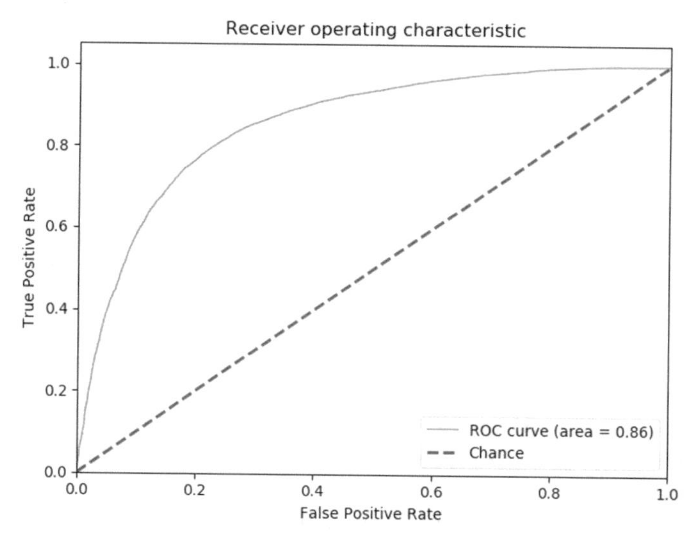

Fig. 9 ROC curve for keystroke dynamics recognition tool

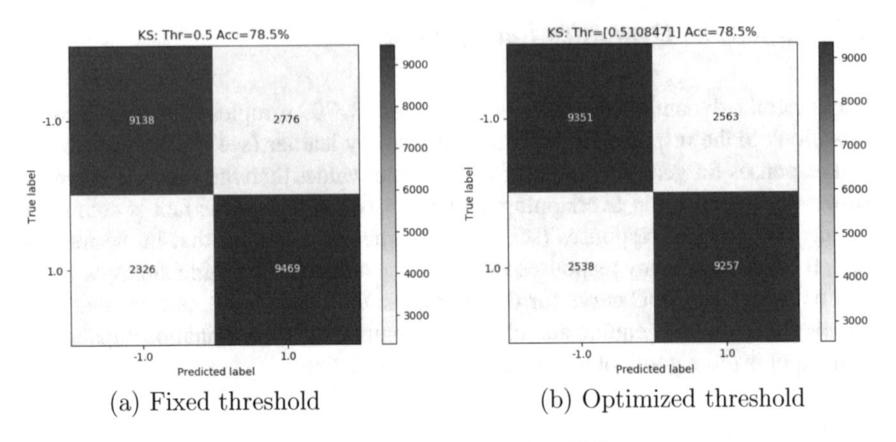

(a) Fixed threshold (b) Optimized threshold

Fig. 10 Confusion matrix for keystroke dynamics recognition tool

If we look to the confusion matrix for a threshold value of 0.5, the accuracy of this biometric tool is 78.5%, the same value obtained after threshold optimization (see Fig. 10). Looking at the confusion matrices, it can be seen that failures are balanced in FN and FP.

Keystroke features are sensible to hardware settings, that is, features captured for a learner using the keyboard of a laptop or the one of a desktop computer are slightly different. As TeSLA system captures the keystrokes from the VLE web interface, it has no access to the base hardware, and this information cannot be used to filter or adapt the features. Therefore, the biometric tool needs to deal with such changes, and this affects its performance. Nevertheless, this tool provides a good accuracy, allowing to detect cheating attempts during assessment activities.

4.3 Voice Recognition

For the voice recognition biometric tool we have analyzed 2, 905 samples from 195 learners. Analyzing the distribution of the data (see Fig. 11a) we can identify two different patterns. On the first half of the learners the genuine and cheating responses are centered on 0.5, while on the second half are centered below 0.4. In addition, on the second half responses for genuine and cheating responses are mixed. Looking at the histogram of the responses (see Fig. 11b) we see a huge overlapping between both histograms.

Looking at the ROC curve (see Fig. 12), is clear that this biometric tool is not able to classify learners between genuine and cheating. The AUC is 0.51, close to the random decision.

Looking at the predictions made by this tool using a threshold of 0.5 (see Fig. 13) we can see that most of the genuine samples are detected as cheating attempts, obtaining a final accuracy of only 38.6%. Optimizing the threshold, we found that

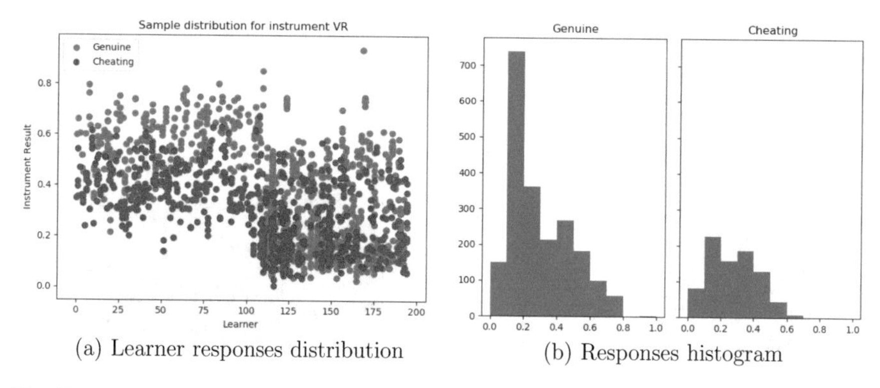

(a) Learner responses distribution (b) Responses histogram

Fig. 11 Visualization of the voice recognition tool responses

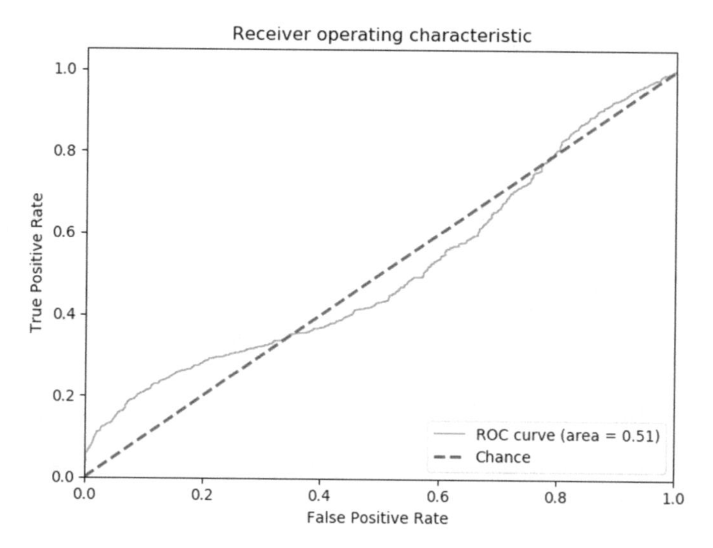

Fig. 12 ROC for voice recognition tool

the best threshold is 0.25, with an accuracy of 46.6%. The threshold optimization try to correct the classification of the genuine samples, but since the response of genuine and cheating attempts are overlapped, the final value is under the random decision, and therefore is not a useful prediction.

With an analysis of the data, we discovered that the two patterns in the biometric tool responses (see Fig. 11a), correspond to learners from an on-line institution and from a blended institution. From the original dataset, we divided the data into two different datasets, one for the on-line institution with 107 learners and 616 samples, and the other with 83 learners and 2, 186 samples. The data distribution for the two datasets is shown in Fig. 14.

Ploting the ROC curve for both datasets in a separate way (see Fig. 15), in the case of on-line institution the biometric tool have a good predictive value, with a AUC

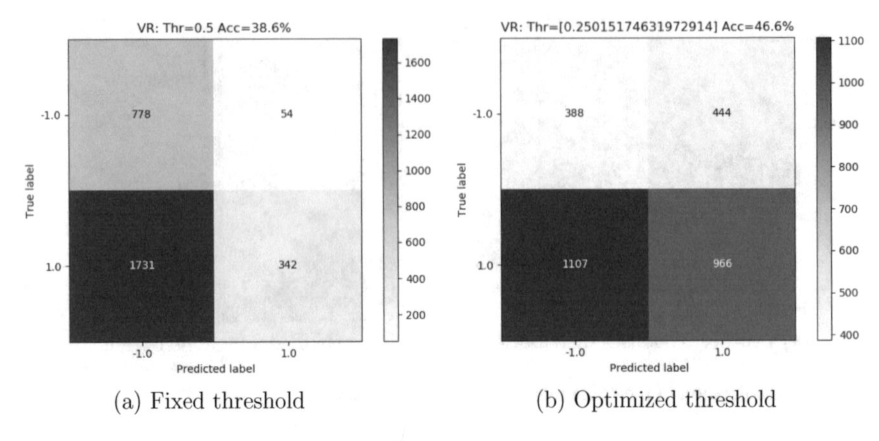

Fig. 13 Confusion matrix for voice recognition tool

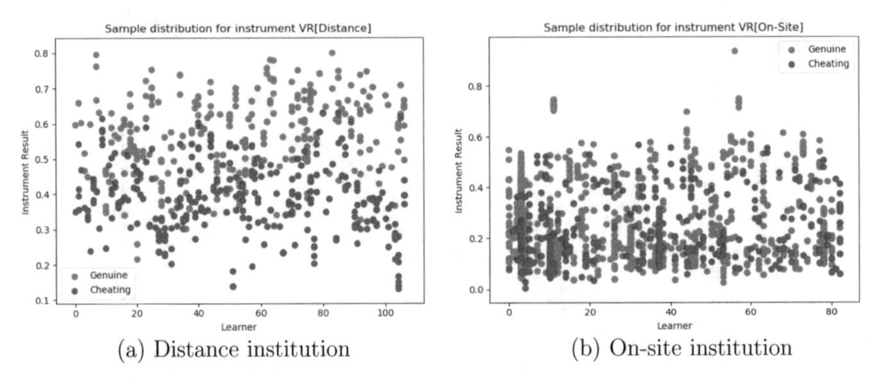

Fig. 14 Voice recognition results distribution for genuine data and cheating attempts taking into account the institution type

value of 0.88. In the case of the blended institution, although the AUC value of 0.58 is better than the previous one, it is still close to the random decision.

If we look to the decisions of this tool for both datasets, we can see that for on-line institution the biometric tool achieve an accuracy of 78.9% with threshold 0.5 and 80% with threshold optimization. However, for face to face institution most of the genuine samples are classified as cheating attempts. In this second case, the maximum accuracy achieved after threshold optimization is 53%, close to the random decision. We randomly selected 10 samples of the on-site dataset and listen the captured audios. We realized that in all of them there is more than one person speaking, as the activities were performed in a classroom, with all learners performing the activity. The conclusion is that the biometric tool is working as expected, since genuine samples are not actually genuine as more than one people is speaking and therefore to reject those samples is a correct behaviour (Fig. 16).

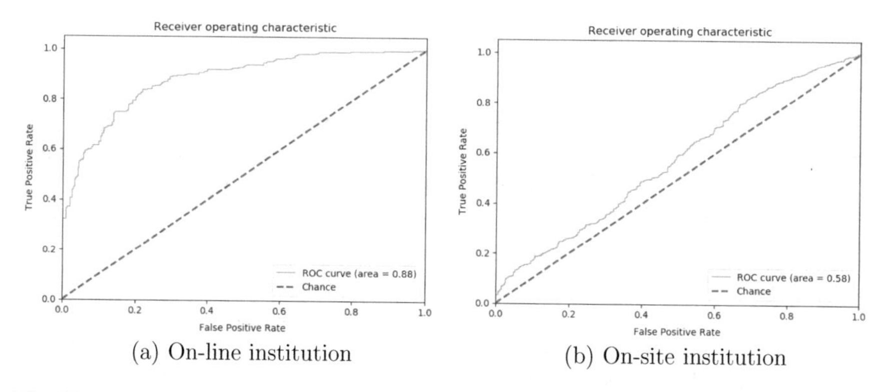

Fig. 15 ROC for voice recognition tool taking into account the institution type

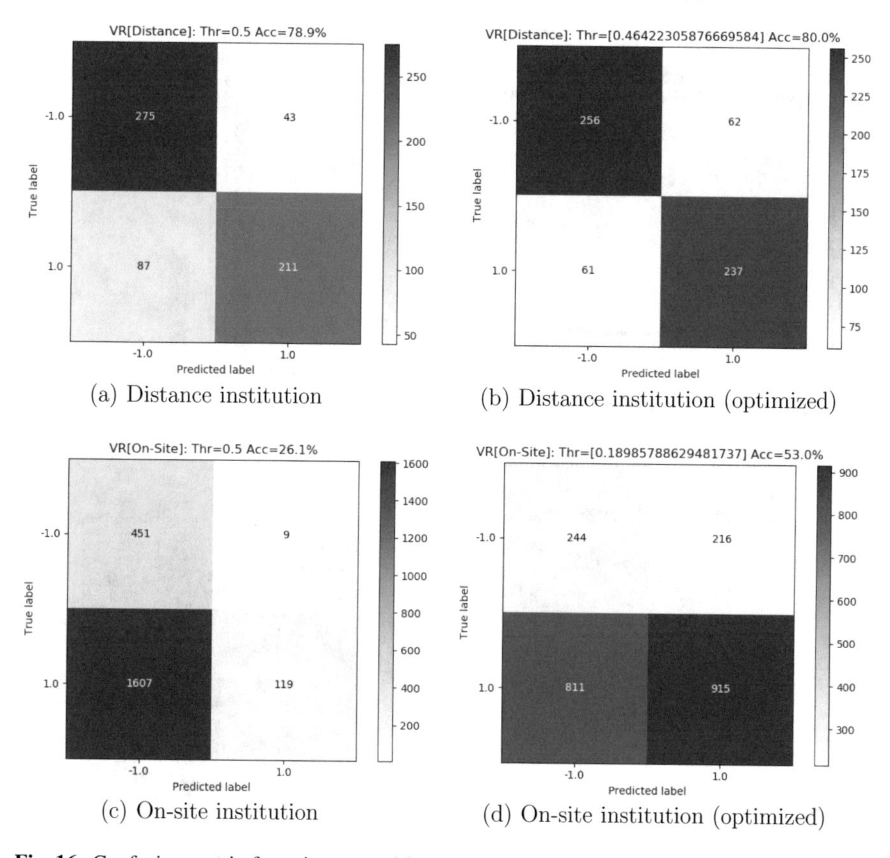

Fig. 16 Confusion matrix for voice recognition tool taking into account the institution type

4.4 Forensic Analysis

For this biometric tool we analyze 606 samples from 67 learners. In contrast with the previous tools, in this case the response of the tool is not a value between 0 and 1, but a binary decision {0, 1}. The response of this tool for all the learners is shown in Fig. 17. We can see that in general genuine documents receive a high value, while many cheating attempts are not detected.

Looking at the ROC curve (see Fig. 18a), we see that this biometric tool have an AUC of 0.71, lower than the previous tools, but over the random decision. As the output of this tool is binary, the threshold is fixed at 0.5 and the confusion matrix is shown in Fig. 18b. As stated in the histogram of the responses, this tool tends to consider documents as genuine. The final accuracy of this tool is 75.2%.

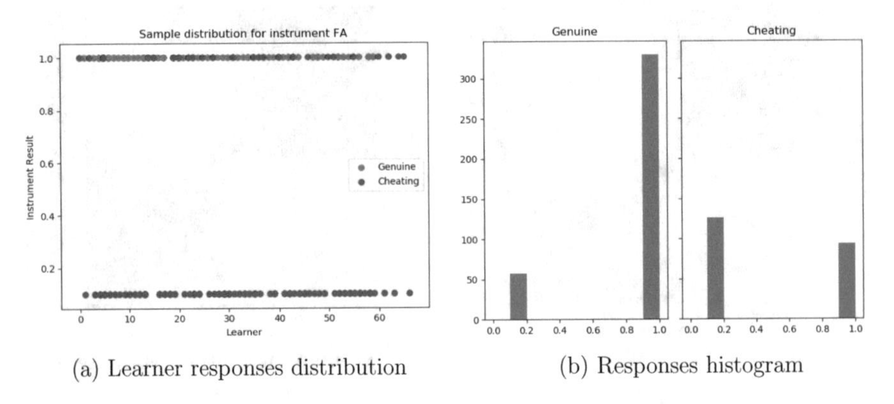

(a) Learner responses distribution (b) Responses histogram

Fig. 17 Visualization of the forensic analysis tool responses

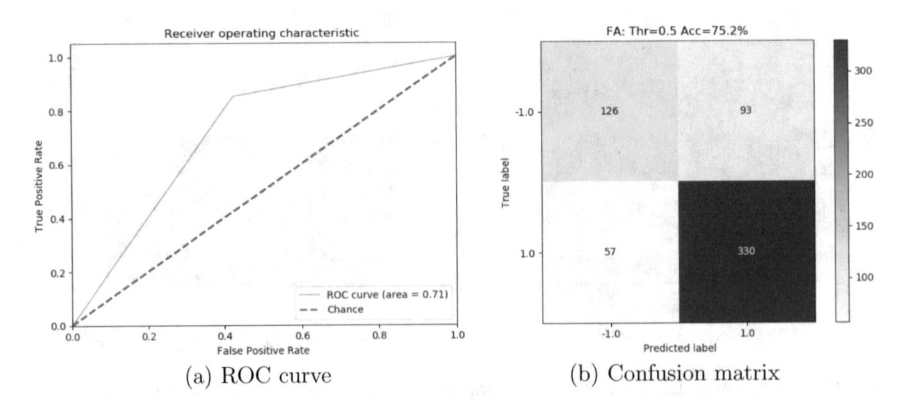

(a) ROC curve (b) Confusion matrix

Fig. 18 ROC curve and confusion matrix for forensic analysis tool

Table 3 Summary of the metrics for all biometric tools

Instrument	AUC	Accuracy (%)	Accuracy Opt. (%)
Face recognition	1	97.7	98.2
Keystroke dynamics recognition	0.86	78.5	78.5
Voice recognition	0.51	38.6	46.6
Voice recognition (Distance)	0.88	78.9	80
Voice recognition (On-site)	0.58	26.1	53
Forensic analysis	0.71	75.2	75.2

5 Conclusions

In this chapter we have analyzed the results of the instruments used as part of the TeSLA system with real data captured during the pilots. Biometric systems are mainly designed to work in security applications, usually with a controlled environment. Porting those tools to an educational environment has been a challenge, and require to adapt the basic philosophy of the instruments. In a security scenario, those tools must ensure the identity and are optimized in order to minimize the False Positive ratio. When those tools are used in an educational application like the TeSLA system, those tools need to deal with uncontrolled scenarios, and must optimize the False Negative ratio. That is, while in a security application is accepted to deny the access if we are not sure of the identity, in the case of an educational application we prefer to miss a cheating attempt in front of classifying as cheating a genuine activity.

The numeric results are summarized in Table 3. We can state that in all the cases the accuracy is over the 75%, which has been shown to be an acceptable value for this application, specially if multiple biometric tools are used simultaneously. In the case of Voice Recognition we found that this tool cannot be used in classrooms or public spaces where multiple people are speaking.

Acknowledgements This research was supported by RTI2018-095232-B-C22 grant from the Spanish Ministry of Science, Innovation and Universities (FEDER funds), and NVIDIA Hardware grant program.

References

1. Asha S, Chellappan C (2008) Authentication of e-learners using multimodal biometric technology. In: International symposium on biometrics and security technologies, 2008. ISBAST 2008, pp 1–6
2. Dalal N, Triggs B (2005) Histograms of oriented gradients for human detection. In: Proceedings of the 2005 IEEE Computer Society conference on computer vision and pattern recognition

(CVPR'05), vol 1', CVPR '05. IEEE Computer Society, Washington, DC, USA, pp 886–893. https://doi.org/10.1109/CVPR.2005.177

3. Dehak N, Kenny PJ, Dehak R, Dumouchel P, Ouellet P (2011) Front-end factor analysis for speaker verification. Trans Audio Speech Lang Proc 19(4):788–798. https://doi.org/10.1109/TASL.2010.2064307
4. Fawcett T (2006) An introduction to ROC analysis. Pattern Recogn Lett 27(8):861–874 (ROC Analysis in Pattern Recognition) http://www.sciencedirect.com/science/article/pii/S016786550500303X
5. Ferreira J, Santos H, Patrao B (2011) Intrusion detection through keystroke dynamics. In: Communications and multimedia security. Springer, Berlin, Heidelberg, pp 81–90
6. Frantzeskou G, Stamatatos E, Gritzalis S, Chaski C, Stephen Howald B (2007) Identifying authorship by byte-level n-grams: the source code author profile (SCAP) method. In: IJDE, vol 6
7. Ivanova M, Bhattacharjee S, Marcel S, Rozeva A, Durcheva M (2018) Enhancing trust in eAssessment—the TeSLA system solution. In: Technology enhanced assessment conference
8. Kambourakis G, Damopoulos D (2013) A competent post-authentication and non-repudiation biometric-based scheme for m-learning. In: Proceedings of the 10th IASTED international conference on web-based education (WBE 2013). ACTA Press, Innsbruck, Austria, pp 821–827
9. Kazemi V, Sullivan J (2014) One millisecond face alignment with an ensemble of regression trees. In: Proceedings of the IEEE conference on computer vision and pattern recognition, pp 1867–1874
10. Khoury E, Kinnunen T, Sizov A, Wu Z, Marcel S (2014) Introducing I-vectors for joint anti-spoofing and speaker verification. In: The 15th annual conference of the International Speech Communication Association
11. Kutafina E, Laukamp D, Bettermann R, Schroeder U, Jonas SM (2012) Wearable sensors for elearning of manual tasks: using forearm EMG in hand hygiene training. Sensors (Basel) 16(8):1221
12. López-Monroy AP, Montes-y Gómez M, Escalante HJ, Villaseñor-Pineda L, Stamatatos E (2015) Discriminative subprofile-specific representations for author profiling in social media. Knowl-Based Syst 89:134–147
13. Pastor López-Monroy A, Montes-y Gómez M, Escalante HJ, Villaseñor-Pineda L, Stamatatos E (2015) Discriminative subprofile-specific representations for author profiling in social media. Knowl-Based Syst 89:134–147
14. Monaco J, Bakelman N, Cha S-H, Tappert C (2013) Recent advances in the development of a long-text-input keystroke biometric authentication system for arbitrary text input. In: Intelligence and security informatics conference (EISIC), 2013 European, pp 60–66
15. Peralta D, Galar M, Triguero I, Paternain D, García S, Barrenechea E, Benítez JM, Bustince H, Herrera F (2015) A survey on fingerprint minutiae-based local matching for verification and identification: taxonomy and experimental evaluation. Inf Sci 315:67–87
16. Pinto P, Patrão B, Santos H (2014) Free typed text using keystroke dynamics for continuous authentication. In: De Decker B, Zúquete A (eds) Communications and multimedia security. Springer, Berlin, Heidelberg, pp 33–45
17. Pirbhulal S, Zhang H, Mukhopadhyay SC, Li C, Wang Y, Li G, Wu W, Zhang Y-T (2015) An efficient biometric-based algorithm using heart rate variability for securing body sensor networks. Sensors (Basel) 15(7):15067–15089
18. Prasad NV, Umesh S (2013) Improved cepstral mean and variance normalization using Bayesian framework. In: 2013 IEEE workshop on automatic speech recognition and understanding, pp 156–161
19. Rabiner LR, Schafer RW (2007) Introduction to digital speech processing. Found Trends Signal Process 1(1):1–194. https://doi.org/10.1561/2000000001
20. Rabuzin K, Baca M, Sajko M (2006) E-learning: biometrics as a security factor. In: 2006 international multi-conference on computing in the global information technology - (ICCGI'06), pp 64–64

21. Sapijaszko GI, Mikhael WB (2012) An overview of recent window based feature extraction algorithms for speaker recognition. In: 2012 IEEE 55th international midwest symposium on circuits and systems (MWSCAS). IEEE, New York, pp 880–883
22. Schroff F, Kalenichenko D, Philbin J (2015) Facenet: a unified embedding for face recognition and clustering. In: 2015 IEEE conference on computer vision and pattern recognition (CVPR), pp 815–823
23. Sumithra M, Devika A (2012) A study on feature extraction techniques for text independent speaker identification. In: 2012 international conference on computer communication and informatics (ICCCI), pp 1–5
24. Tata S, Patel JM (2007) Estimating the selectivity of TF-IDF based cosine similarity predicates. SIGMOD Rec 36(2):7–12. https://doi.org/10.1145/1328854.1328855
25. Wagner A, Wright J, Ganesh A, Zhou Z, Mobahi H, Ma Y (2012) Toward a practical face recognition system: robust alignment and illumination by sparse representation. IEEE Trans Pattern Anal Mach Intell 34(2):372–386
26. Wechsler H, Phillips JP, Bruce V, Soulie FF, Huang TS (2012) Face recognition: from theory to applications, vol 163. Springer Science & Business Media, Berlin
27. Lee Y, Chen J, Tseng CW, Lai S-H (2016) Accurate and robust face recognition from RGB-D images with a deep learning approach. In: Richard ERH, Wilson C, Smith WAP (eds) Proceedings of the British machine vision conference (BMVC), pp 123.1–123.14
28. Zhong Y, Deng Y (2015) A survey on keystroke dynamics biometrics: approaches, advances, and evaluations. Recent advances in user authentication using keystroke dynamics biometrics. Science Gate Publishing

Engineering Cloud-Based Technological Infrastructure

Josep Prieto-Blazquez and David Gañan

Abstract In the last decade cloud development has grown exponentially and increasingly companies and institutions decide to make the leap and provide their solutions into a cloud platform. This is mainly because of the evident benefits of using cloud-based solutions like high availability, scalability or redundancy, which are impossible or too much costly to be achieved on-premises. Cloud platforms may reduce considerably the costs and difficulty of maintenance, while offering flexible solutions to each need (Talukder and Zimmerman in Computing. computer communications and networks. Springer, London, 2010 [1]). Enterprises have a lot of services available to their needs, some of them very basic like a database, a web site or storage, but also many other advanced services for business intelligence, big data, artificial intelligence and other emerging technologies. However, from the architectural point of view, cloud-based solutions suppose some challenges not existing in traditional on-premises solutions that must be tackled during the engineering process. Whether a monolithic application may fit well on-premises, cloud environments work better with modularization and separation of services. This separation and distribution of services requires additional work to define communications between components and provide security techniques to guarantee data protection against potential risks like data loss, privacy breaks, unwanted access, etc. (Ghaffari et al. in 2016 8th international symposium on telecommunications (IST). IEEE, pp 105–110, 2016 [2]). Performance is also critical in cloud-based solutions, especially when services are located in different nodes on the internet. All these concerns suppose a quite big learning curve to the technical team that prevents some enterprises to move to cloud, especially when they try to migrate already existing solutions. This chapter discusses about the ins and outs of engineering a cloud-based solution applied to a real use case.

Keywords Engineering · Cloud · Architecture · Security · Performance

---

J. Prieto-Blazquez (✉) · D. Gañan
EIMT-UOC Rambla Poblenou, 156, 08018 Barcelona, Spain
e-mail: jprieto@uoc.edu

D. Gañan
e-mail: dganan@uoc.edu

D. Baneres et al. (eds.), *Engineering Data-Driven Adaptive Trust-based e-Assessment Systems*, Lecture Notes on Data Engineering and Communications Technologies 34,
https://doi.org/10.1007/978-3-030-29326-0_4

1 Introduction

Cloud platform major benefits are related with the easiness of setting up an infrastructure in the cloud comparing with setting up on-premises. In a cloud environment you can deploy a new machine or service in question of minutes, while doing the same on-premises can take considerably more time (purchasing the hardware, installing it physically, setup operating systems, etc.), and it can be considerably cheaper in cloud [1].

Furthermore, cloud environments enable vertical and horizontal scaling when you require it, in just few clicks and return back to the previous state with the same easiness, while doing the same thing with owned hardware will be unthinkable.

Additionally to the ease of use, cloud platforms also offer many other services that can help in the accomplishment of the business requirements or even enable new business lines and opportunities.

However, deploying an application in the cloud requires some additional considerations, the most important related with security issues. This is why some companies are reticent to migrate their solutions to the cloud, especially if those solutions are monolithic systems that do not scale well. Another common reason to not migrate to cloud environments is the perception of the loss of control over data.

As another reason that prevents companies to migrate from monolithic infrastructure on-premises to a cloud environment is that it supposes generally an investment in development in order to implement secure communication protocols, data encryption, and separation of modules and making them interoperable, enable modules to work in parallel with other instances, etc. This is why starting a new project oriented for the cloud is easier than migrating an existing one.

The rest of the chapter is structured as follows. Section 2 explores the current bibliography about engineering and architecting solutions for the cloud. Section 3 describes TeSLA system (Trust-based authentication and authorship e-assessment analysis) as a use case of a system developed for cloud environments, and Sect. 4 discusses about the particularities of the system because of the decision of deploying it in the cloud. Section 5 summarizes some conclusions and Sect. 6 closes the chapter with some further work of the presented use case.

2 Background

Cloud platforms offer multiple services of different types that can be categorized in many ways, but in general there are three distinguished layers or service models (XaaS services) [3, 4]:

SaaS: Software as a Service

In this model the software that typically is installed on client machines is provided from cloud infrastructures. Some well-known examples of services in this layer are Google Gmail and other tools, Microsoft Office 365, or Dropbox among many others.

PaaS: Platform as a Service

Cloud platforms offer all the tools for developers to create applications without the need to worry about infrastructure, like databases, web hosting, storage or even more complex services like IA processing, big data analysis, biometric authentication or cognitive services.

IaaS: Infrastructure as a Service

This model enables companies to create a hardware infrastructure in the cloud, and allows creating virtual machines (either shared or dedicated), connections, storage, etc.

Some contributions differentiate some services into more detail and talk about security as a service, storage as a service or testing as a service among others [5].

More or less all the reviewed contributions [3–6] coincide in identifying the main advantages and issues of cloud platforms, the most important are the easy of management and the possibility of reducing costs. The scalability of components, the uninterrupted service or the characteristics for disaster management are also mentioned as the main advantages of the cloud.

On the other hand, the most important issues mentioned are related with security and privacy. These issues are covered in detail in many contributions like [7] which makes an exhaustive review of literature in cloud security issues and classifies them into different categories like software, storage and computing, virtualization, etc. [8] includes some of these limitations and proposes some ideas to mitigate them. Ghaffari et al. [2] also talks about security issues and proposes a reference model to control security in cloud environments aligned with some security standards.

In [9, 10] the aspects related to the migration of existing solutions to cloud environments are discussed. Both contributions define some refactorings or changes required to adapt existing architectures for the cloud.

Finally, there are many contributions presenting examples of cloud architectures or framework proposals, and application case studies. As an example, in [11] the authors define a generic framework for enterprise applications in the cloud and [12] describes an architecture for the distribution of multimedia content to mobile devices. There are many other examples of contributions about cloud applications in multiple fields like education or medicine, as an example, the authors in [13] present the architecture of an e-learning system deployed on the cloud, and in [14] the authors explain a framework based on cloud for monitoring student's health.

3 Approach

The chosen methodology to present the specific details of a cloud-oriented architecture consists in the definition and design of a real scenario. The chosen example is the TeSLA system (Trust-based authentication and authorship e-assessment analysis) [15], a framework aimed to detect cheating in online education environments.

The following sections describe the system requirements, the architecture schema, components and interactions with each other, and discuss the specific details as a cloud-architecture.

3.1 System Specifications

The main goal of the system to be constructed is to analyze student's activities in online education in order to detect cheating, like plagiarism or that the activity is done by someone other than the student. The system must support a large number of concurrent students and teachers, provided that online institutions usually have a big community with tens of thousands of members.

There are different tools called instruments that allow the identification of people. Some identification mechanisms of the instrument can be transferable such as a card and an access code; or can be not transferable such as biometric data. Biometric instruments are valid to check if it is the student who really makes the activity, including facial recognition, voice recognition, keyboard pattern recognition or analysis of the style of writing. In terms of copy detection there are also different tools known as anti-plagiarism.

Most of these identification instruments require an initial training process or enrolment that each student must carry out and that allows capturing their characteristic features for later identification. During the process of analyzing the data captured during the activity (called verification), the instruments compare the gathered data with the initial model captured during enrollment and return a probability value about if the student is who is doing the activity.

The main use cases of the system are the following:

- Enrolment of student. The system captures the student's biometric features.
- Student verification. The system captures information of the student during the process of realization of the activity and analyzes if they fit with the data collected during the enrollment process.
- Configuration of activities. The teacher configures the activities to be carried out and the instruments that will be used to identify students.
- Query results. The teacher visualizes the results of the instruments for the different activities and students.

3.2 Global Architecture

The proposed system is composed by the following components (represented in Fig. 1)

- TeSLA Portal (or Portal for short): handles the licensing/enrolment, deployment and statistics.

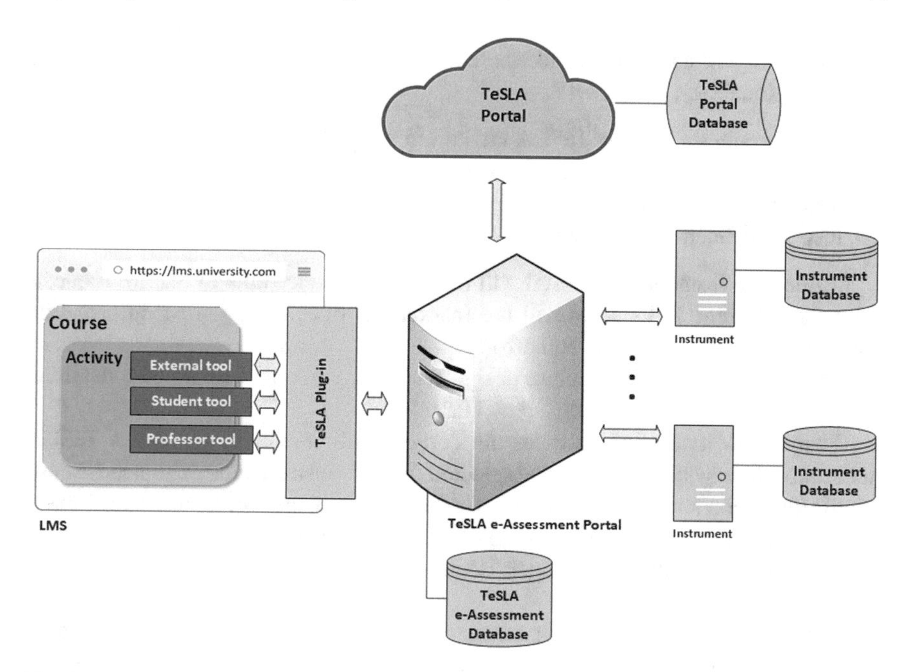

Fig. 1 System global architecturesd°

- TeSLA E-Assessment Portal (TEP for short): acts as a service broker, receiving requests from the Virtual Learning Environment (VLE) plug-ins and forwarding that request to the appropriate instruments and/or the Portal.
- TeSLA Instruments: An instrument evaluates a given sample in order to access the Learner identity/authorship.
- TeSLA Plug-in: generates TeSLA ID used for anonymization, gathers information about the VLE and communicates with TEP.
- External tool: collects data samples to provide to the instruments (e.g. video, sound or keystrokes).
- Learner tool: interacts with the learner using TeSLA.
- Instructor tool: interacts with the instructor using TeSLA.

Next subsections describe each component with more detail.

TeSLA Portal

This component will be responsible of license management and package deployment. For this reason this component is independent from any institution, only one instance will manage all the specific deployments of each institution using TeSLA.

TeSLA Portal will also gather evaluation information and produce statistics with it (instrument calls, responses and results). This information will be useful for the monitoring of the system and to give feedback about the instruments behavior and performance.

The Portal will have a frontend where a user may interact with it, providing three main functionalities:

1. Accept requests for new TeSLA institution setup/deployment;
2. Provide access to statistical information about instruments to instrument developers and provide access to log errors from all TeSLA systems;
3. Institution license management.

Besides the frontend, the Portal will have a web service able to receive requests from the different TEPs across all the Institutions in order to collect information about the evaluation requests and errors.

Figure 2 shows the architecture design for TeSLA Portal, with its own database and communication with the different TEPs.

When a new institution accesses the Portal it will be able to generate a TeSLA license for their systems (VLEs and selected features/instruments). The license contains information about each VLE and, for each of them, the corresponding information about each instrument available to be used by it. This license may be changed at any time, by the Institution manager, issuing a new license, either to add or remove VLEs/instruments.

After a license is created, two packages will be made available to be deployed by the Institution manager: a VLE Package, with a set of components to be integrated within the different VLEs and a Service Package, with a set of components to be installed in the Institution servers.

The license will also contain digital keys, for signing and protection, and an expiration date. This enables each request from the TEP to the instruments and the requests from TEP to the Portal to be validated using the license key. The license is updated by the Institution using a new license provided in TeSLA portal and replacing the old license by the new one in TEP. Figure 3 represents a simple institution deployment overview.

TeSLA Portal will require an authentication method in order to have users securely authenticating in the system. We propose a standard authentication protocol (e.g.

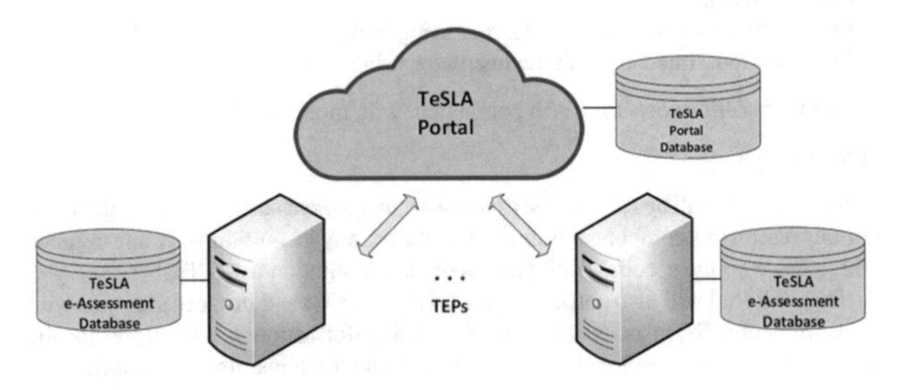

Fig. 2 TeSLA Portal architecture design

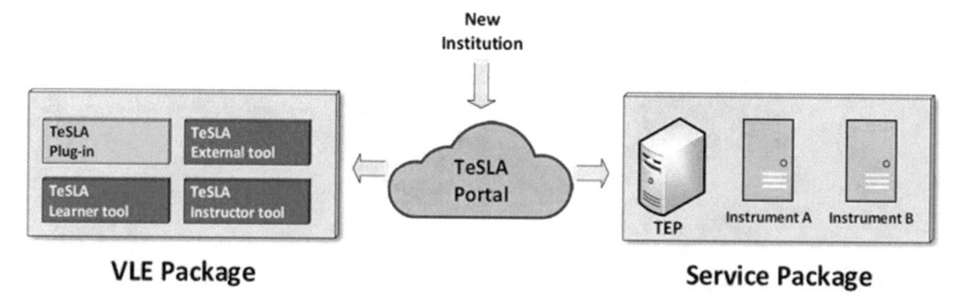

Fig. 3 Institution deployment overview

OpenID with Google, Yahoo, BlogSpot, Flickr, Myspace, Mozilla or WordPress, etc. as an identity provider or SAML with Shibboleth as identity provider).

The use of these authentication protocols with the respective identity providers (http://openid.net/, http://wiki.oasis-open.org/security) will allow the partner responsible for TeSLA Portal to focus its development time in the main Portal responsibilities, removing the effort to support the user account management. Also, these authentication protocols will allow users to use their already existing accounts and take advantage of data protection mechanisms of the large-scale operators as well as the authentication levels that they might offer (e.g. mobile authentication).

E-Assessment Portal (TEP)

TEP will work as a central component of TeSLA System. It will act as a broker in the system, receiving requests from the different Institution's VLEs and forwarding them to their respective instruments. During this process the TEP will validate its license, checking the legitimacy of the request source and if that resource can use the required instrument.

It must be possible to have multiple TEP instances sharing the same data storage and handling requests in order to ensure scalability. This may be accomplished using a load balancer to distribute the requests through those multiple instances. Figure 4 shows the TEP design using this approach, with a load balancer and multiple TEP instances.

TEP will receive requests from the different Institution's VLEs and from the instruments. After receiving a request from the VLEs, TEP will:

1. Validate the Institution associated license(s);
2. If the VLE has permission to use TeSLA and if the requested instrument is enabled for that VLE then it will save the request information in its own database;
3. After validating the license(s), it will send an evaluation request to the respective instrument;
4. Receive a reply from the instrument (valid/invalid sample/error code) and save the evaluation status in the database (in progress/error code).

After an evaluation request from the VLE, TEP will send to TeSLA Portal information about the request. This request information will be used by the Portal to generate statistics about the instruments.

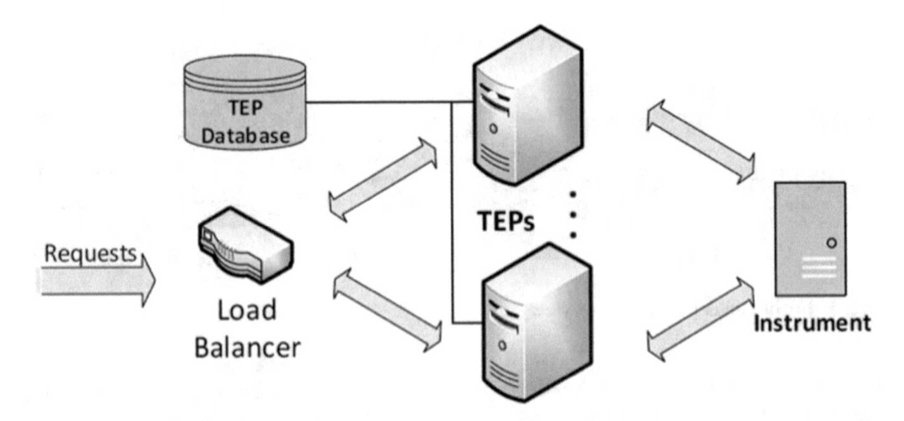

Fig. 4 TEP design

The TEP should also be able to receive evaluation results and update requests from the instruments. This update should be sent to TeSLA Portal with a masked TeSLA ID (original TeSLA ID encrypted using a secret key chosen by the Institution).

Like TeSLA Portal, TEP should be ready to provide information about the system status, usage and result information in order to allow the Institutions to better monitor it. For that purpose, this feature, both in TEP and TeSLA Portal, should be developed as close as possible, given the similarities between them.

The communications between TEP and TeSLA Portal, the different instruments and VLEs are done using RESTful web services to ensure interoperability.

The TEP should be able to adapt to multiple instances of the same instrument, acting as a load balancer for the instruments, e.g. if two instances of the same instrument are listening to requests, then the TEP should split requests between them. If for some reason the workload in an instrument is too high and the Institution decides to add a new instance for that instrument the load balancer should be able to start working with it.

The TEP will handle the learner TeSLA agreement and enrolment; accept new evaluation setups (enable instruments per activity); receive requests for the evaluation results. The TEP should be also prepared to handle with SEND (Special Educational Needs and Disabilities) learners.

At the beginning of an activity, TEP will check if the learner has any SEND and will filter the enabled instruments with the ones that he cannot use. When the instructor requests the learner TeSLA evaluation results he will be informed about the learner SEND, if necessary, and which instruments where disabled in the activity for that SEND learner.

For that purpose, a SEND learner will have a SEND code to describe the SEND and a privacy level that will define if the instructor can or cannot have access to the SEND detailed information. With this, a SEND learner may be treated with special attention by TeSLA System.

Each SEND will have a code, a description and a list of instruments that cannot be used by learners with that SEND.

Instruments

The Instrument will be responsible for all the evaluations requested by TEP and each instrument has three main responsibilities:

1. Receive requests and store them,
2. The instrument must have a thread pool selecting evaluation requests from the database in order to produce a result for each of them,
3. After evaluating them, the result must be sent to the TEP.

A Web Service will be responsible for receiving the requests and stored them in a database ordered by date. The instrument will have at least one thread selecting the oldest evaluation request from the database and processing it. This process is equivalent to a FIFO queue. After the evaluation is completed, the request state is updated and the result must be sent to the TEP.

If required, multiple instances of the same instrument with one or more threads may be running at the same time in order to better serve a high load of evaluation requests. Figure 5 shows the design for the instrument server, along with the web service with multiple threads, services and a database.

When a new request arrives, the instrument must be able to detect if the sample is valid or not for evaluation and reply to the TEP that information, i.e. if the received data is good enough to produce an evaluation (e.g. invalid sample size, no face detected, no sound detected, etc.). This should be done by the instrument as quick as possible (simple algorithm) in order to avoid possible high overload on the system.

Even if the evaluation algorithm does not support multi-threading the web service must be able to receive multiple evaluation requests at the same time. With a web

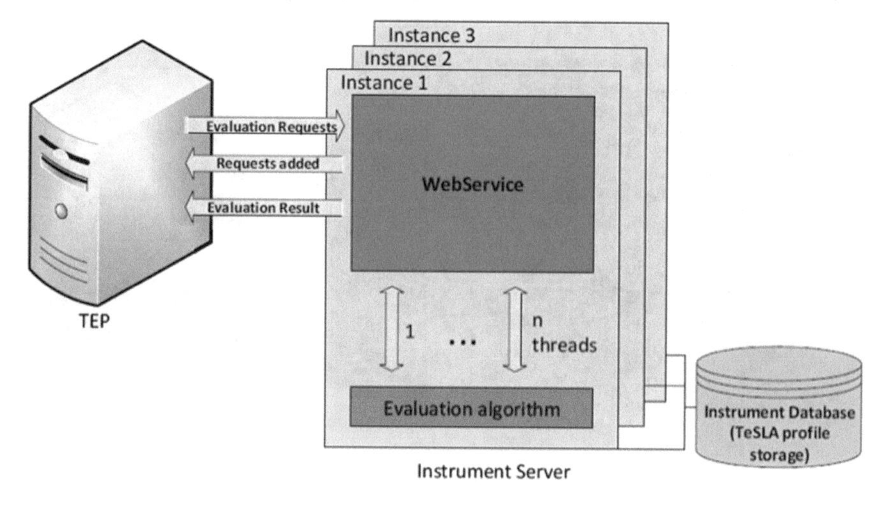

Fig. 5 Instrument server design

service receiving the requests this should not be a problem since it is prepared to handle multiple requests.

If the instrument is able to do multiple evaluations at the same time, a pool of threads should be running, selecting requests from the database and doing the evaluations.

After processing each request and after sending the result to TEP (the score and audit data), the request should be deleted from the request table. The only information that should be kept stored on the Instruments database is the learner profile information from the enrolment.

If the requests are too numerous for the instrument server to handle, then TeSLA System should be prepared to accept a new instrument instance. Multiple instances for the same instrument must share the same data storage and each instrument worker thread must be prepared to evaluate a sample and report the result to the TEP.

Plugin/tools

The plugins/tools are the components to be integrated in the different VLEs in order for them to support TeSLA and communicate with TEP for sending enrolment and verification requests.

Each tool will have its own responsibility and is to be used in different user interactions in the VLEs (Fig. 6 shows the architecture for the Plugins/tools).

TeSLA Plug-in is responsible for all the communication between the VLE and the TEP, so the External, Learner and Instructor Tools must send their requests to the TEP through TeSLA plug-in.

One of the most important things about this Tools and Plug-in is the fact that they need to be supported by the different VLEs. For convenience purposes, this plug-in and tools will be manually integrated in the VLE. In the final version of TeSLA System, all calls from the course scope will be able to be performed through LTI. For calls performed in the activity scope a build-in plug-in allowing LTI calls from

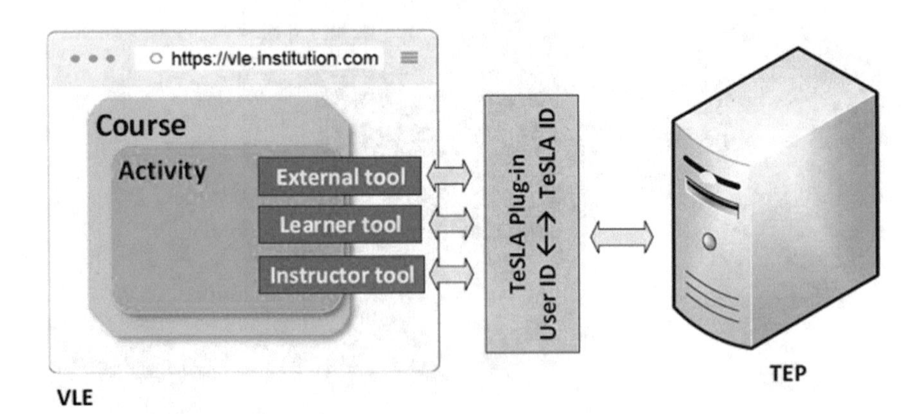

Fig. 6 Plugins/tools design

activity is required and has to be implemented as a VLE plug-in. The LTI support will guarantee a good system lifespan.

As mentioned before, the TeSLA plug-in will be responsible for communicating with TEP. Besides that, the TeSLA plug-in will also be responsible for providing information about the VLE to the tools (e.g. learner and Institution identification, course information, activity information or the VLE type).

The plug-in must also create the TeSLA ID, applying a sha256 hash to the original VLE user ID [16], and store it on TEP database. The TeSLA ID will be the learner identification used by the TEP and the Instruments. It should be created right after the agreement acceptance and right before starting the TeSLA enrolment phase.

External tools

The external tool will be responsible for collecting sample data from the user, pre-process that sample data and send it to the TEP using the TeSLA plug-in. This data gathering will be minimal and non-intrusive and the only thing the learner will receive is feedback if something went wrong (e.g. an audio icon in the page corner when the TeSLA receives an invalid sound or photo sample).

The data gathering will only begin once the learner accesses to the online activity. The learner is always informed, at the beginning of an activity with TeSLA, about what data will be collected and will have to the option to accept the agreement if not accepted/rejected yet.

Instructor tool

The instructor tool will have the responsibility to configure an activity with TeSLA and display the TeSLA evaluations of a learner's activity. During the configuration the instructor may choose which instruments he wants to use in the specific activity.

The instructor tool also allows the instructor to see the TeSLA evaluation results obtained from the instruments. The instructor will be able to navigate between courses and activities, and view for each activity a list of the students, and the corresponding evaluation results for each one. These results are not a true or false result but a confidence rating, the biggest the better. For some instruments the instructor might be able to see the audit data (e.g. the photo used in the evaluation).

Learner tool

The learner tool main responsibility is to inform the learner about the use of TeSLA System in an activity. The learner will be informed about which instruments are enabled for that activity and might see information about these instruments. The learner may accept to use TeSLA or not when accepting or rejecting the agreement. This acceptance will only happen once per agreement version. The learner tool will also handle the learner enrolment process.

At the beginning of an activity a request for learner agreement and TeSLA activity configuration information is sent to the TEP. Then, the TEP will check if the activity has TeSLA enabled and if the learner has already accepted the TeSLA agreement.

The learner information will be sent to the learner tools as response to the first request and two things might happen, either the learner already accepted the agreement for that activity or not. If the learner accepted, the external tool will start gathering data. If the learner had not accepted, then he will be prompted to accept it.

At the end two things may happen, either the learner accepts the agreement being a request sent to the TEP to update the learner agreement and the external tool then starts gathering data or the learner does not accept the agreement and TeSLA will use the alternative defined by the instructor to that activity (e.g. leave the activity or be informed that he will need to take an extra assessment in the institution facilities, etc.).

The gathering will take into account the fact that the learner may have special needs. If an instrument is mandatory but the learner should not use it (e.g. collecting voice from voiceless people) then the instrument will be automatically disabled. This information will be reported to the instructor on the moment the he requests the evaluation results.

4 Discussion

There are some well-known non-functional requirements, already discussed in the background section that should be taken into account when designing the architecture of the system, but the decisions taken to ensure them may vary when the system will be deployed into the cloud. This section describes the decisions made in the case of study for support these non-functional requirements.

Modularization

The correct modularization of a system affects to other non-functional requirements like extensibility, performance or operability. The whole TeSLA system was entirely designed as a highly-modular system. One of the most impacting decisions in the architecture of the system was to enforce modularization with the use of containerization, which is the encapsulation of each module or component inside a container that is one minimal unit of deployment and operation and can be scaled separately. The ecosystem of multiple containers is commonly known as microservices architecture, and the most well-known example of containerization platform is Docker [17].

Robustness/reliability

One of the first decisions made in the project was to stablish the required mechanisms to ensure the robustness of the system. The number of partners in the project was quite big and each one was in charge of developing different components of the system, so the first decision in order to ensure robustness was to adopt some DevOps tools, concretely using GitHub as a central code repository, using one repo for each component and enforcing continuous integration and delivery.

Another of the decisions was to increase the testability of the system, so each component should have its own unitary tests, and acceptance tests, and also a set of integration tests were defined to ensure the correct integration between components.

In order to enhance integrability of components, which also favors a robust system, some well-defined interfaces were defined for each component, and exposed through REST endpoints accepting JSON format.

Extensibility

TeSLA system was designed to be extensible in some different ways, and the modularization of components, the correct definition of interfaces and the interoperability between components were the key to success.

On one hand, TeSLA should be able to include other identification instruments from other parties in the future, so the instrument interface were defined independently of the type or the source of the instrument. This way, any instrument that would be integrated into the system, only should define a thin layer of integration between the instrument and the TEP.

On the other hand, the system should be able to work with different VLEs (at the end of the project it is compatible with Moodle out-of-box). The plugin component (see Fig. 6) can be replaced in each VLE with a specific implementation that enables to communicate with the corresponding VLE, maintaining the interfaces defined with tools and the TEP.

Performance

The number of potential students and instructors to be using the TeSLA system is quite big, more than 20,000, so it should be prepared to support a big number of concurrent requests. This is not only achieved by developing high performant components, but also by scaling some parts of the system.

Scalability can be addressed in two ways: either by scaling vertically (increase machine resources) or horizontally (increase the number of machines supporting a service). The vertical scaling does not affect in general to the behavior of a component, but horizontal scaling requires that multiple instances of the same component behave correctly when running concurrently.

Both vertical and horizontal scaling requires a lot of operational efforts from the management team when the deployment is done on-premises. On the contrary, if the deployment is done in cloud, scaling a component is just a few clicks operation on the administration panel of the cloud infrastructure.

Security

There are many things to consider about the system security. The most obvious are related with communications security using encryption to avoid man-in-the-middle attacks, or the authentication and authorization mechanisms for each of the interfaces of the system (either visual interfaces or endpoints). This is especially important to ensure in cloud systems because data is moved from the on-premises installation to the exterior.

In order to increase even more the security in the communications between components, each component uses its own client certificate emitted by a trusted certifying authority (CA) which is another component of the system. The use of a CA also enables the possibility of handling license verification, using the expiration time of the certificate.

Another aspect of the security of the system is the data privacy. The TeSLA system is able to manage informed consents from students and only enable the use of the system if they agreed to that consent. Additionally, the system does not store personal information about students; the way to identify them is using a unique TeSLA ID which is generated for each student. The translation between the TeSLA ID and the corresponding student is stored in the Plugin component which is installed always in the VLE side (this means in the institution premises).

Operability

Operability of the system includes deployment and maintainability. The containerization of the components helps in the deployment of the system, which is deployable through an installation script adaptable to each institution. The maintainability is enhanced by the use of the Portal component tools.

The Portal includes tools for administrators, they can deploy and manage components (including instruments and instances), and centralizes the monitoring of all the components of the system (status, health, logs, etc.).

Operability also implies sometimes making changes on the infrastructure. Here is where the cloud facilitates the work, not only because setting up the infrastructure is nearly transparent to administrators, but also because it saves all the costs of purchasing and maintaining hardware and infrastructures.

Usability

This requirement is not directly related with architecture, but it is very important each time more. A system can be fully functional, performant, robust, secure, but if it is not usable it will not succeed.

The development of TeSLA system took into account some premises (finally accomplished or not) about the usability of the system:

- The system would be easy to use for students; it should not be intrusive or impede them to work normally if there is some failure. The enrolment phase would be balanced between the requirements of the instruments and the time of efforts required from the students to finish it.
- The system would be easy to use for instructors; it should be easy to setup activities and the configuration of the system should be understandable for non-technical people. The results shown should be easily interpretable.
- The system would be easy to be managed at admin level. Operations should be understandable and easily reachable.
- The visual interfaces of the system should have accessibility into account.
- The system has to be aware of SEND students, and offer alternatives (for example a dumb person cannot use a voice recognition system, another alternative should be offered like face recognition or keystroke).

5 Conclusions

Engineering applications that will be deployed in a cloud infrastructure does not differ dramatically from engineering distributed applications deployed on-premises. However some specific considerations should be taken into account when using a cloud infrastructure, overall those regarding to security and data privacy because data is not anymore stored and managed on-premises but in a remote server. Performance and availability is also an important point to design so moving to the cloud does not affect the response time of the application.

Furthermore, cloud technologies offer very interesting features to distributed applications, mainly at easing the setup and maintenance of infrastructure because almost every operation or maintenance in the cloud can be done just with a couple of mouse clicks, when the same operation on-premises would take some hours or days and cost considerable more. As an example, when the performance decreases, cloud infrastructure allows to scale either vertically or horizontally in just few minutes and even is able to automate the process when some performance alert is thrown. This reduces costs and the time required to solve the problem, when on-premises would take a lot of effort or maybe it would be feasible to perform.

This chapter discussed about engineering applications for the cloud and described the most important points that distinguish typical on-premises application architectures from that cloud based. As an example of a successful use case of architecture engineering for cloud applications, an application for detecting cheating in online assessment called TeSLA system is presented. The TeSLA system and the instruments were tested at various stages of their development in three consecutive pilots over three years, involving over 23,000 students from seven universities.

One of the most interesting points that can be highlighted about the architecture of the example is the use of containerization (specifically Docker containers) for modularizing thus isolating components and easing communications between them. At the same time, containerization allows for scalability options and also eases deployment. Is also worth to mention the security measures defined in the infrastructure to either ward off external access to de system without the required permissions, but also to protect sensitive data and avoid incorrect usage of each module.

Finally, but not less important, is worth to mention the special effort when defining the architecture in order to enhance the reusability and extensibility of the system.

6 Future Work

The TeSLA system was designed to accomplish the functional and non-functional requirements mentioned in previous sections, however, as is common in the major part of the projects, not all the requirements were able to be finished completely, and some problems or issues had raised from the initial design. Some of these issues will be faced until the end of the project, but others will remain as further work for the exploitation of the system.

One of the most relevant issues found during the tests of the system was the bottleneck provoked by the TEP component. As it is a central component receiving and producing multiple communication and coordination between components, any delay on this component blocks the ability to process new requests. The solution proposed for solving this issue is to delegate TEP responsibilities into a thin API component that uses a queue management system to coordinate the rest of components requests using tasks.

Some of the instruments used in the TeSLA system during the project are commercial, that means they will not be available out of the project, so it will be necessary to provide new instruments to cover them in the different types of identification.

Finally, the deployment of the TeSLA system is currently based on a deployment script that must be customized for each institution where it should be installed, and the sources of components are stored into a GitHub repository. In order to be able to exploit the product the deployment has to be simplified, even the option to exploit TeSLA as a SaaS service should be considered.

Acknowledgements (Blind for Review) This work is partly funded by the Spanish Government through grant TIN2014-57364-C2-2-R "SMARTGLACIS".

References

1. Talukder AK, Zimmerman L, APH (2010) Cloud economics: principles, costs, and benefits. In: Antonopoulos N, Gillam L (eds) Cloud computing. computer communications and networks. Springer, London
2. Ghaffari F, Gharaee H, Forouzandehdoust MR (2016 Sept) Security considerations and requirements for cloud computing. In: 2016 8th international symposium on Telecommunications (IST). IEEE, pp 105–110
3. Rimal BP, Choi E, Lumb I (2009 Aug) A taxonomy and survey of cloud computing systems. In: Fifth international joint conference on INC, IMS and IDC, 2009. NCM'09. IEEE, pp 44–51
4. Jadeja Y, Modi K (2012 Mar) Cloud computing-concepts, architecture and challenges. In: 2012 international conference on computing, electronics and electrical technologies (ICCEET). IEEE, pp 877–880
5. Rimal BP, Jukan A, Katsaros D, Goeleven Y (2011) Architectural requirements for cloud computing systems: an enterprise cloud approach. J Grid Comput 9(1):3–26
6. Tsai WT, Sun X, Balasooriya J (2010 Apr) Service-oriented cloud computing architecture. In: 2010 seventh international conference on information technology: new generations (ITNG). IEEE, pp 684–689
7. Fernandes DA, Soares LF, Gomes JV, Freire MM, Inácio PR (2014) Security issues in cloud environments: a survey. Int J Inf Secur 13(2):113–170
8. Tripathi A, Mishra A (2011 Sept) Cloud computing security considerations. In: 2011 IEEE international conference on Signal Processing, Communications and Computing (ICSPCC). IEEE, pp 1–5
9. Andrikopoulos V, Binz T, Leymann F, Strauch S (2013) How to adapt applications for the cloud environment. Computing 95(6):493–535

10. Zimmermann O (2017) Architectural refactoring for the cloud: a decision-centric view on cloud migration. Computing 99(2):129–145
11. Karim B, Tan Q, El Emary I, Alyoubi BA, Costa RS (2016) A proposed novel enterprise cloud development application model. Memet Comput 8(4):287–306
12. Felemban M, Basalamah S, Ghafoor A (2013) A distributed cloud architecture for mobile multimedia services. IEEE Netw 27(5):20–27
13. Masud MAH, Huang X (2012) An e-learning system architecture based on cloud computing. World Acad Sci Eng Technol Int J Comput Electr Autom Control Inf Eng 6(2):255–259
14. Verma P, Sood SK, Kalra S (2018) Cloud-centric IoT based student healthcare monitoring framework. J Ambient Intell Humaniz Comput 9(5):1293–1309
15. https://tesla-project.eu/
16. Dadda L, Macchetti M, Owen J (2004) The design of a high speed ASIC unit for the hash function SHA-256. IEEE Computer Society (384, 512), 70–75
17. https://www.docker.com/

Security and Privacy in the TeSLA Architecture

Christophe Kiennert, Malinka Ivanova, Anna Rozeva and Joaquin Garcia-Alfaro

Abstract In this chapter, we address security and privacy aspects in TeSLA, from a technical standpoint. The chapter is structured in three main parts. Firstly, we outline the main concepts underlying security in TeSLA, with regards to the protection of learners' data and the architecture itself. Secondly, we provide an empirical analysis of a specific deployment in one of the members of the consortium. Some representative aspects such as security levels in terms of storage, processing and transfer are analyzed in the deployment of TeSLA at the Technical University of Sofia. In the third part, we address identity management issues and outline additional efforts we consider worth exploring.

Keywords Authentication · Authorship · Security · Public key infrastructures · X.509 certificates · Anonymity · Privacy · GDPR

Acronyms

CA	Certification Authority
GDPR	General Data Protection Regulation
JWT	JSON Web Token
LMS	Learning Management System

C. Kiennert · J. Garcia-Alfaro (✉)
Institut Mines-Telecom, Institut Polytechnique de Paris, Paris, France
e-mail: jgalfaro@ieee.org; garcia_a@telecom-sudparis.eu

C. Kiennert
e-mail: christophe.kiennert@telecom-sudparis.eu; kiennert@telecom-sudparis.eu

M. Ivanova
College of Energy and Electronics, Technical University of Sofia, Sofia, Bulgaria
e-mail: m_ivanova@tu-sofia.bg

A. Rozeva
Department of Informatics, Technical University of Sofia, Sofia, Bulgaria
e-mail: arozeva@tu-sofia.bg

D. Baneres et al. (eds.), *Engineering Data-Driven Adaptive Trust-based e-Assessment Systems*, Lecture Notes on Data Engineering and Communications Technologies 34,
https://doi.org/10.1007/978-3-030-29326-0_5

LTI	Learning Tools Interoperability
OCSP	Online Certificate Status Protocol
PKI	Public Key Infrastructure
RSA	A public-key cryptosystem
SAML	Security Assertion Markup Language
TLS protocol	Transport Layer Security protocol
TTP	Trusted Third Party
TUS	Technical University of Sofia
UUID	Universally Unique Identifier
VLE	Virtual Learning Environment

1 Introduction

TeSLA aims at providing learners with an innovative environment, which allows them to take assessments remotely, thus avoiding mandatory attendance constraints. From a technical standpoint, TeSLA has been designed as a flexible architecture, in which traditional learning management systems and virtual learning environments are seen as the main entry points of an educational platform. The architecture itself is comprised of several entities, some of them located at the institution side, establishing communications with the learning environments or with external tools embedded into the learners' browsers; others belonging to a separate domain independent of the institution. Securing such architecture consisted in expressing the security needs regarding sensitive and personal data on the one hand, and analyzing threats both on hosts and network on the other. Moreover, the choices made on security measures ensured that TeSLA is compliant with existing technical standards and recommendations [1–4], and also with legal requirements, such as the European General Data Protection Regulation [5].

From a security perspective, the main properties ensured by TeSLA are authentication, authorship, confidentiality and integrity. Authentication aims at proving an entity's identity to another party; authorship consists in proving the identity of the creator of a piece of work; confidentiality consists in encrypting data to prevent information disclosure to unauthorized parties; and integrity aims at preventing fraudulent data alteration. Over the network, the most convenient way to implement these three traditional security properties was to deploy the well-known Transport Layer Security (TLS) protocol [6], which allows entities to authenticate to each other and creates a secure tunnel with data encryption and integrity check.

Authentication in TLS does not rely on passwords, but on X.509 certificates. These certificates rely on asymmetric cryptography, and create an association between a public key and an identity. Any entity can authenticate itself via the certificate, as long as it owns the associated private key, which is never transmitted over the network. The certificate management requires a Public Key Infrastructure (PKI), in which specific trusted entities, called Certification Authorities, are in charge of the certificate delivery.

As we will see in the first part of this chapter, TeSLA comes with a PKI to manage X.509 certificates within the TeSLA domain on the one hand, and within the institution domain on the other. This way, communication between the various entities of the TeSLA architecture can be entirely secured, in all the three dimensions aforementioned (i.e., authentication, confidentiality, and integrity). Other aspects that have been carefully taken into account in TeSLA are in terms of data protection from a privacy perspective. In fact, the identity of the learners is never disclosed within the TeSLA domain. The architecture has been conceived to respect sensitive data of the learners, for both legal and ethical reasons. In a nutshell, the enforcement of privacy in TeSLA consists in minimizing the personal information retrieved from learners during their interactions with the system, as well as encrypting and anonymizing the data exchanged or stored in the databases whenever required. In the end, the TeSLA architecture provides the learner with pseudo-identifiers, which hide the learner's genuine identity when taking e-assessment activities.

In this chapter, we elaborate further on all the aforementioned security and privacy aspects, and provide additional elements for further research perspectives and discussions. More specifically, Sect. 2 outlines the main concepts underlying the TeSLA architecture from a technical perspective. Section 3 reports a practical hands-on analysis conducted by members of the consortium during the pilot phases of the project. Section 4 discusses additional enhancements towards enhanced privacy features in future TeSLA releases. Finally, Sect. 5 concludes the chapter. Some parts of this chapter have been previously published in, TeSLA [7].

2 Technical Security Features in TeSLA

2.1 Security Problems and Use Cases

Specific security problems related to e-assessment systems in the eLearning literature point out the following question [8, 9]: "*How can learners and educators be confident that the e-assessment system can be trusted so that it can detect cheating attempts?*" To properly answer the above question, we shall first raise concerns about situations in which a precise e-assessment system can be misused.

From an educational point of view, several situations can source the problems. The main one is often referred to as the recognition of the learner's identity. In case learner's identity is being used by someone else, rather than the real learner to be assessed, we can refer to identity misuse, which leads to the following use cases:

- E-assessment could occur in a controlled environment, such as a university building under educator supervision, and it is a common case in universities with blended-learning. In this situation, the misuse of the system is possible when an educator is responsible for a big number of learners and the educator does not recognize their faces. Then, additionally the educator must check, e.g., the learner's national card, to be sure of their real identity.

– It is also possible that the e-assessment process gets performed in uncontrolled environment outside the university building where the educator does not have any control on learners' identity. This is the typical situation for online learning environments, in which the educator must be sure that the assessed learner is the same as the one from the declared personal data.

In the two aforementioned cases, a fair e-assessment process can be compromised if learner's identity changes. We consider the problem of assuring identity authentication as the main challenging problem to address by the e-assessment process. By applying a suitable authentication mechanism [10], the educator can ensure the identity of the assessed learner no matter where the assessment is located, hence avoiding the necessity of checking the identity periodically, as this can be time consuming.

If during the e-assessment process, private or sensitive data are transmitted, a second major problem arises. This is related to the disclosure of sensitive information to unauthorized parties. Regarding this issue, the following two use cases are defined:

– During an e-assessment process, learners may share more data than needed. Here, the role of the educator is very important, because he has to design the assessment scenarios in a way that will collect only the data needed to ensure a successful assessment process. The learners should not have to provide information that does not concern either the educator, or the improvement of the teaching and learning process, or the formation of the final mark. For example, if the educator starts a forum topic that is part of e-assessment scenario, this must exclude problems for discussion by learners that will reveal more private or sensitive data. The collection of any additional data will foster options for information disclosure.
– Learners' or educator's information can be stolen in result of the internal or external intervention of a malicious user and the e-assessment might be compromised. The loss of information of learners' achievements in this case will not allow the educator to form the final learners' marks. As a result, learners may have to take the assessment activities again and the educator has to mark them again. Before that, the educator has to prepare new variants of the same assessment activities. It is time consuming and is an overload task for learners and educators. Of course, learners' data can be potentially stolen in traditional assessment environments, but in online assessment the information is much more vulnerable.

The two aforementioned use cases make the possibility for information disclosure very high, especially when data is transmitted from one system component to another, or from one organization to another. This can cause difficulties during the e-assessment process. This second problem concerns data confidentiality. It requires data protection and access control, to avoid the disclosure of data to unauthorized parties. Since the e-assessment data are stored in records and databases, fraudulent alteration of data must be addressed. Data modification leads to serious e-assessment problems for learners and educators. The following use cases are identified:

- An adversary (learner, faculty, university staff, etc.) can gain unauthorized access to the educational records and databases, in order to modify private or sensitive information (e.g., the outcomes of a quiz activity). This leads to a confusing situation and unclear picture for the educator.
- An adversary with unauthorized access to the e-assessment tasks before they are assigned to learners, could modify or distribute them to learners (e.g., for financial profit). The e-assessment process may lose its meaning which is to evaluate and measure the real learners' knowledge and skills.
- It may also be possible for the adversary to corrupt or delete part or the whole assessment information. This creates difficulties for the learners and the educator.

In all those aforementioned cases, the main challenging problem concerns data integrity, i.e., how to assure that data is secured in the case of fraudulent data alteration.

2.2 *Integration of Security Measures in the TeSLA Architecture*

The TeSLA architecture[1] is comprised of several entities (see Fig. 1), some of them located in a cloud infrastructure, and shared among several institutions; some others deployed individually at an institutional level (e.g., one per university). Regardless of the location of each entity, TeSLA must secure the establishment of communications between entities such as Learning Management Systems (LMS) and Virtual Learning Environment (VLE), as well as with external tools embedded into the learners' browsers. Securing such architecture is a complex task. It consists in expressing all the security needs regarding sensitive and personal data on one hand, and analyzing threats and security levels on both hosts and network elements on the other. The choices made on the underlying security properties of the TeSLA architecture must also follow requirements in terms of learners' privacy, as those expressed by General Data Protection Regulation (GDPR) directives for all individuals within the European Union and the European Economic Area [5]. In a nutshell, this requires that the architecture guarantees (1) the ability to ensure the confidentiality and integrity of system communications and related services; and (2) ability to guarantee proper pseudonymization process of all user identities; and (3) ability to guarantee an appropriate protection of all the personal data stored or processed by the system as well.

Consequently, the security services provided by the TeSLA architecture concern the enforcement of authentication and protection of both communications and data storage. Authentication aims at proving an entity's identity to another party, leading to providing enough guarantees in terms of confidentiality and integrity. In turn, confidentiality consists in protecting data to prevent information disclosure to unauthorized parties. Integrity aims at preventing fraudulent data alteration. Over the

[1]More detailed information related to the TeSLA architecture can be found in Chap. 4: Engineering Cloud-based Technological Infrastructure to Enforce Trustworthiness.

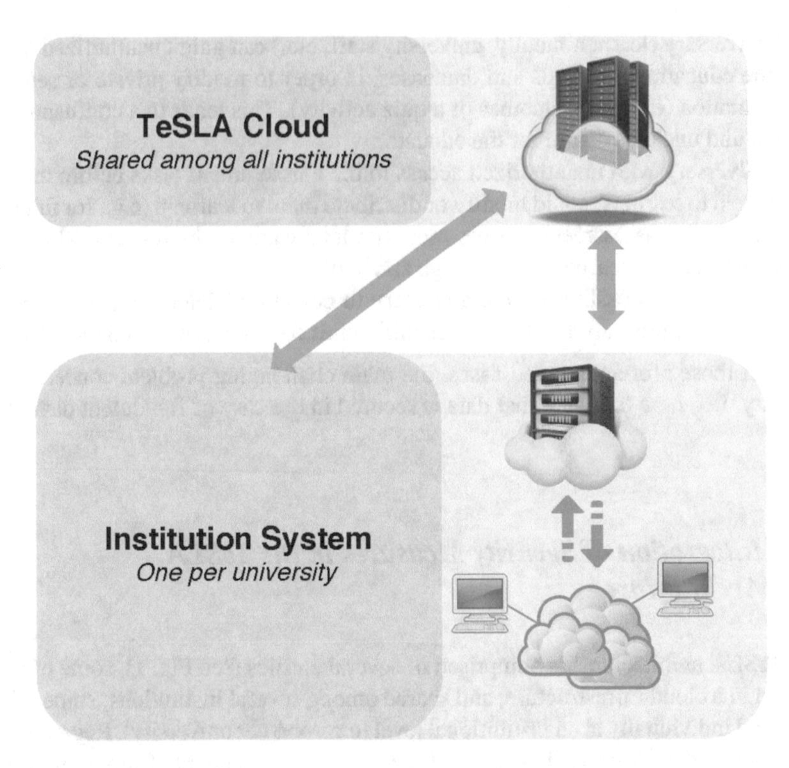

Fig. 1 The TeSLA architecture

network, the most convenient way to implement these security services is to use the TLS (Transport Layer Security) protocol [6], which allows entities to authenticate to each other and creates a secure tunnel with data encryption and integrity checks.

Authentication in TLS relies on the use of X.509 certificates [11], which, in turn, use asymmetric cryptography, and create an association between a public key and an identity. Any entity can authenticate itself via its certificate, as long as it owns the associated private key, which is never transmitted over the network. The certificate management requires a Public Key Infrastructure (PKI) [12], in which specific trusted entities, called Certification Authorities (CA), are in charge of certificate delivery. The TeSLA architecture has its own PKI, to manage the certificates within the TeSLA domain on one hand; and within the institution domain on the other. This way, the communications between the various entities of the architecture can be entirely secured.

Some of the aforementioned elements and mechanisms are elaborated further in the following sections as a summary of the main actions and guidelines followed during the design phase of the TeSLA architecture. Such actions and guidelines are the result of a careful analysis conducted by the technical members of the TeSLA project, to guarantee that the resulting architecture follows generic best practices and well-established security standards (see [1–4] and citations thereof, for further details).

2.3 *TLS and PKI-Based Communication*

The TeSLA architecture guarantees that traditional information security properties such as confidentiality, integrity and authentication are always respected. This is achieved as follows: (1) use of TLS to secure all the exchanges between components of the architecture; (2) deployment of a PKI associated to the TeSLA architecture; (3) enforcement of mutual authentication between all the components of the architecture.

The TLS protocol ensures confidentiality, integrity, authentication and non-repudiation altogether for two communicating entities. The protocol consists of two phases: the handshake, during which the security parameters are negotiated (in particular, cipher and hash algorithms [13]. The communicating entities are hence authenticated (either mutually or one-way). In the second phase, a secure tunnel is established between the two communicating entities, ensuring that all data are properly encrypted and cannot be modified by an attacker during transmission. Symmetric keys are used to encrypt all the TLS exchanges. The keys are automatically and dynamically generated during the initial handshake of the TLS protocol.

TLS-based authentication requires X.509 digital certificates, which are managed by the PKI. The principle of a certificate is to assess the link between an entity and its public key, through a TTP (Trusted Third Party) called a Certificate Authority (CA). The CA digitally signs certificates itself, or delegates the signature activity to intermediate entities. The validation of a certificate during the authentication process includes the following steps: (i) check the expiration date of the certificate; (ii) verify the signature of the certificate; (iii) check if the signing CA is recognized as a trusted CA; (iv) check if the certificate has not been revoked.

The PKI model proposed for the TeSLA architecture is available in [14]. The PKI secures, for instance, all the exchanges with biometric instruments, i.e., those TeSLA components in charge of evaluating data to assess learner's identity and authorship (see Sect. 2.1, second use case: "*the educator must be sure that the assessed learner is the same as the one from the declared personal data*"). More information about the use of PKI certificates is provided in the following sections.

2.3.1 Certificate Management

The PKI model proposed in [14] for the TeSLA architecture identifies the following four representative certificate authorities: (i) TeSLA CA; (ii) TeSLA Intermediate CA; (iii) Institution CA; and (iv) Institution Intermediate CA.

Firstly, the TeSLA CA is the top certificate authority regarding the TeSLA PKI. Basically, this certificate authority is only used once, to sign the TeSLA intermediate CA signature request. It is recommended to use this certificate as scarcely as possible

[3, 6]. Then, the TeSLA intermediate CA signs the Institution CAs certificates (one for each institution) and delivers the client and server certificates for the TeSLA components. The Institution CA is the top certificate authority regarding the institution based TeSLA components. For security purposes, the use of the certificate associated to the Institution CA is minimized, e.g., it is used only once, to sign the Institution intermediate CA signature request. As for the TeSLA CA certificate, it is also recommended to use it as scarcely as possible. Finally, the Institution Intermediate CA is used to deliver client and server certificates of the architecture components (e.g., backend components of TeSLA system and its corresponding databases).

2.3.2 Revocation Lists

The TeSLA PKI shall maintain, update and provide secure access to two main revocation lists [15]: (i) the revocation list associated to the TeSLA CA; and (ii) the revocation list associated to the TeSLA intermediate CA. Each institution using its corresponding CAs has to manage, update and provide secure access to two revocation lists: the revocation list associated to the Institution CA, and the revocation list associated to the Institution intermediate CA.

With respect to the secure connections between the TeSLA components, the certificate validity must be checked with respect to their revocation lists. A certificate may indeed be valid (i.e., not expired and with a correct signature), but marked as revoked. Finally, the use of the Online Certificate Status Protocol (OCSP) [16] has also been included at the core PKI functionality of TeSLA, for obtaining the revocation status of X.509 digital certificates.

2.3.3 Cryptographic Keys

We conclude this section with a quick overview regarding the security procedures that have to be applied when a private key is disclosed, as suggested in Barker [17], ANSSI [1]. Possible incidents are classified in terms of critical levels (in which zero represents the most critical one, i.e., the one with the highest priority).

- Level 0—If the TeSLA CA private key has been compromised, then the whole system is compromised. The whole TeSLA PKI has to be recreated, and all the certificates and CAs that had previously been generated must be revoked.
- Level 1—If the TeSLA Intermediate CA private key has been compromised, then the TeSLA CA has to revoke this certificate. All the certificates that were signed by the TeSLA Intermediate CA have to be revoked as well. On the other hand, if a client/server private key associated to a certificate signed by the TeSLA Intermediate CA has been compromised, then the TeSLA Intermediate CA has to revoke this certificate.

- Level 2—If the Institution CA private key has been compromised, then the TeSLA Intermediate CA has to revoke this certificate. All the certificates that were signed by the Institution CA have to be revoked as well.
- Level 3—If an Institution Intermediate CA private key has been compromised, then the TeSLA Institution CA has to revoke this certificate. All the certificates that were signed by the Institution Intermediate CA have to be revoked as well. Likewise, if a client/server private key associated to a certificate signed by the Institution Intermediate CA has been compromised, then the Institution Intermediate CA has to revoke this certificate.

Finally, the CA certificates use RSA keys with a modulus of at least 4096 bits [1, 13, 17]. The validity is fixed to ten years maximum (also limited by the TeSLA license validity period). Client and server certificates use RSA keys with a modulus of at least 2048 bits [1, 17]. The validity is fixed to one year. Attention must be paid to certificate management to prevent architecture malfunctioning, e.g., new certificate emission to clients and servers before their actual certificates expire.

3 Security Evaluation of a Representative TeSLA Deployment

In this section, we provide an empirical security hands-on analysis of the deployment of TeSLA at the Technical University of Sofia (TUS for short) [18]. The analysis builds upon an existing methodology [19], provided by TUS for the verification of secure web services and applications. The analysis focuses on the e-assessment environment at TUS, involving Moodle [20], complemented by the TeSLA release deployed at TUS for the pilots of the TeSLA project. The criteria, parameters and underlying tools for the evaluation used during the analysis support multi-criteria decision-making and are based on fuzzy set theory and fuzzy logic.

3.1 Aims and Background

3.1.1 Fuzzy Sets and Fuzzy Logic

An online environment implementing variety of web services is always vulnerable to various threats and attacks. The correct evaluation of its security status is a complex task, due to difficulties in considering and proper identification of all the factors that influence it. Decision-making and tasks concerning security issues have to be performed based on incomplete information provided by experts. Such information is mainly in form of natural language statements involving linguistic variables, i.e. “High”, “Medium”, “Low”, rather than numbers. Cases like these require dealing with approximations of numbers, which are close to a given real number, rather than with crisp real numbers and crisp intervals. Classic mathematical description and formalization turns out to be inapplicable to such knowledge, expressed in natural language statements, referred to as fuzzy statements.

An approach providing for finding optimal decision by such expert systems, which has been adopted for the evaluation of the security status of TeSLA online environment is based on the fuzzy set theory. It states the conceptualization of fuzzy statements by appropriate fuzzy sets in R (real numbers set) in order to treat them as fuzzy numbers. It has been widely adopted in decision support, knowledge based and expert systems. It has become the background of methodology designed at conceptual level [19], which addresses managers, technical professionals and other authorities involved in securing the online environment in an organization. It has been implemented for performing empirical analysis of the security level of the deployment of TeSLA in the online virtual learning environment at TUS.

Fuzzy set [21] $\tilde{A}$ of set X is defined by its membership function $\mu\tilde{A}: X \rightarrow [0, 1]$ as $x \rightarrow \mu\tilde{A}(x) \in [0, 1]$. The fuzzy membership function $\mu\tilde{A}(x)$ indicates the degree of belonging of $x \in X$ to $\tilde{A}$. Value 0 corresponds to absolute non-membership and value 1 to full membership. For set X with elements $x_1, x_2, \ldots, x_n$, the fuzzy set $\tilde{A} = \{(x_1, \mu(x_1)), (x_2, \mu(x_2)), \ldots, (x_n, \mu(x_n))\}$. Fuzzy sets can be represented by different kinds of membership functions. The most popular variant of membership function is the "triangle" one, as it provides for the simplification of calculations performed upon fuzzy sets. The triangular membership function [22], depends on three scalar parameters a, c (lower and upper bound) and b (mean value) and is defined as (1):

$$\mu_A(x) = \begin{cases} \frac{x-a}{b-a}, & a \leq x \leq b \\ \frac{c-x}{c-b}, & b < x \leq c \\ 0, & a < x, x > c \end{cases} \tag{1}$$

Fuzzy numbers, defined by triangular membership function are referred to as triangular fuzzy numbers. A fuzzy number is denoted by a triplet (a, b, c).

The basic operations on triangular fuzzy numbers $\tilde{A}_1 (a_1, b_1, c_1)$ and $\tilde{A}_2 (a_2, b_2, c_2)$ are defined as follows:

Addition +: $\tilde{A}_1 \oplus \tilde{A}_2 = (a_1 + a_2, b_1 + b_2, c_1 + c_2)$,
Subtraction −: $\tilde{A}_1 \ominus \tilde{A}_2 = (a_1 - c_2, b_1 - b_2, c_1 - a_2)$,
Multiplication ×: $\tilde{A}_1 \otimes \tilde{A}_2 = (\min(a_1a_2, a_1c_2, c_1a_2, c_1c_2), b_1b_2, \max(a_1a_2, a_1c_2, c_1a_2, c_1c_2))$.
Comparison ≤: $\tilde{A}_1 \leq \tilde{A}_2 \rightarrow (a_1, b_1, c_1) \leq (a_2, b_2, c_2)$.
Mean $m(\tilde{A}) = 1/3(a + b + c)$,
Variance $\sigma(\tilde{A}) = 1/18(a^2 + b^2 + c^2 - ab - ac - bc)$.

Fuzzy numbers are implemented in fuzzy logic, which is an approach for computing, based on partial, i.e. truth to a certain degree and not on the absolute truth (true/false). This is close to real life cases where a lot of imprecise data are generated. It provides for representation of generalized human cognitive abilities in software solutions. Fuzzy logic involves building fuzzy IF-THEN rules as a way for formalizing human natural language and facilitating decision-making. Sample fuzzy rule is IF x is A THEN y is B, where A and B are linguistic values defined by fuzzy sets, x and y are input and output variables. The input to the rule is a crisp value x, which is transformed into fuzzy set by applying a specialized function. The procedure is referred to as fuzzification. The output to the rule is a fuzzy set assigned to the output variable y, which is turned into a crisp number after performing the reverse transformation called defuzzification.

The process involving fuzzy logic based inference on expert knowledge simulates human reasoning by implementing relatively simple mathematical concepts.

3.1.2 Fuzzy Logic Algorithm for Empirical Security Analysis

The main goal of the algorithm is the identification of the influence of selected criteria and parameters that are most relevant for assuring high security level of the virtual learning environment with TeSLA deployment at TUS. Performance ratings and importance weights for them are defined by using fuzzy numbers. Fuzzy logic is implemented in the decision-making process, as suggested in [23]. Fuzzy performance-importance index of each criterion and parameter is calculated to facilitate the improvement of the security level of the TeSLA e-Assessment environment.

A. Criteria definition
   Criteria that are relevant to the security degree of the services related to TeSLA deployment at TUS are defined. Figure 2 presents the hierarchical relationship of a global criterion (GC), criteria (C_i) and their parameters (P_i).
B. Definition of security level scale with triangular fuzzy numbers for criteria boundaries.
C. Definition of linguistic variables with triangular fuzzy numbers for rating the performance and weighing the importance of the security criteria and parameters.
D. Security evaluation by three experts and aggregation of the obtained fuzzy numbers for performance-ratings and importance-weights (Eq. 2).

$$a = \sum_{i=1}^{3} a_i/3,\ b = \sum_{i=1}^{3} b_i/3,\ c = \sum_{i=1}^{3} c_i/3 \tag{2}$$

E. Calculation of the fuzzy indexes of criteria C_i and their parameters.
   The fuzzy indexes of each criterion C_i are calculated by Eq. (3):

$$WR = \sum_{i,j=1}^{n} W_{ij}R_{ij} / \sum_{i,j=1}^{n} W_{ij}, \tag{3}$$

where W_{ij} are the importance-weights of parameters, and R_{ij} are the performance-ratings.

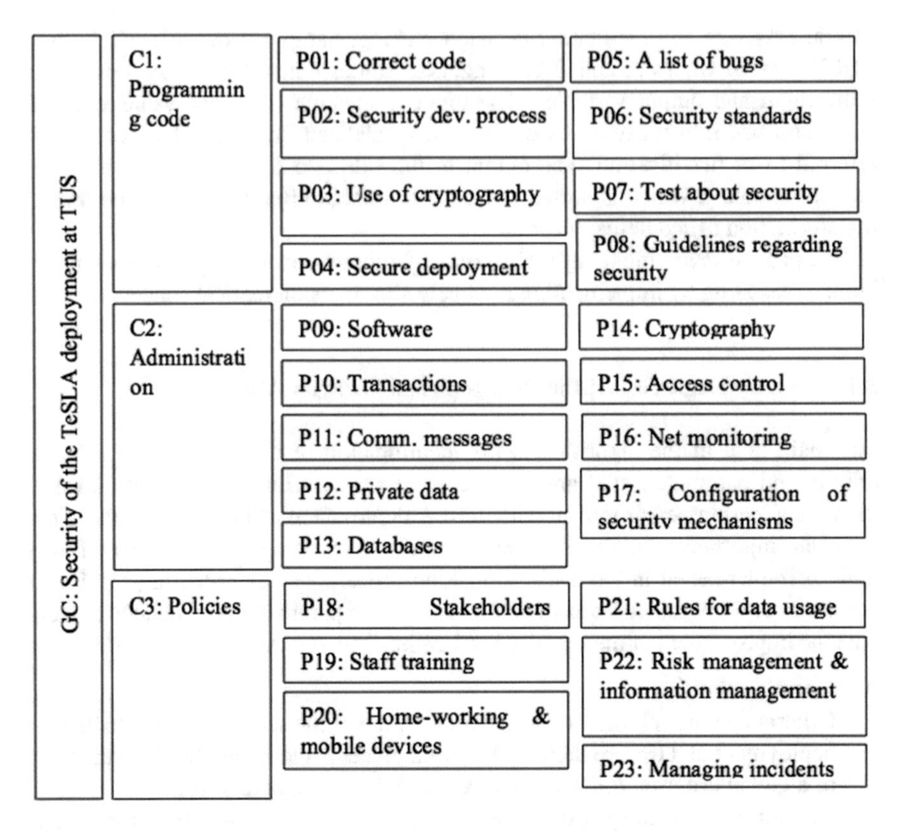

Fig. 2 Criteria and parameters for security evaluation of TeSLA deployment

F. Calculation of the fuzzy security index for the global criterion. Match the result obtained as triangular fuzzy number to the linguistic variable corresponding to the respective security level in the scale from **B**.
G. Validation of the security level determined in **F** by the Euclidean distance of the global criterion to each security level.
The Euclidean distance is the distance between two triangular fuzzy numbers X (x_1, x_2, x_3) and Y (y_1, y_2, y_3). It is calculated by Eq. (4) (cf. [CHK08] and citations thereof, for further details):

$$D(X, Y) = \sqrt{\frac{1}{6}\left[(x_1 - y_1)^2 + 4(x_2 - y_2)^2 + (x_3 - y_3)^2\right]} \quad (4)$$

The calculated security level corresponds to the minimal value of the Euclidean distance and its linguistic value is obtained from the scale, defined in **B**.
H. Calculation of the performance-importance indices and rating scores of all parameters of the security criteria for suggesting the ones that could be improved. Performance-importance indices FPII are calculated by Eq. (5):

$$FPII = \left[(1, 1, 1) \ominus W'_{ij}\right] \otimes PRI, \tag{5}$$

where W'_{ij} is the importance-weight fuzzy number of all the parameters with reversed places of fuzzy values and PRI is their matched performance-rating fuzzy number. FPII numbers are then matched to crisp rating scores by using Eq. (6):

$$RS = (x_1 + 2x_2 + x_3)/4, \tag{6}$$

The minimal values of the rating scores indicate which parameters must be addressed as vulnerabilities that could compromise the security status.

3.2 Results of Empirical Security Analysis

B → The security status of the TeSLA deployment environment at TUS is evaluated at five levels. The linguistic variables and the fuzzy numbers that have been chosen to define the boundaries of each security level are shown in Table 1.

C → The security criteria and parameters' rating of performance and weighing of importance are performed with linguistic variables and fuzzy numbers defined in Table 2. A scale from 0 to 5 is chosen for each fuzzy set definition. The importance-weights range from 0 to 1.

D → The rating of performance **R** and weighing the importance **W** of the security criteria and parameters have been obtained by surveying three experts involved in the

Table 1 Linguistic variables and fuzzy numbers of defined security levels

	Security levels				
Linguistic variable	NS (non-secure)	PS (poorly secure)	MS (moderately secure)	S (secure)	VS (very secure)
Fuzzy set	(0, 0.6, 1.2)	(1, 1.6, 2.2)	(2, 2.6, 3.2)	(3, 3.6, 4.2)	(4, 4.5, 5)

Table 2 Performance-ratings and importance-weights of the criteria and parameters

R (Performance rating)		W (Importance weighting)	
Linguistic variable	Fuzzy set	Linguistic variable	Fuzzy set
P (Poor)	(0, 0.6, 1.2)	Very low (VL)	(0, 0.12, 0.24)
F (Fair)	(1, 1.6, 2.2)	Low (L)	(0.2, 0.32, 0.44)
G (Good)	(2, 2.6, 3.2)	Medium (M)	(0.4, 0.52, 0.64)
VG (Very good)	(3, 3.6, 4.2)	High (H)	(0.6, 0.72, 0.84)
E (Excellent)	(4, 4.5, 5)	Very High (VH)	(0.8, 0.9, 1)

TeSLA system pilots at TUS as software and security professionals. Sample expert evaluation in terms of the linguistic variables is shown in Table 3.

Excerpt of the three votes for criterion C_1 and parameters P01 and P02 represented as fuzzy numbers is presented in Table 4.

The votes for all criteria and parameters have been aggregated (Eq. 2). The definitions of the obtained ratings **R** and weighs **W** for all parameters are presented in Table 5.

E → The fuzzy indices of each criterion C_i are calculated (Eq. 3) by considering the influence of the respective parameters.

The fuzzy numbers of criterion C_1 using the values in Table 5 are obtained as follows:

$$C_1 = [5 \otimes (0.8, 0.9, 1) \otimes (4, 4.5, 5) \oplus 2 \otimes (0.8, 0.9, 1) \\ \otimes (3, 3.6, 4.2) \oplus (0.6, 0.72, 0.84) \otimes (3, 3.6, 4.2)] \\ /[7 \otimes (0.8, 0.9, 1) \oplus (0.6, 0.72, 0.84)] = (3.65, 4.18, 4.71)$$

For C_2 and C_3, the following fuzzy indices are obtained:

$C_2 = (3.78, 4.3, 4.82)$, from the aggregated values for parameters P09 to P17;
$C_3 = (3.35, 3.91, 4.47)$, from the aggregated values for parameters P18 to P23.

Table 3 Performance-ratings and importance- weights expert vote

Criterion C_i	Weight W_i	Parameter P_{ij}	Weight W_{ij}	Rating R_{ij}
C_1	VH	P01	VH	E
		P02	VH	E
		P03	VH	E
		P04	VH	E
		P05	H	VG
		P06	VH	VG
		P07	VH	E
		P08	VH	VG

Table 4 Excerpt of votes for performance and importance of criteria and parameters

Criterion C_i	Weight W_i	Parameter P_{ij}	Weight W_{ij}	Rating R_{ij}
C_1	(0.7, 0.85, 1) (0.7, 0.85, 1) (0.7, 0.85, 1)	P01	(0.5,0.7, 0.85) (0.7,0.85, 1) (0.7,0.85, 1)	(3, 4, 5) (4, 5, 6) (4, 5, 6)
		P02	(0.5,0.7, 0.85) (0.5,0.7, 0.85) (0.7, 0.85, 1)	(4, 5, 6) (3, 4, 5) (4, 5, 6)
		…	…	…

Table 5 Aggregated fuzzy sets for ratings and weights of the parameters

C_1			C_2			C_3		
P_{ij}	R_{ij}	W_{ij}	P_{ij}	R_{ij}	W_{ij}	P_{ij}	R_{ij}	W_{ij}
P01	(4, 4.5, 5)	(0.8, 0.9, 1)	P09	(4, 4.5, 5)	(0.8, 0.9, 1)	P18	(4, 4.5, 5)	(0.8, 0.9, 1)
P02	(4, 4.5, 5)	(0.8, 0.9, 1)	P10	(4, 4.5, 5)	(0.8, 0.9, 1)	P19	(4, 4.5, 5)	(0.8, 0.9, 1)
P03	(4, 4.5, 5)	(0.8, 0.9, 1)	P11	(4, 4.5, 5)	(0.8, 0.9, 1)	P20	(3, 3.6, 4.2)	(0.8, 0.9, 1)
P04	(4, 4.5, 5)	(0.8, 0.9, 1)	P12	(4, 4.5, 5)	(0.8, 0.9, 1)	P21	(3, 3.6, 4.2)	(0.8, 0.9, 1)
P05	(3, 3.6, 4.2)	(0.6,0.7,0.8)	P13	(4, 4.5, 5)	(0.8, 0.9, 1)	P22	(3, 3.6, 4.2)	(0.6, 0.7, 0.8)
P06	(3, 3.6, 4.2)	(0.8, 0.9, 1)	P14	(4, 4.5, 5)	(0.8, 0.9, 1)	P23	(3, 3.6, 4.2)	(0.8, 0.9, 1)
P07	(4, 4.5, 5)	(0.8, 0.9, 1)	P15	(4, 4.5, 5)	(0.8, 0.9, 1)			
P08	(3, 3.6, 4.2)	(0.8, 0.9, 1)	P16	(3, 3.6, 4.2)	(0.8, 0.9, 1)			
			P17	(3, 3.6, 4.2)	(0.8, 0.9, 1)			

F → The global criterion index is obtained in a similar way as the indices of C_1, C_2 and C_3, i.e., $GC = (3.59, 4.13, 4.7)$. It summarizes the influence of the criteria and their parameters with respect to security of the TeSLA system deployment at TUS. The obtained result in terms of fuzzy values is matched to the corresponding linguistic variable of the security level described in Table 1, and it is characterized as *very secure* deployment.

G → The Euclidean distance of the global criterion GC to each security level (Eq. 4) is presented in Table 6. The linguistic value of the security level (cf. Table 1) is determined by the minimal value of the calculated Euclidean distance. The minimal value is D(GC, VS) = 0.038, which corresponds to security level denoted as *very secure*. This result validates the security level, obtained in **F**.

H → Identification of criteria and parameters that could be improved for enhancing the security level. For this purpose, Fuzzy Performance-Importance Indices are calculated (Eq. 5). FPII of parameter P01 is obtained as follows:

$$FPII = [(1, 1, 1) \ominus (1, 0.9, 0.8)] \otimes (4, 4.5, 5) = (0, 0.45, 1).$$

Table 6 Euclidean distance from global criterion to each security level

D(GC, VS)	D(GC, S)	D(GC, MS)	D(GC, PS)	D(GC, NS)
0.038	0.383	1.187	1.796	2.503

Table 7 Fuzzy performance-importance indices and rating scores of parameters

Parameter	FPII (Fuzzy-performance importance index)	RS (Rating score)
P01	(0, 0.45, 1)	0.475
P02	(0, 0.45, 1)	0.475
P03	(0, 0.45, 1)	0.475
P04	(0, 0.45, 1)	0.475
P05	(0.48, 1.01, 1.68)	1.04
P06	(0, 0.36, 0.84)	0.39
P07	(0, 0.45, 1)	0.475
P08	(0, 0.36, 0.84)	0.39
P09	(0, 0.45, 1)	0.475
P10	(0, 0.45, 1)	0.475
P11	(0, 0.45, 1)	0.475
P12	(0, 0.45, 1)	0.475
P13	(0, 0.45, 1)	0.475
P14	(0, 0.45, 1)	0.475
P15	(0, 0.45, 1)	0.475
P16	(0, 0.36, 0.84)	0.39
P17	(0, 0.36, 0.84)	0.39
P18	(0, 0.45, 1)	0.475
P19	(0, 0.45, 1)	0.475
P20	(0, 0.36, 0.84)	0.39
P21	(0, 0.36, 0.84)	0.39
P22	(0.48, 1.01, 1.68)	1.04
P23	(0, 0.36, 0.84)	0.39

The calculated value (Eq. 6) is the rating score. The RS index of P01 is:

$$RS = \frac{0 + 2.0.45 + 1}{4} = 0.475.$$

The obtained Fuzzy performance-importance indices and rating scores of all parameters are presented in Table 7.

3.3 *Evaluation and Recommendations*

The minimal values of the rating scores indicate which parameters must be addressed in terms of potential vulnerability for compromising the security status of the TeSLA deployment at TUS. Table 7 shows that the minimal value of rating score is 0.39, which is bounded to parameters P06, P08, P16, P17, P20, P21, P23. This means that special attention must be paid to the following criteria and specific parameters, in order to assure proper maintenance of the security level obtained by the deployment of TeSLA at TUS:

- P06 (applying standards for security) and P08 (guidelines regarding security) referring to criterion C_1 (programming code);
- P16 (network monitoring) and P17 (security configuration) referring to criterion C_2 (administration);
- P20 (home-working and mobile devices), P21 (rules for data usage) and P23 (managing incidents) referring to criterion C_3 (policies).

In the context of TeSLA deployment in the virtual learning environment at TUS P08 implies the need for requirements and recommendations for the secure elaboration, adoption, administration and use of the TeSLA web services. P16 refers to the configuration of a firewall; traffic monitoring and understanding web services' proper functioning. In case of recognized problems, vulnerabilities and possible threats and attacks should be detected. P17 implies proper certificate authorization management; efficient access control of different user categories, i.e. students, teachers, managers, administrative staff.

The obtained and validated result "very secure" of the fuzzy security index after performing fuzzy logic based empirical analysis of security issues indicated that the deployment of TeSLA at TUS is at sufficient security level. Another analysis result suggested some criteria and parameters subjected to further improvement in order to guarantee its maintenance at the required security level.

4 Identities and Privacy Management

As pointed out in Sect. 2 of this chapter, TeSLA provides numerous protection features to properly secure the exchange of learners' data from institutional and cloud domains, to third party services. To guarantee the high degree of security evaluated in Sect. 3, the learner identity managed by the TeSLA components must be seamlessly and securely linked to the identity in use in each context.

Notice that a single learner may have several digital identities. For instance, the identity provided by the university where the learner is registered. Usually, this identity is linked with learner data such as first and last names, date of birth, and visual data (e.g., learners' photography). When the learner accesses the new services provided by TeSLA, direct mapping is performed between login and personal data, if the connection is within the institutional domain. However, whenever the services refer to other domains (e.g., cloud domain or third party agents), the learner may have to make use of other digital identities needed to authenticate to these services. Since these identities all refer to the same learner, they cannot be decorrelated.

Several solutions exist to reduce the need for various credentials and refer to only one identity. The first one consists in adding third party services as plugins to the learning environments, and forwarding the necessary personal data to the services, based on the identity in use in the institutional domain. For instance, standards like LTI (Learning Tools Interoperability) make these interactions possible [24]. A trust relationship is established between domains, e.g., using a shared secret that ensures the security of all the remaining exchanges based on the OAuth 1.0 standard [25].

The second solution, which is adapted to the case of third party services built independently from TeSLA, consists in relying on identity federation, implemented in several standards such as OpenID, SAML (*Security Assertion Markup Language*), or Shibboleth. Identity federation consists in delegating authentication to an identity provider. The user who wishes to access the service provider is redirected to the identity provider for authentication, where authorization token is generated to certify the authentication success. This token is then transmitted to the service provider, which allows the user to be regarded as authenticated. This way, in case of deploying TeSLA as a standalone service, the enforcement of authentication can be managed using traditional identity federation services, i.e., by simply allowing institutional domains to act as identity providers.

4.1 Use of Pseudonymous Identification in TeSLA

While the above standards provide the technological basis to link learner's identity with TeSLA, specific attention should be paid to privacy issues. As already anticipated in Sect. 1, the personal data associated to one's identity must be carefully managed during the association of entities and domains, e.g., to avoid the unauthorized disclosure of private information. In order to make it possible for the learner to take assessments without disclosing personal information, TeSLA provides pseudonymity management. Learners are authenticated and authorized in TeSLA without allowing TeSLA to know the real identity of the learner.

Anonymity is only partial in this context, since links between the TeSLA identifiers (hereinafter denoted as TesLA ID) and the learners' true identities remain available at institutional level (e.g., at the university domain). The precise approach, from a technical standpoint, is as follows. Each institution (e.g., university) generates a series of randomized Universally Unique IDentifiers (UUIDs) (TeSLA uses version four of the UUID standard [26] for each learner. As such, the institution generating the UUIDs is the only entity able to make the link between a precise UUID and the records of the learner. Using public information, such as learner's e-mail address to generate a UUID using version 3 or version 5 of the UUID standard, is excluded for operational deployments, since it allows attackers to compute all the possible TeSLA IDs from the learner's directory, and deduce the link between learners' names and TeSLA IDs.

The UUIDs are stored in databases that are shared between all the remaining components of the TeSLA architecture (see Sect. 2). A dedicated component, i.e., the TeSLA Identity Provider, is mapped to the database in order to receive requests from the remaining TeSLA components. This provider is issued with learners' true identities, and replies with the corresponding UUID. The communication between TeSLA components (e.g., TeSLA plugins) and the identity provider is mutually authenticated with TLS (see Sect. 2). In case learners' authentication is certificate-based, e.g., learners interacting with TeSLA through a series of plugins, the learner only needs to authenticate once. The certificate used for the authentication of each component is associated to learner's true identity. Then, when a request is sent to TeSLA, the system retrieves the TeSLA ID associated to learner's identity from the identity provider, and eventually communicates with other TeSLA components, while guaranteeing the pseudonymity of the learner.

Some external tools, embedded as JavaScript code within the learner's web browser, also need to communicate to the TeSLA system without revealing learner's true identity, nor retrieving the TeSLA ID either. A session token mechanism, based on JWT (JSON Web Tokens) [27], is used for this purpose. When a TeSLA plugin authenticates to the identity provider and retrieves the TeSLA ID, tokens are created and provided to the external tools, using public key cryptography to secure the signature of the tokens.

With respect to the protection of learners' data outside their respective institution data centers, no traceability features are implemented. Apart from the TeSLA ID association, stored at the identity provider (within the learner's institution domain), all the remainder personal data of learners, such as the IP address or similar data, which could be used to map different sessions of the same learner, are omitted. As a result, the architecture presented in Sect. 2 provides full pseudonymity for learners. Learner's identity remains only known within the institution, while never transmitted to other components.

4.2 Future Directions Towards Extended Privacy Functionalities

As already stated in Sects. 2 and 3, an e-assessment system like TeSLA has to be properly secured with classical measures, such as authentication, data ciphering and integrity checks, in order to mitigate cyber-attacks that might lead to disastrous consequences, such as data leakage or identity theft. In addition, and to meet the GDPR recommendations [5], it is also necessary to ensure a reasonable level of security in the system (cf. Sect. 3).

Security and privacy are two very close domains, and yet important differences have to be highlighted, since it is possible to build a very secure system that fails to ensure any privacy properties. Security, from a technological standpoint, consists in guaranteeing specific requirements at different levels of the architecture, such as confidentiality, integrity or authentication. It mainly targets the exchange and storage of data, which, in the case of TeSLA, may contain some traces of learner's biometric data, learner's assessment results, and other sensitive information. In contrast with security, privacy consists in preventing the exploitation of metadata to ensure that no personal information leakage will occur. However, it always remains mandatory to comply with legal constraints, which may prevent full anonymization of the communications. Therefore, the main objective of privacy, from a technological perspective, is to reveal the least possible information about user's identity, and to prevent any undesired traceability, which is often complex to achieve.

In the context of TeSLA, several privacy technological filters are already included in the underlying design of the architecture. The randomized TeSLA identifier associated to each learner (cf. Sect. 4.1) is a proper example. This identifier is used each time the learner must access TeSLA, hence ensuring pseudonymous identification of learners—full anonymity not being an option in TeSLA for legal reasons. Yet, a randomized identifier alone cannot protect learners against more complex threats such as unwanted traceability. The system can still be able to link two different sessions of the same learner. A technical solution to handle such issues, which is proposed as potential extension of the TeSLA PKI architecture [28, 29], is the use of anonymous certification.

Anonymous certification allows users to prove they are authorized to access a resource without revealing more than they need about their identity. For example, users can be issued with certified attributes that may be required by the system verifier, such as "Older than 18", "studies at IMT", or "lives in France". When the users want to prove that they own the right set of attributes, they perform a digital signature based on the required attributes, allowing the system verifier to check if a precise user is authorized, sometimes without even knowing precisely which attributes were used.

Such an approach could be integrated in several points of the TeSLA architecture where it is not necessary to identify the learner. For example, to access course material on a learning environment, it should be enough to prove that the learner comes from an allowed institution and is registered for this course. That way, it becomes impossible for the learning environment to follow the studying activity of each learner, while still letting learners access the course material. Similarly, when a learner has taken an assessment, the learner's work can be anonymously sent to anti-cheating tools (such as anti-plagiarism). With anonymous certification, each tool might receive a request for the same work without being able to know which learner wrote it, but also without being able to correlate the requests and decide whether they were issued by the same learner.

We argue that anonymous certification might prove to be a solid and innovative asset to enhance privacy in TeSLA, and to prevent traceability of learners whenever traceability is not required. Other approaches might also be added, following the same direction for privacy enhancement. One of them consists in mixing together the data stored in a database in order to make it impossible to associate the various attributes of a table entry, hence offering anonymous data storage. Should such a technique be integrated to TeSLA, it would guarantee that even a data leak from a sensitive database will not provide any certain information to anyone—as long as the leaked data do not contain secrets such as private keys or passwords.

In terms of trust, enhanced features beyond learners' privacy can also be added to future releases of the architecture. A system like TeSLA, where learners have to take e-assessments under strict anti-cheating countermeasures, requires a high degree of trust from learners in order to be widely deployed and accepted as a legitimate assessment tool. TeSLA should provide public guarantees that its claims regarding privacy and security are met, meaning that TeSLA is as transparent as possible with respect to personal data management processes. Though it is not directly related to security and privacy, TeSLA should also ensure transparency regarding the anti-cheating decision processes, and let learners know how these decisions are made while informing them of possible resorts at their disposal in case of false positive detection.

Pushing the analysis further requires an overall look on the global architecture, and on the fundamental choices that led to its design. Among these choices, relying on biometry for learners' authentication there is one that particularly stands out in terms of privacy. Contrary to a password, which would authenticate learners using what they know, biometric samples authenticate them using what they are. The data transmitted over the network, from the learner's computer to the TeSLA instruments, are parts of learners' identity and as such, are much more sensitive than mere passwords, which can be changed at will. With encrypted data exchanges over the network, TeSLA ensures that these biometric samples cannot be retrieved by an attacker. The anonymous treatment of samples by the TeSLA instruments strongly limits the risks in terms of unwanted access and exploitation of personal data.

The choice of biometric-based authentication for learners who are taking e-assessments entails other issues. Firstly, the biometric samples are collected from the learner's computer, which by definition has no guarantee whatsoever regarding security. Even if the samples are not meant to be stored on the learner's computer, the risk of personal data theft at this point is independent from the TeSLA architecture, but is induced by the choice to rely on biometry. As such, it should be taken into account for further improvement of the TeSLA system. Secondly, even though the biometric samples are anonymized before they are sent to the TeSLA instruments, it may be better not to send such sensitive data to TeSLA at all, and decentralize Trusted Third Parties (TTPs) as much as possible. The role of TeSLA is to offer a specific service, namely the possibility to take e-assessments. It does not, and could not act as a TTP. In the current configuration, what happens to the biometric samples depends on how TeSLA is managed. With a TTP, which would have no specific connection to TeSLA or to the academic institutions, there would be a dedicated entity whose explicit role would be to guarantee the treatment of these sensitive data, independently of the current TeSLA policy. Notice that anonymous certification will benefit of such a TTP-decentralization, as well.

To sum up, we consider that improving the privacy in TeSLA requires further decentralization of its fundamental choices, in order to offer the best privacy guarantees. Even if the use of biometry is maintained as it is, extending current TTP elements, such as the TeSLA Public Key Infrastructure (PKI) , and the underlying Certification Authorities (CAs) , would be a significant step in this direction.

5 Conclusion

On top of addressing numerous challenges, we have seen in this chapter that the TeSLA architecture has been designed with security guarantees in terms of communication exchanges, as well as in terms of learners' data protection with respect to their privacy, in compliance with the GDPR requirements of the European Union. Ensuring privacy in TeSLA consists in minimizing the personal information retrieved by the system during its interactions with learners. While obviously securing the access to databases, TeSLA makes sure to anonymize every sensitive data collected from the learner. This process applies to e-assessments, which are taken by learners with an anonymized identifier, but also to the biometric samples required to authenticate the user. These biometric samples are anonymized in the same way before reaching the TeSLA system, where they are dispatched to various instruments that will analyze them accordingly, and return the results to TeSLA.

In the first part of this chapter, we have addressed possible security risks in different learning scenarios implemented in an e-assessment process from learners' and educators' perspectives. It highlights the recognition and verification of learner's identity, the disclosure of information to unauthorized parties and the fraudulent data alteration as the most challenging ones. Technical solutions, guidance and actions implemented as security services in the architecture of the TeSLA e-assessment system, are outlined and discussed. The presented solution is mainly based on TLS protection via authorized certificates and public key infrastructures. Certificate management and security procedures to apply in case of private key disclosure were also explained. It has been shown that TeSLA guarantees the required security level concerning confidentiality and integrity of system communications of the e-assessment process in different learning scenarios, which respects the European regulations for the appropriate protection of all personal data referring to user identities.

In the second part, we have performed a methodological verification of the underlying services of TeSLA. The evaluation has been performed in the technical context of the TeSLA deployment at TUS (the Technical University of Sofia), as a representative member of the TeSLA consortium where the pilots of the TeSLA project were conducted. The evaluation concerned the deployment environment and experience gained by TUS during the execution of the three pilots of the project. The methodology used for the execution of our evaluation is based on the use of Fuzzy Set Theory and Multi-Criteria Decision Making Methodologies. Both disciplines have been applied to analyze the current security level of the TeSLA environment at TUS, in order to provide information to responsible professionals, interested bodies as well as to end users about the security level of the TeSLA deployment at TUS and suggest issues for its proper maintenance.

In the third part, we have presented the precise approach for pseudonymization of learners in TeSLA. We have also provided some ideas for future research directions, towards extended privacy functionalities. Among them, we have highlighted enhancements such as adding anonymous certification, and improving the level of transparency of the whole system. We have argued that the technical completion of the TeSLA platform, as well as its seamless integration to usual educational activities, are probably the two most obvious factors that one can name. TeSLA must succeed in convincing learners that they can trust the system as a legitimate examination module that is devoid of any serious risk for their personal data. Ensuring privacy and transparency not only allows TeSLA to meet the requirements of the GDPR; it will greatly help TeSLA to obtain learners' trust, even more than achieving legal and ethical considerations.

References

1. ANSSI (2016) Best practices. Available from: https://www.ssi.gouv.fr/administration/bonnes-pratiques/ (20 Oct 2016)
2. IEEE Standards (2018) 29148—2018—ISO/IEC/IEEE international standard—systems and software engineering—life cycle processes—requirements engineering. Available from: https://ieeexplore.ieee.org/document/8559686. (30 Nov 2018)
3. ISO (2013) ISO/IEC 27001:2013—information technology—security techniques—information security management systems—requirements. Available from: https://www.iso.org/standard/54534.html (1 Oct 2013)
4. OWASP (2013) OWASP top 10 most critical web application security risks… Available from: https://www.owasp.org/index.php/Category:OWASP_Top_Ten_Project (1 Nov 2016)
5. EUR-Lex (2016) Regulation (Eu) 2016/679 of the European Parliament and of the council of 27 April 2016 on the protection of natural persons with regard to the processing of personal data and on the free movement of such data, and repealing Directive 95/46/EC (General data protection regulation), 2016. Available from: https://eur-lex.europa.eu/eli/reg/2016/679/oj (27 Apr 2016)
6. Dierks T, Rescorla E (2008) The transport layer security (TLS) protocol. Available from: https://tools.ietf.org/html/rfc5246 (11 Nov 2018)
7. TeSLA (2016) TeSLA home page. Anonymous certification in TeSLA. Available from: https://tesla-project.eu/anonymous-certification-tesla/ (20 Jul 2017)
8. Apampa KM, Wills G, Argles D (2009) Towards security goals in summative E-assessment security. In: 2009 international conference for internet technology and secured transactions (ICITST), pp 1–5. Available from IEEE Xplore Digital Library (29 Jan 2010)
9. Thamadharan K, Maarop N (2015) The acceptance of E-assessment considering security perspective: work in progress. World Acad Sci Eng Technol Int J Comput Inf Eng 9(3):874–879
10. Laurent M, Bouzefrane S (eds) (2015) Digital identity management. ISTE, London
11. ITU (2016) X.509: information technology—open systems interconnection—the directory: public-key and attribute certificate frameworks. Available from: https://www.itu.int/rec/T-REC-X.509-201610-P/en (14 Oct 2016)
12. Cooper M, Dzambasow Y, Hesse P, Joseph S, Nicholas R (2005) Internet X.509 public key infrastructure. Certification path building. Available from: https://tools.ietf.org/html/rfc4158 (11 Nov 2018)
13. Menezes AJ, van Oorschot PC, Vanstone SA (2011) Handbook of applied cryptography. CRC Press, US
14. Kiennert C, Rocher PO, Ivanova M, Rozeva A, Durcheva M, Garcia-Alfaro J (2017) Security challenges in e-assessment and technical solutions. In: 8th international workshop on interactive environments and emerging technologies for eLearning, 21st international conference on information visualization, London, UK, pp 366–371. Available from IEEE Xplore Digital Library (16 Nov 2017)
15. Cooper D, Santesson S, Farrell S, Boeyen S, Housley R, Polk W (2008) Internet X.509 public key infrastructure certificate and certificate revocation list profile. Available from: https://tools.ietf.org/html/rfc5280 (11 Nov 2018)

16. Santesson S, Myers M, Ankney R, Malpani A, Adams C (2013) X.509 internet public key infrastructure online certificate status protocol—OCSP. Available from: http://www.rfc-editor.org/info/rfc6960 (11 Nov 2018)
17. Barker E (2016) Recommendation for key management, part I: general. Available from: https://csrc.nist.gov/publications/detail/sp/800-57-part-1/rev-4/final (12 Feb 2019)
18. Baró-Solé X, Guerrero-Roldan AE, Prieto-Blázquez J, Rozeva A, Marinov O, Kiennert C, Rocher PO, Garcia-Alfaro J (2018) Integration of an adaptive trust-based E-assessment system into virtual learning environments—the TeSLA project experience. Internet technology letters. Available from: https://doi.org/10.1002/itl2.56 (09 June 2018)
19. Ivanova M, Rozeva A (2017) Methodology for realization of secure web services. In: Proceedings of academics world international conference, Edinburgh, UK, pp 16–21
20. Kumar S, Dutta K (2011) Investigation on security in LMS MOODLE. Int J Inf Technol Knowl Manage 4(1):233–238
21. Zadeh L (1965) Fuzzy sets. Inf Control 8:338–353
22. Porebski S, Straszecka E (2016) Membership functions for fuzzy focal elements. Arch Control Sci 26(3):395–427
23. Ansari S, Mittal P, Chandna R (2010) Multi-criteria decision making using fuzzy logic approach for evaluating the manufacturing flexibility. J Eng Technol Res 2(12):237–244
24. Durand G, Downes S (2009) Toward simple learning design 2.0. In: 2009 4th international conference on computer science & education, pp 894–897. Available from IEEE Xplore Digital Library (01 Sept 2009)
25. Leiba B (2012) OAuth web authorization protocol. IEEE Internet Comput 16(1):74–77. Available from. https://www.computer.org/csdl/magazine/ic/2012/01/mic2012010074/13rRUxjyX0o (20 Feb 2012)
26. Leach P, Mealling M, Salz R (2005) A universally unique identifier (UUID) URN namespace. Available from: https://tools.ietf.org/html/rfc4122
27. Jones M, Bradley J, Sakimura N (2015) JSON Web Token (JWT). Available from: http://www.rfc-editor.org/info/rfc7519 (19 Jan 2019)
28. Kiennert C, Kaaniche N, Laurent M, Rocher PO, Garcia-Alfaro J (2017) Anonymous certification for an e-assessment framework. In: Proceedings of 22nd Nordic conference on secure IT systems (NordSec 2017), Tartu, Estonia, pp 70–85
29. Kaaniche N, Laurent M, Rocher PO, Kiennert C, Garcia-Alfaro J (2017) PCS, a privacy-preserving certification scheme. In: 22nd ESORICS symposium 12th international workshop on data privacy management (DPM 2017), Oslo, Norway, pp 239–256

Design and Implementation of Dashboards to Support Teachers Decision-Making Process in e-Assessment Systems

Isabel Guitart Hormigo, M. Elena Rodríguez and Xavier Baró

Abstract The growing number of universities adopting some form of e-learning in recent years has raised some concerns about how to ensure students' authentication, and the authorship of the assessment activities they deliver. There are several strategies and market tools that can help teachers in these tasks. While the usage of plagiarism detection tools for checking authorship is common practice (above all in fully online universities), the use of biometric instruments for ensuring students' identity is less extended. Although all these tools collect a large amount and variety of data, there is a lack of software systems that can integrate such data, and show the information that may be extracted from these data in a visual and meaningful way that fits the teachers' needs. Precisely, the objective of this chapter is to present a set of dashboards that integrate data collected by different kinds of authentication and authorship instruments, oriented to assist the decision-making process of teachers, above all in case of suspicion of students' dishonest academic behavior. Although these dashboards have been designed and implemented in the context of TeSLA project, the experience and conclusions provided here are of interest to researchers and practitioners aiming to develop dashboards with learning analytics purposes at higher education. For this reason, this chapter also provides a discussion and review of the most prominent analytical efforts in universities.

Keywords e-assessment systems · Authentication · Authorship · Decision-making · Learning analytics · Key performance indicator · Dashboards · Audit data

---

I. Guitart Hormigo (✉) · M. E. Rodríguez · X. Baró
Universitat Oberta de Catalunya, Rambla del Poblenou, 156, 08018 Barcelona, Spain
e-mail: iguitarth@uoc.edu

M. E. Rodríguez
e-mail: mrodriguezgo@uoc.edu

X. Baró
e-mail: xbaro@uoc.edu

D. Baneres et al. (eds.), *Engineering Data-Driven Adaptive Trust-based e-Assessment Systems*, Lecture Notes on Data Engineering and Communications Technologies 34,
https://doi.org/10.1007/978-3-030-29326-0_6

Acronyms

API	Application Programming Interface
ICT	Information and Communication Technologies
KPI	Key Performance Indicators
LMS	Learning Management System
MOOC	Massive Open Online Courses

1 Introduction

The university is experiencing a paradigm shift driven by the transformation of the knowledge society, and the specific demands of an increasingly competitive university environment. In such a context, improving the management of the university, and achieving better levels of quality (in teaching, innovation and research) have become a priority in an economically sustainable context [23]. One of the ways in which the university may achieve previous goals is to make objective decisions with the help of analytical tools. By using them, the university will be able to make decisions based on data and evidences, thus reducing the decision-making processes based on beliefs and intuitions [24]. Unfortunately, the university is not exploiting all its analytical potential [6, 47]. A large amount of data, coming from different and heterogeneous data sources, is generated within the university, but data are neither systematically collected nor analyzed for decision-making purposes [15]. For this reason, several authors classify the university in the lower stages of the models of analytical maturity [25, 47].

The innovation carried out in the teaching and learning processes driven by the irruption of the Information and Communication Technologies (ICT) has modified the traditional university. ICT have made possible new types of learning, such as e-learning. Currently, most of the traditional universities incorporate different forms of e-learning in their teaching and learning processes, as a way of complementing traditional face-to-face classes (i.e. blended learning). In addition, ICT have enabled the emergence of fully online (or virtual) universities. Blended and online universities use Learning Management System (LMS) platforms, such as Moodle[1] and Blackboard[2] for e-learning purposes. Frequently, online universities develop their own LMS by adding new additional services, addressed to support all the university stakeholders and processes, giving rise eventually to online (or virtual) campus. The major features of e-learning [24, 28, 43, 47] include flexibility and openness in the learning process, thus creating new opportunities for students that have special educational needs (as it would be the case of students who suffer some disability), as well as the increase of teaching quality and students' satisfaction. Innovation in

[1] https://moodle.org/.

[2] https://www.blackboard.com/index.html.

e-learning has also led to the creation of a new type of data, the digital trace. Each student's interaction within the LMS generates a digital trace composed of data, such as navigational data, textual data, the patterns a student follows when he or she accesses to the available learning resources [31], etc.

As consequence, there is a growing and genuine interest of the teachers in extracting new knowledge based on the analysis of academic data and the digital traces of students, which are stored in the databases and logs of the LMS and virtual campus [33]. Analyzing these rich data sets (in quantity and diversity) means knowing the students better (e.g. the concepts they learn more easily; or conversely, the concepts they are facing more difficulties, the interests they have, the most appropriate learning resources for each student profile, etc.). The analysis also facilitates the improvement of the LMS and campus services (for example, enhancements in the structure and navigation functionalities). In summary, teachers can know what is really happening in their courses, and they can be able to make decisions based on data analysis. During this analysis process, monitoring and provision of feedback to the students are also essential elements. Feedback can help students to achieve competences acquisition, i.e. the data analysis may allow a comparison between the student' expected learning goals achievement and his or her present state [27]. This analytical discipline focused on the university domain is referred in the literature as *learning analytics* [45]. Most of the current analytical initiatives and research works in this field have as main objective to improve the students' academic success, through the creation of tools aiming to personalize and facilitate the learning process, the reduction of dropout rate, and the enhancement of students' performance.

Inherently, e-learning involves the realization of assessment activities that students perform at home during the course, without the supervision of the teaching staff. Probably, this is one of the major weaknesses or criticisms regarding e-learning, and there is a certain concern for ensuring the authorship and authentication of the assessment activities submitted by the students. For these reasons, most universities (including fully online universities) still maintain traditional face-to-face examinations at the end of each academic term. In spite of this, there is a diversity of tools that can help teachers to find similarities among the assessment activities submitted by the students, or the inappropriate use of information sources available at Internet [1, 38]. For example, plagiarism tools as Urkund,[3] or forensic analysis techniques. Other strategy consists of conducting online assessment activities within proctored settings coupled with biometrics instruments as face and voice recognition or keystroke dynamics [18]. Although the use of previous mechanisms can help to prevent or detect cheating, the literature suggests a third strategy that lies in modifying students' attitudes toward cheating. Such strategies include, among others, informing students or requiring agreement with an honor code [8]. Data about authorship and authentication, used according to the national legal framework and ethical code of the university, can complement the data sets used in the learning analytics field. Data sets can serve to enhance the knowledge teachers have about their students. More-

[3]https://www.urkund.com/about-urkund.

over, they are valuable assets in the decision-making process in case of students' dishonest behavior.

In spite the research efforts in the learning analytics field, the reality is that few teachers really benefit from the opportunities that the analysis of such huge amount of data brings, because there is a lack of analytical tools that support teachers in their decision-making process to improve teaching quality [23]. Although learning analytics is the new framework for posing questions related to teaching and learning processes, the answers should be turned into new tools and dashboards that help teachers, students, and managers to better perform and understand their tasks within an educational institution [33]. Furthermore, these analytical tools must be integrated within the LMS in order the teachers (as well as other stakeholders) can have an efficient and effective access to the exploitation and visualization of the analysis results.

The goal of this chapter is to discuss a set of dashboards of an analytical tool specifically designed to support the decision-making process regarding the authorship and authentication of the e-assessment systems, in the context of TeSLA project. Although the chapter focuses on TeSLA dashboards, most of the discussion and conclusions provided are also of interest for practitioners and researchers interested in the development of dashboards for supporting decision-making processes in the learning domain. The overall objective of TeSLA project is to define and develop an e-assessment system, which provides an unambiguous proof of students' academic progression during the whole learning process, while avoiding the time and physical space limitations imposed by face-to-face examinations [43].

The chapter is structured as follows: Sect. 2 discusses the most relevant analytic techniques that can be applied in e-learning, and how they can be used to improve the university analytical potential, as well as the large volume of data that is generated in e-learning environments. Following this, Sect. 3 presents the visual component of the analytical tools, the dashboards. Subsequently, the data sources captured in TeSLA, how they are analyzed and presented in dashboards are analyzed in Sect. 4. These dashboards are intended to support teachers in decision-making processes about the authorship and authentication of the assessment activities carried out by students in e-learning environments. Finally, Sect. 5 provides the conclusions and future work.

2 Analytical Efforts in Higher Education

This section reviews the analytical systems available at the university, and how they have evolved along the time, as well as the large amount and variety of data generated within the university. Among the wide range of systems, we focus on analytical systems and data generated in online teaching-learning environments.

2.1 Learning Analytics

The interest in analytical systems in academic and institutional environments is not new. The objective of the first analytical systems was efficient planning and management of the university [19, 35, 49]. The institutional initiatives are concerned with improving organizational processes, such as personnel management or resource allocation, improving efficiency within the university, and also measuring and monitoring external indicators defined by the third evaluation parties (government or external quality agencies), and internal indicators related to the strategic objectives of the university [26, 29, 37]. Decades later, the area of *learning analytics* emerges. Siemens and Long [45] defined learning analytics as "the measurement, collection, analysis and reporting of data about students and their contexts, for purposes of understanding and optimizing learning and the environments in which it occurs". The learning analytics systems may use analytical techniques to deal with different aspects of education, such as understanding the learning process of students [17, 50], finding out the students at risk of dropping out [40], or calculating the satisfaction of students at institutional and departmental level [7, 47]. In publications as [39, 51] there are examples of learning analytics tools, whereas several case studies are published in [47].

The analysis of academic data and the digital traces generated in the LMS or other smart devices (as tablets or mobile phones) in educational contexts has boosted other analytic disciplines, as it would be the case of the *educational data mining* field. Whereas educational data mining looks for new patterns in data, and concentrates more on the automated discovery of intelligence and information from data, learning analytics applies known predictive models, placing a considerable focus on leveraging human judgment through the presentation of meaningful information extracted from large data sets [12, 46]. This chapter focuses on the learning analytics field.

2.2 Data in Higher Education

As aforesaid, universities have large volumes of data, but little analytical culture [20]. Then they cannot accumulate great amounts of knowledge from such data. Universities usually provide simple and routine reports to their teachers, obtained from their academic management information systems. However, these reports do not tend to provide thorough analysis of data, being of little relevance for taking academic decisions. An analytical system requires great volumes of data to extract knowledge. The key factor for having success with these systems is not to have large quantities of data, but to know how to analyze and organize the data correctly to answer the analytical questions of each user [24].

The learning and teaching processes performed in LMS (and virtual campus) generate digital traces of the tasks performed for all LMS users, generating quickly huge amounts of data, very heterogeneous and of different type: structured, semi-structured and unstructured [17, 23]. In those systems, all possible communications among students, and between students and teachers are done within the LMS, and therefore can be stored. Furthermore, LMS can gather navigational data, describing the way students learn and the way teachers teach. The stored data do not contain personal information about students and teachers, but about the way they navigate through the LMS, the services they use, the resources they consume, etc. In addition, any data should be anonymized. Thus, interactions performed within the LMS generate a great amount of data, allowing the use analytical techniques [17, 24] to extract, manipulate and analyze data. Such analysis may have several benefits to universities, such as providing a whole picture of the university or its departments and faculties, detecting and identifying students at risk of dropping out, estimating student satisfaction or detecting improvements in teaching.

3 The Importance of Dashboards in Analytical Environments

In this section, we firstly discuss dashboard tools created in business environments, and how their success has helped to transfer them to educational contexts. Secondly, we analyze how dashboards can be used in the field of learning analytics in higher education.

3.1 Dashboards in Organizations

Dashboards are the key visualization components of analytical tools. A dashboard shows the analysis results by means of friendly, clear, concise and intuitive visual interfaces, and it is easy to manipulate [9]. Eckerson [13] defined a dashboard as an "information delivery system that parcels out information, insights, and alerts to users on demand". Few [16] defined dashboards as a "visual display of the most important information needed to achieve one or more objectives; consolidated and arranged on a single screen so the information can be monitored at a glance". A dashboard is a "performance management system that allows users to track and respond to organizational or institutional activities based on key goals-based and objectives-based metrics or indicators" [34].

An effective dashboard has a set of characteristics [14] that enables users to: monitor performance against corporate strategy metrics, monitor processes that drive daily business, and monitor the progress toward organizational goals; analyze data across dimensions and hierarchies to determine the causes of problems; and manage collaboration and decision making by keeping up-to-date performance data at the fingertips of managers.

Dashboards have gained popularity as a method for executives, managers, and employees to easily keep an eye on key metrics and move to obtain the insight they need to resolve issues quickly, efficiently, and effectively [13]. They help to proactively improve decisions. Therefore, dashboards need to be concise and complete reports with a simple and accessible structure that shows not only the progress of a set of key indicators, but also all the relevant data related to the selected indicators. The visualization of data is considered crucial, because it converts the abstract and complex data to the concrete and visible by amplifying human cognition [27, 39].

In a dashboard, the information is represented through a combination of diagrams, tables, figures, maps, and images, among other elements. The objective is to make the information more understandable. Dashboards must be intuitive and emphasize visual simplicity instead of excessive graphics plenty of colors [16]. An effective dashboard has to communicate the fundamentals concepts, thus simplifying complex interactions. It also has to prevent from the risk that users accidentally change data, and it has to allow that users can do sophisticated analysis, without the help of specialists. Users should be able of detecting patterns, and relationships in large amounts of data, enabling the evaluation of different what-if scenarios, controlling the strategic indicators and, in general, thus facilitating the understanding of the results of the analytical tools. These advantages contribute to increasing the number and diversity of inexperienced users of the organization in technical and analytical aspects.

In the design process of a dashboard, it is as important to know what information is relevant, as to know what data are viable and available [16]. Dashboards are made from a set of metrics, called Key Performance Indicators (KPI). These metrics define the performance measures that represent the strategy of the organization.

3.2 Dashboards in Higher Education

The first dashboards on the higher education appeared in the late 1990s [34], and continue to be implemented due to pressures for accountability and strategic planning of increasingly scarce financial resources [11, 34, 41]. Areas related to academic affairs have recently increased the use of analytical systems [34] and dashboards [37, 48]. As consequence, dashboards are becoming increasingly popular as evaluation and performance management tools in the university [34]. These dashboards are relevant to define, monitor and evaluate the strategy and management [22] of a university campus [37], and/or the university at the national level [26, 29]. These dashboards support the decision-making processes of the university at different levels [24], for example, from the executive council [5] to the departments' directions [37], or the decisions made in the management of academic programs, such as offering a new academic program or a new course [32, 52]. There are also dashboards oriented to define, monitor and review the research strategy of the university [10, 22]. KPI included in this type of dashboards allow measuring the level of achievement of the strategic objectives, such as budgets, student enrollment, student performance and graduates of an academic program [37, 48].

The use of dashboards to analyze the data generated in the learning process and with a teaching purpose is more recent [44]. In some cases, dashboards with academic indicators are beginning to replace classical performance reports. Results suggest that these dashboards increase efficiency in academic making-decisions processes [36], because they enable that teachers make academic decisions based on data and evidences, thus improving teaching quality. For example, and regarding the learning process, dashboards can help to identify trends in student needs, or students' success and dropout patterns. It worth noting that dashboards (with the required adaptations) can also provide personalized feedback to the students, for example, dashboards can show students information about their learning progress at different levels (e.g. assessment activities, use of learning resources, and enrolled courses). The information provided by all these dashboards is mainly built upon the analysis of the data generated within the LMS. Frequently, dashboards are complemented with warning systems [3, 4] that trigger notifications of possible problems (both teachers and students).

Schwendimann et al. [44] present a review on dashboards in the fields of learning analytics and educational data mining. Learning dashboards are becoming popular due to the increased use of educational technologies, such as LMS and Massive Open Online Courses (MOOC). There are synonymous terms for learning dashboards, such as "educational dashboard", "learning analytics dashboard", "student dashboard" and "web dashboard" [27, 42, 44]. Schwendimann et al. [44] define learning dashboard as "a single display that aggregates different indicators about student(s), learning process(es) and/or learning context(s) into one or multiple visualizations". Park and Jo [39] define learning dashboard as "an interactive, historical personalized, and analytical monitoring display that reflects students' learning patterns, status, performance and interactions". Thereby, learning dashboards are an effective tool to support student motivation and the self-regulation process [27]. Dashboards provide feedback to increase students' self-reflection and self-awareness and improve learning strategies and learning outcomes and retention [27]. Learning dashboards can be classified into three types [39]:

1. Dashboards for teachers only: they inform about the student learning status and help in class management, the provision of feedback and evaluation, in a scalable way. For example, the LOCO-Analysis tool is focused on providing feedback on students' learning activities and performance [2].
2. Dashboards both teachers and students: they present learning patterns of students and help improving learning strategies and learning outcomes. For example, the GLASS tool [30] provides a visualization of learning performance with a comparison whole class group, while the SAM tool [21] enables students' self-reflection and awareness of what and how they are doing.
3. Dashboards students only: they provide feedback in regard to their learning performance, such as the Course Signal tool [3].

As aforementioned, the dashboards are based on data coming from different data sources. The main data sources used to obtain data for the dashboards are classified into six types [44]: (1) Logs used to track computer-mediated user activity; (2) Learning artifacts used or produced by the users; (3) Information asked directly from the users for analytics purposes; (4) Institutional database records; (5) Physical user activity; and (6) Application Programming Interfaces (API) for collecting data from external platforms. In spite of this, most of the dashboards relied on a single data source or combined between two or three data sources.

Concerning KPI, the dashboards usually show KPI that can be grouped in the next six types [44]:

1. Student-related: KPI that provide information about "Who are the students?" such as age, prior education, competences, or university entrance grade.
2. Action-related: KPI that provide information about "What do students do while learning?" such as number of page visits, number of file downloads, or time spent on tasks.
3. Content-related: KPI that provide information about "What is the content involved in students' learning?" such as sentiment of the messages, or topics covered and omitted in the report.
4. Result-related: KPI that provide information about "What is the result of students' learning?" such as average grade, or grades distribution in a group.
5. Context-related: KPI that provide information about in "Which context does the learning take place?" such as location of students around a table top, or geographical location.
6. Social-related: provide information about "How do students interact with others during learning?" such as network showing communication with others in a group forum, or direction of interaction in a group around a table top.

Dashboards use several visual techniques and the most popular representation are [44]: bar charts, line graphs, tables, pie charts, network graphs, tag cloud, signal lights, bubble chart, scatterplot, text and physical map.

The dashboards designed and implemented in the TeSLA project belong to the type of dashboards only for teachers. The KPI belong to the type of action-related, content-related, and result-related. TeSLA dashboards use a variety of visual techniques, such as signal lights. The data used, analyzed and displayed on the dashboards comes from a combination of different types of data, specifically the data captured by the instruments used in the project (authentication and authorship instruments). In the literature review, no learning dashboards with similar purposes were found.

4 Design and Implementation of TeSLA Dashboards

In this section, we present the raw data generated by TeSLA instruments, how these data are analyzed and, finally, the different dashboards designed to present the information derived from the analysis to the teachers.

4.1 Data Typology

In the context of TeSLA project, different types of data are captured and stored, depending on the instruments under consideration. The instruments tested are face recognition, voice recognition, keystroke dynamics, forensic analysis and plagiarism. Face recognition uses web camera and generates data samples with the student's face. Voice recognition aims to record student's voice by creating a set of audio samples. Keystroke dynamics is based on student's typing on the computer keyboard and recognizes two key features: the time for key pressing and the time between pressing two different keys. The forensic analysis compares the writing style of different text typed by the same student and verifies that he/she is the author. Plagiarism checks whether the submitted documents by a student are his/her original work and they are not copy-pasted from other works.

On the one hand, face recognition, voice recognition and keystroke dynamics allow students' authentication based on the analysis of captured images, audio and typing data while the students perform an assessment activity. In the case of face and voice recognition, authentication can also be checked over assessment activities submitted by the students (for example, video/audio recordings). On the other hand, forensic analysis checks authentication and authorship based on the analysis of text documents provided by the same student (in fact, authorship is inferred from the fact the student is who has written the text document), while plagiarism detects similarities among text documents delivered by different students ensuring thus authorship.

The authentication instruments require learning a model for the user (i.e. a biometric profile of the student needs to be built). This model is used as a reference for subsequent checking. In addition to these instruments, other instruments that address security aspects are provided. These instruments are face and voice anti-spoofing. These instruments increase confidence in the results provided by face and voice recognition, avoiding common cheating practices in the use of such instruments (for example, in the case of face recognition, the anti-spoofing detects pre-recorded images o still pictures of the student).

4.2 Data Analysis and Visualization

One of the key aspects of TeSLA is the ability to combine multiple instruments to analyze student's data and documents. The huge variability and amount of data collected and analyzed by the different instruments represent a challenge on the visualization of the results in a clear, intuitive and useful manner. The goal of the dashboards is to minimize the time teachers devote to the analysis of the TeSLA results, providing graphical representations based on icons, plots and natural language descriptions of the computed indicators (i.e. the KPI).

4.2.1 Motivation

For each assessment activity, TeSLA system collects many samples for the different instruments enabled for the assessment activity. The different instruments analyze such samples and return a numeric value, having multiple results for each instrument and student in the context of the assessment activity (the reader can find additional information regarding that, in Chaps. 2 and 3 included in this volume). In Fig. 1, we can see the results visualization (when instruments are integrated in Moodle) for a given assessment activity with two active instruments (face recognition and keystroke dynamics). At left side we can see an obfuscated view of the students' name, and selecting a student, we see the results provided by selected instrument for this student in the assessment activity (specifically, in Fig. 1, the selected tab corresponds to face recognition). Colors are based on predefined threshold levels fixed by institution for each instrument. There are three colors: green, orange and red. Red color implies that TeSLA system detected possible cheating. Orange color implies that TeSLA system detected a possible cheating, but it is not clear. Green color implies that TeSLA system did not detect cheating.

Although we can add some semantic information on the results, like a color legend helping the teacher understanding the values provided by instruments, the teacher needs to go through students and instruments to decide about the legitimacy of student's assessment activity. When the assessment activities are long in time, and/or the number of students is large, the time teachers require spending in this task increases up to the point that makes results totally useless. This is especially dramatic when instruments, like face recognition, which analyze many samples during the

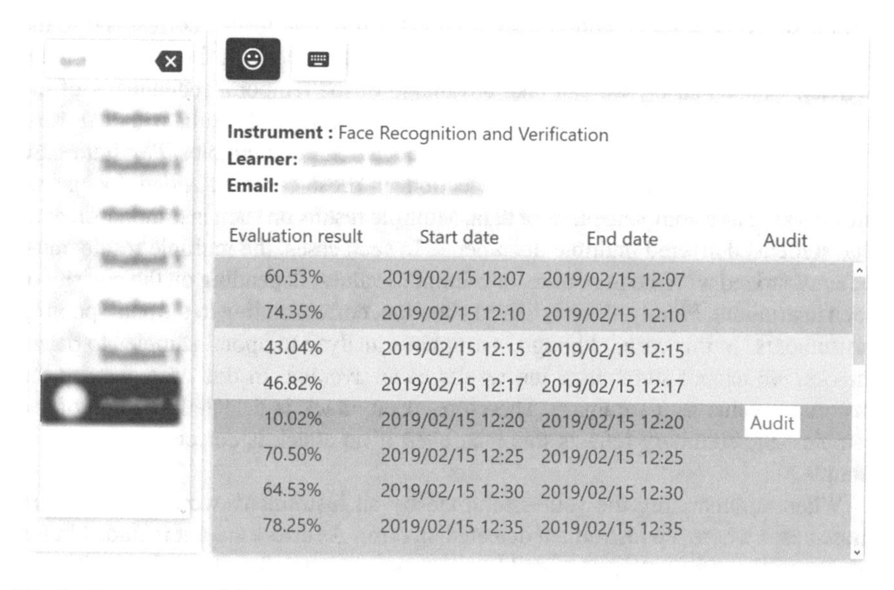

Fig. 1 Assessment activity's results visualization implemented in Moodle

assessment activity are used. These concerns were confirmed during the pilots (which consisted of the experimental and gradual integration of the TeSLA system in 7 consortium member universities) where teachers had the opportunity of seeing this visualization and providing improvement suggestions (for additional information, see Chap. 12 in this volume). For all these reasons, thereafter we explain the computed indicators used for building a set of dashboards allowing teachers to analyze the results in a simpler way. We introduce the indicators, and how the results are combined (or aggregated) to get a final summary. Those indicators are the foundation of a set of three dashboards that allow a coarse-to-fine navigation of the results.

4.2.2 Indicators Generation

Following the standard defined in the TeSLA project, instruments return real values between zero and one. The interpretation of instrument results depends on each instrument, but in all cases except plagiarism and anti-spoofing instruments, licit behaviors have results near one, while cheating ones are near zero. In the case of plagiarism, the result is the percentage of similarity between a document and other documents, and therefore, good results are near zero. Anti-spoofing instruments provide values near to one when an attack is detected. The property of the instrument that defines this fact is known as polarity. In addition, instruments can return an error value when some issue is detected, or the provided data are not valid due to capture error or not supported formats.

As aforementioned, for one assessment activity, we use to have multiple results for the same instrument and student. In the case of face recognition, voice recognition and keystroke dynamics recognition instruments, the different results correspond to the analysis of the samples captured at different temporal moments during the assessment activity. In such cases, we can take advantage of the temporal redundancy of the values, and summarize the instrument results as the average of results without errors. In other cases, the data are analyzed as whole, not as temporal samples. This is the case of forensic analysis and plagiarism detection. In those cases, provided documents are analyzed as a complete piece of data. Multiple results on such instruments means that students delivered multiple documents. In such cases, the multiple results must be summarized with the maximum or minimum values, depending on the polarity of each instrument. Finally, there is a third situation, corresponding to the anti-spoofing instruments. In this case, although instruments analyze temporal samples to detect attacks, we cannot summarize the results as an average. In this case, we use the maximum value as a summary. Therefore, if an attack is detected in some of the samples, the summary reflects this fact, even if no attack is detected in any other sample.

When summarizing the values returned by an instrument, we need to discard those cases where the instrument detected an error. As errors are not included in the summary, the information provided by the summary (and, therefore, the confidence we have on this value) depends on how many samples are correctly evaluated. Let's define the result of an instrument I and its confidence as:

$$result(X_I) = \begin{cases} \frac{1}{|V(X_I)|} \sum_{k \in V(X_I)} k \; if \; I \in \{FR, VR, KD\} \\ \max V(X_I) \quad if \; I \in \{FRA, VRA, PL\} \\ \min V(X_I) \quad if \; I \in \{FA\} \end{cases}$$

$$confidence(X_I) = \frac{|V(X_I)|}{|V(X_I)| + |E(X_I)|}$$

where:

- X_I is the set of results provided by the instrument *I*.
- Instrument *I* can be face recognition (*FR*), voice recognition (*VR*), keystroke dynamics (*KS*), face recognition anti-spoofing (*FRA*), voice recognition anti-spoofing (*VRA*), plagiarism detection (*PL*) and forensic analysis (*FA*).
- $V(X_I)$ is the subset of X_I containing valid results.
- $E(X_I)$ is the subset of X_I containing errors.

After computing this compact representation of the results, it is possible to simplify the results visualization of Fig. 1 to a single flat table, providing for each student just the result and the confidence for each used instrument, removing the teacher has the need to select a student and instrument. Threshold values used in Fig. 1 can be applied to the result value as before, providing a color legend. The use of threshold values allows providing a color legend of the values to help teachers. However, those values are computed statistically to maximize the accuracy with an acceptable false-alarm rate. That is, values are chosen to detect as much fraudulent assessment activities as possible, while required teacher time is minimized. The use of fixed thresholds is based on global optimizations, without considering the student and assessment activity particularities. To reduce the number of false alarms, in the dashboards, we use adaptive thresholds, which consider the historic information of the student for each instrument, and how this instrument behaves with other students in the same assessment activity.

For computational reasons, the analysis of the instrument behaviors for the student and assessment activity is done using histograms. We use 10 bins for the histograms, where each bin has the number of times the instrument reported a valid result in the range of the bin (see Table 1).

Table 1 Range of values included in each histogram bin

Bin	0	1	2	3	4
Range	[0, 0.1)	[0.1, 0.2)	[0.2, 0.3)	[0.3, 0.4)	[0.4, 0.5)
Bin	5	6	7	8	9
Range	[0.5, 0.6)	[0.6, 0.7)	[0.7, 0.8)	[0.8, 0.9)	[0.9, 1.0]

From the histogram, we compute the following statistical estimators:

$$P_{hist}(x) = \begin{cases} \dfrac{hist(bin(0)) + \frac{hist(bin(1))}{2}}{\sum_{i\in[0,9]} hist(i)} & if\ x < 0.1 \\ \dfrac{\frac{hist(bin(x)-1)}{2} + hist(bin(x)) + \frac{hist(bin(x)+1)}{2}}{\sum_{i\in[0,9]} hist(i)} & if\ 0.1 \le x < 0.9 \\ \dfrac{\frac{hist(bin(8))}{2} + hist(bin(9))}{\sum_{i\in[0,9]} hist(i)} & if\ x \ge 0.9 \end{cases}$$

$$P^L_{hist}(x) = \begin{cases} 0 & if\ x < 0.1 \\ \frac{\sum_{i\in[0,bin(x))} hist(i)}{\sum_{i\in[0,9]} hist(i)} & if\ x \ge 0.1 \end{cases}$$

where:

- $hist(i)$ is the value of i-th bin of the histogram *hist*.
- $bin(x)$ is bin corresponding to the value x.

Using those statistical estimators, a set of indicators is computed over different histograms. We compute the context histogram H^I_C over $V(X_I)$, and H^I_L over $V\left(X^L_I\right)$, where X^L_I are the results provided by instrument I for student L. Using the previous definition P_{hist}, and using those new histograms, we use an indicator of the goodness of a certain value x using an estimation of its quartile, both taking into account the student history and the contexts. Quartiles are estimated as:

$$Q^I_L(x) = \begin{cases} 1\ if\ P^L_{H^I_L}(x) \le 0.25 \\ 2\ if\ 0.25 < P^L_{H^I_L}(x) \le 0.5 \\ 3\ if\ 0.5 < P^L_{H^I_L}(x) \le 0.75 \\ 4\ if\ P^L_{H^I_L}(x) > 0.75 \end{cases}$$

$$Q^I_C(x) = \begin{cases} 1\ if\ P^L_{H^I_C}(x) \le 0.25 \\ 2\ if\ 0.25 < P^L_{H^I_C}(x) \le 0.5 \\ 3\ if\ 0.5 < P^L_{H^I_C}(x) \le 0.75 \\ 4\ if\ P^L_{H^I_C}(x) > 0.75 \end{cases}$$

Once the indicators are defined, the final alert level for an instrument I is computed as follows in Table 2, and each alert level is transformed in a color (see Table 3).

4.3 TeSLA Dashboards

Although TeSLA system includes other dashboards (for example, technical administrators), in this chapter we present those addressed to provide teachers with the

Table 2 Conditions used to generate the alert level

Conditions					Level
$\lvert X_I \rvert = 0$					N/A
$\lvert X_I \rvert > 0$	$Confidence(X_I) < 0.5$				W
	$Confidence(X_I) \geq 0.5$	$Polarity(I) = 1$	$Q_L^I(x) \leq 2$	$Q_C^I(x) \geq 3$	Ok
				$Q_C^I(x) < 2$	A
			$Q_L^I(x) = 3$		W
			$Q_L^I(x) = 4$		Ok
		$Polarity(I) = -1$	$Q_L^I(x) \geq 3$	$Q_C^I(x) \leq 2$	Ok
				$Q_C^I(x) > 3$	A
			$Q_L^I(x) = 2$		W
			$Q_L^I(x) = 1$		Ok

Table 3 Relationship between alert level and color

Level	Color
N/A	Gray
W: Warning	Orange
A: Alert	Red
OK	Green

evidences of TeSLA instruments for a given assessment activity carried out by a set of students. The different dashboards we present correspond to coarse-to-fine of results details level, ranging from the assessment activity level, to the details for a specific instrument for a concrete student.

4.3.1 Global Results Visualization Dashboard

The objective of this dashboard (see Fig. 2) is to identify the students that the TeSLA system has detected as possible cheaters on an assessment activity. The legend of the icons as well as the colors used on the dashboards is depicted in Fig. 3.

This dashboard presents a student level summary of the results obtained within the assessment activity (see the *summary* column in the left part of Fig. 2). This column has three indicators, where each one informs about one aspect of the student assessment activity: identity (i.e. authentication), authorship, and security (anti-spoofing instruments), as depicted in Fig. 3. The teacher should further investigate any student that has any warning (orange) or alert (red) indicators in each column. These indicators summary is generated from each instrument summary. In the case of the example shown in Fig. 2, the value associated to the summary column about the student identity combines the individual summary results obtained by the student by

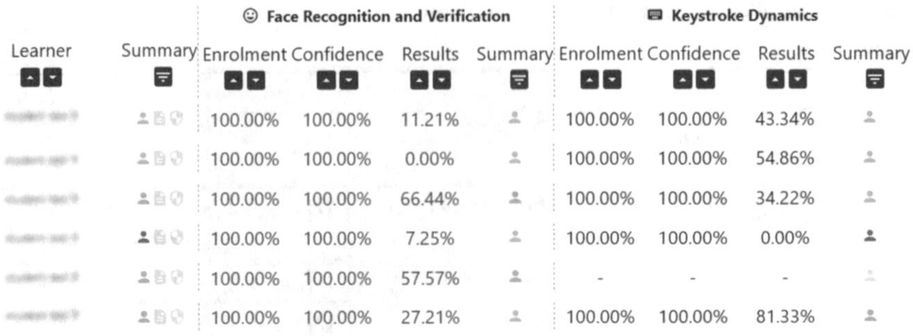

Learner	Summary	Face Recognition and Verification: Enrolment	Confidence	Results	Summary	Keystroke Dynamics: Enrolment	Confidence	Results	Summary
[illegible]		100.00%	100.00%	11.21%		100.00%	100.00%	43.34%	
[illegible]		100.00%	100.00%	0.00%		100.00%	100.00%	54.86%	
[illegible]		100.00%	100.00%	66.44%		100.00%	100.00%	34.22%	
[illegible]		100.00%	100.00%	7.25%		100.00%	100.00%	0.00%	
[illegible]		100.00%	100.00%	57.57%		-	-	-	
[illegible]		100.00%	100.00%	27.21%		100.00%	100.00%	81.33%	

Fig. 2 Global dashboard with the summary results for all students on an assessment activity

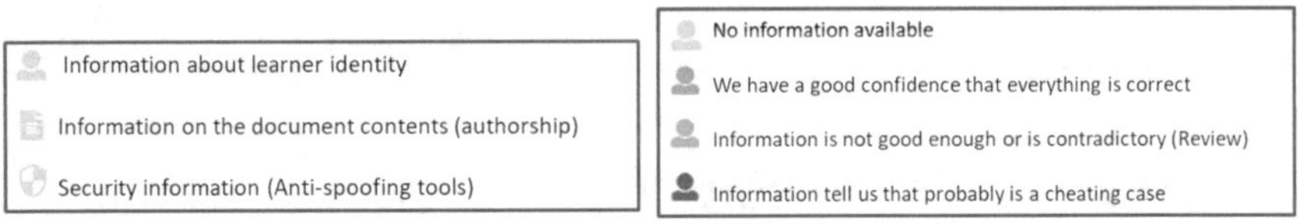

Information about learner identity
Information on the document contents (authorship)
Security information (Anti-spoofing tools)

No information available
We have a good confidence that everything is correct
Information is not good enough or is contradictory (Review)
Information tell us that probably is a cheating case

Fig. 3 Legend of the icons and colors used on the dashboards

face and keystroke recognition instruments (as shown in the middle and right part of Fig. 2). The color of each indicator in the summary column is decided as shown in Table 4. If teachers want to see only some indicators, and certain values of these indicators, they can use filters (see Fig. 4).

This global dashboard also shows individual information for each enabled instrument in the assessment activity, after the summary column. In case of the dashboard depicted in Fig. 2, there are two instruments active: face recognition and keystroke dynamics. The information associated to each instrument informs the teacher about: *enrolment percentage*, *confidence*, *results*, and *summary*. The enrolment percentage is the information relative to enrolment for this student for this instrument. It should be 100% to get good results for the instrument. The confidence value expresses the confidence of the result computed for the instrument (denoted as results in Fig. 2),

Table 4 Indicator color decision

Color	Color	Result
Green	Green	Green
Green	Red	Red
Green	Orange	Green
Orange	Red	Red
Gray	Green	Green
Gray	Orange	Orange
Gray	Red	Red

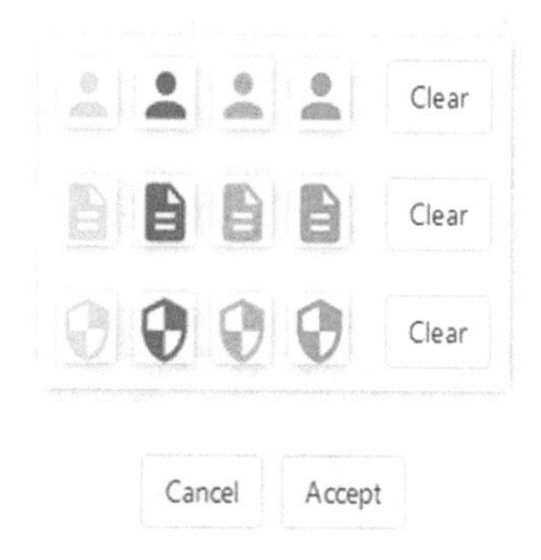

Fig. 4 Column filter

which in turn is the summary result of the instrument for the student in the assessment activity. As explained previously, in the case of instruments as face and keystroke dynamics instruments, this result is computed as consequence of the analysis of the samples captured at different temporal moments during the assessment activity. Finally, the summary value (represented as a colored icon) is the combination of the summary results and confidence for the instrument about this student in this assessment activity.

4.3.2 Student Details Dashboard

The objective of this dashboard (shown in Fig. 5) is to see in more detail what TeSLA system detects during an assessment activity for a concrete student. The teacher can reach this dashboard by clicking in the student's name (the list of students is available in the global dashboard, as it is depicted in the left part of Fig. 2). The student details dashboard provides two types of information: temporal results and instruments' information.

In the temporal results section, the teacher can see the evolution for each instrument. For example, in Fig. 5 we can see that there are two instruments enabled in the assessment activity: face recognition and keystroke dynamics. On the one hand, the face recognition instrument was working during all the assessment activity (see the blue curve in Fig. 5). That means that the instrument captured face images of the student at different temporal moments, and while the student was doing the assessment activity. On the other hand, the keystroke dynamics instrument only worked at the end of the activity (orange curve in Fig. 5). In this case, it is a correct behavior, because the first part of the assessment activity consisted of a set of multiple-choice

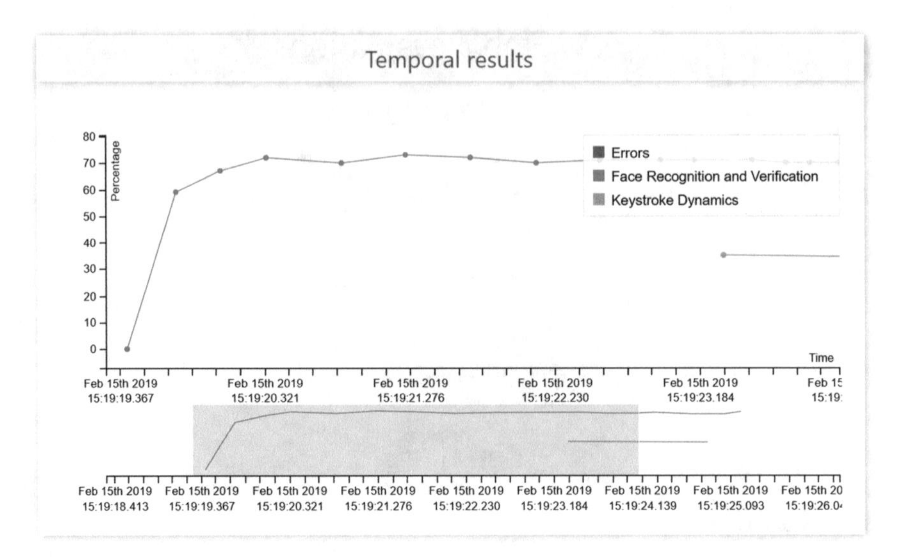

Fig. 5 Student details dashboard: temporal results

questions, while at the end of the assessment activity, there were open questions. It was in this last part of the assessment activity when student typed text (his or her answers to the posed questions), which allowed capturing his or her typing patterns by the keystroke dynamics instrument. It is worth noting that during the samples capture not errors were produced (if this had been the case, it would have been represented with a curve in red in Fig. 5).

In the instruments' information section of the student details dashboard, teachers can contextualize the results with the help of two histograms: the *student histogram* and the *context histogram*. These histograms are computed from statistical estimators and indicators presented in Sect. 4.2.2. Although there is an information section (with its associated histograms) for each instrument enabled in the assessment activity, for the sake of simplicity, Fig. 6 only shows these histograms for the case of face recognition instrument.

In the student histogram (depicted in the left part of Fig. 6), teachers can obtain summarized information of the results obtained by the student for a specific instrument. There are three different kinds of information: all student results with this instrument in all assessment activities supervised by TeSLA (in blue color), current results in this assessment activity (in green), and the better results (in orange) that this student had with this instrument in other assessment activities. Basically, this histogram explains how this student actuated with this instrument in the past and in the current assessment activity, helping the teacher to better contextualize the student's results. For example, if student has a bad result with the instrument in this assessment activity, but the student usually has had bad results with this instrument in the past, then it is possible that the bad result obtained in the current assessment activity is not too bad when put in context. The context histogram allows teachers can

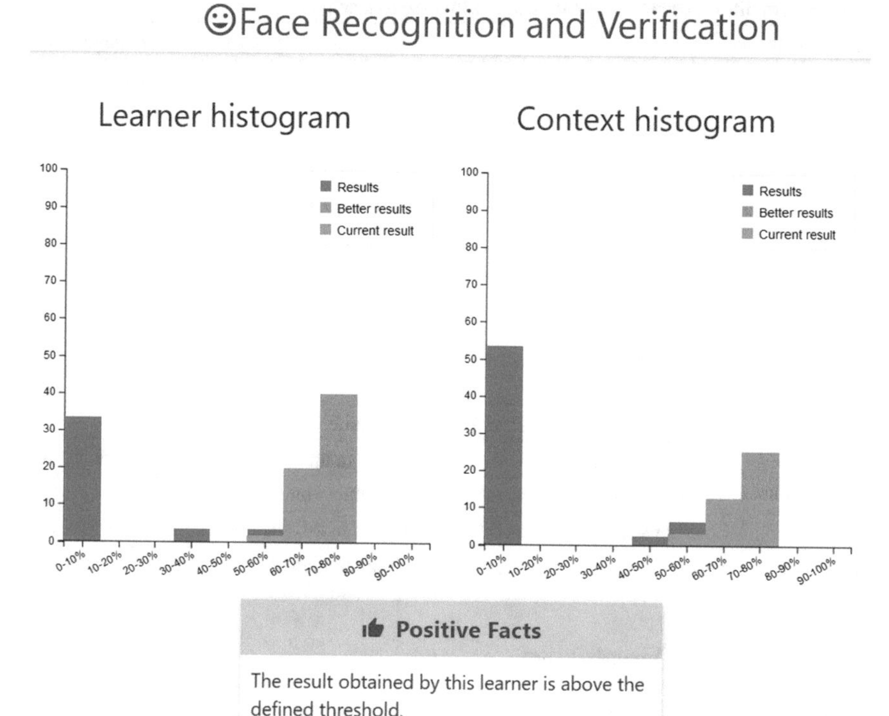

Fig. 6 Student details dashboard: instruments' information

see the results of this student with all other students in this assessment activity. The objective of this histogram is to contextualize the student's results into this assessment activity when compared with the results obtained by all the students that did the same assessment activity. Again, there are three different kinds of information: all students' results in this assessment activity (in blue), current results of this student in this assessment activity (in green), and better results of all students in this assessment activity. The histograms are complemented with natural language descriptions (or feedback messages addressed to the teacher). Tables 5 and 6 show the mechanism to generate these messages.

Table 5 Conditions used to select messages provided to teachers

Conditions			Sign	Message
$\|X_I\| = 0$			N/A	No-info
$\|X_I\| > 0$	$Confidence(X_I) < 0.5$		N/A	No-info
	$Confidence(X_I) \geq 0.5$	$Q_L^I(x) \leq 2$	Pos	Good-lear
		$Q_C^I(x) \leq 3$	Pos	Good-cont
		$Q_L^I(x) \geq 3$	Neg	Bad-lear
		$Q_C^I \geq 2$	Neg	Bad-cont

Table 6 Feedback message table

Message ID	Message
No-info	Available information is not enough to decide
Good-lear	Looking at results obtained for this student in this instrument, this is a good result
Good-cont	This is a good result if we compare with other students in the context
Bad-lear	Looking at results obtained for this student in this instrument, this is a bad result
Bad-cont	This is a bad result if we compare with other students in the context

4.3.3 Audit Data Visualization

Hereby, we explain the last type of information that teachers can examine. This information is related with cases where TeSLA system has enough confidence and evidences about a potential case of dishonest behavior of the student during the assessment activity. This information is named audit data, and its objective is to help teachers to make a final decision. TeSLA system does not make decisions regarding this issue; it is always the teacher who has the responsibility of deciding whether a cheating case took place. Furthermore, audit data are only available for visualization in these cases, in order to comply with data protection and privacy policies (for a deeper discussion about these issues, see Chap. 12 in this volume).

There are two instruments that generate audit data: face recognition and plagiarism. The audit data generated by the plagiarism instrument have been discussed in Chap. 2 in this volume. Therefore, hereby we concentrate on the audit data provided by face recognition instrument. It worth noting that the audit data of the remaining instruments (voice recognition, keystroke dynamics and forensic analysis) are a matter of current research.

Face recognition audit data include two types of information (organized in two different sections): information about the enrolment process (shown in Fig. 7) and image details captured during the assessment activity (depicted in Fig. 8). In the case of Fig. 8 three potentially problematic situations (from left to right) are shown: (1) student not recognized (because there was no person in the image); (2) student not recognized by the instrument (because the person that did the enrolment and the person appearing in the image are different persons); and (3) multiple students

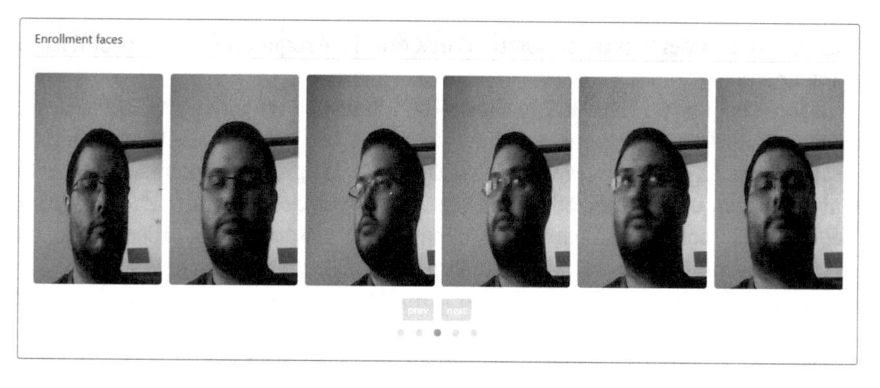

Fig. 7 Audit data detail for face recognition (enrolment)

Fig. 8 Audit data detail for face recognition (assessment activity)

detected (the assessment activity was not done individually). Please note that, in the last case, the instrument recognized the student that did the enrolment (score value of 63.55%).

5 Conclusions and Future Work

The advantages of analytical systems in business environments, as well as the success of dashboards, have helped to transfer them to educational contexts, with some adaptations, in order to provide solutions that fulfill the requirements and peculiarities of the university

In this chapter, we have firstly shown the interest of the university in the development of analytical systems, and how they have been progressively incorporated (Sects. 2 and 3). Secondly, we have discussed the design and implementation of an analytical tool, contextualized within the TeSLA project (Sect. 4). This analytical tool includes a set of dashboards that support the decision-making process of teachers about the authentication and authorship of the assessment activities submitted by the students in fully online and blended learning environments, thus contributing to enhance the e-assessment process.

The TeSLA system has been implemented following the iterative and incremental development methodology for software development. In each iteration, there was a new system version that improved and extended the previous one, which was validated through pilots (i.e. it was tested in real educational settings in the seven universities participating in the project). The opinion of the participants (teachers and students) was collected, with the aim of introducing improvements. One of the elements that was validated was a first visualization of authentication and authorship results (Sect. 4.2.1). The enhancements suggested by the teachers have given rise to the set of final dashboards discussed in Sect. 4.3. Those dashboards have been validated at one of the universities that participated in the pilots. As a future work, it is intended to extend the validation with the teaching staff of the other institutions. Other future lines of work include the development of dashboards that show the audit data for voice recognition, keystroke dynamics and forensic analysis instruments, as well as the possibility of developing dashboards addressed to students.

References

1. Ali AMET, Abdulla HMD, Snasel V (2011) Overview and comparison of plagiarism detection tools. In: DATESO pp 161–172
2. Ali L, Asadi M, Gašević D, Jovanovic J, Hatala M (2013) Factors influencing beliefs for adoption of a learning analytics tool: an empirical study. Comput Educ 62:130–148
3. Arnold KE, Pistilli MD (2012) Course signals at Purdue: using learning analytics to increase student success. In: Proceedings of the 2nd international conference on learning analytics and knowledge. ACM, pp 267–270
4. Baneres D, Rodriguez-Gonzalez ME, Serra M (2019) An early feedback prediction system for learners at-risk within a first-year higher education course. IEEE Trans Learn Technol
5. Baranovic M, Madunic M, Mekterovic I (2003) Data warehouse as a part of the higher education information system in Croatia. In: Proceedings of the 25th international conference on information technology interfaces, 2003. ITI 2003. IEEE, pp 121–126
6. Campbell JP, DeBlois PB, Oblinger DG (2007) Academic analytics: a new tool for a new era. Educause Rev 42(4):40
7. Chatti MA, Dyckhoff AL, Schroeder U, Thüs H (2013) A reference model for learning analytics. Int J Technol Enhanced Learn 4(5–6):318–331
8. Chertok IRA, Barnes ER, Gilleland D (2014) Academic integrity in the online learning environment for health sciences students. Nurse Educ Today 34(10):1324–1329
9. Davenport T, Harris J (2017) Competing on analytics: updated, with a new introduction: the new science of winning. Harvard Business Press

10. Di Tria F, Lefons E, Tangorra F (2012) Research data mart in an academic system. In: 2012 spring congress on engineering and technology. IEEE, pp 1–5
11. Doerfel ML, Ruben BD (2002) Developing more adaptive, innovative, and interactive organizations. New Dir High Educ 2002(118):5–28
12. Elias T (2011) Learning analytics. Learning, 1–22
13. Eckerson WW (2010) Performance dashboards: measuring, monitoring, and managing your business. Wiley, New York
14. Eckerson W, Hammond M (2011) Visual reporting and analysis, TDWI best practices report. TDWI, Chatsworth
15. Ferguson R (2012) Learning analytics: drivers, developments and challenges. Int J Technol Enhanced Learn 4(5/6):304–317
16. Few S (2006) Information dashboard design. O'Reilly
17. Gašević D, Dawson S, Siemens G (2015) Let's not forget: learning analytics are about learning. TechTrends 59(1):64–71
18. Glendinning I (2014) Responses to student plagiarism in higher education across Europe. Int J Educ Integrity 10(1)
19. Glover RH (1986) Designing a decision-support system for enrollment management. Res High Educ 24(1):15–34
20. Glover RH (1993) Developing executive information systems for higher education: new directions for institutional research, vol 77. Jossey-Bass
21. Govaerts S, Verbert K, Duval E, Pardo A (2012) The student activity meter for awareness and self-reflection. In: CHI'12 extended abstracts on human factors in computing systems. ACM, pp 869–884
22. Green J, Rutherford S, Turner T (2009) Best practice in using business intelligence to determine research strategy. Perspectives 13(2):48–55
23. Guitart I, Conesa J (2015) Analytic information systems in the context of higher education: expectations, reality and trends. In: 2015 international conference on intelligent networking and collaborative systems. IEEE, pp 294–300
24. Guitart I, Conesa J (2016) Creating university analytical information systems: a grand challenge for information systems research. In: Formative assessment, learning data analytics and gamification. Academic Press, pp 167–186
25. Guitart I, Conesa J, Casas J (2016) A preliminary study about the analytic maturity of educational organizations. In: 2016 international conference on intelligent networking and collaborative systems (INCoS). IEEE, pp 345–350
26. Ida M (2014) Structure of university database system and data analysis. In: 16th international conference on advanced communication technology. IEEE, pp 553–557
27. Kim J, Jo IH, Park Y (2016) Effects of learning analytics dashboard: analyzing the relations among dashboard utilization, satisfaction, and learning achievement. Asia Pac Educ Rev 17(1):13–24
28. Kitto K, Bakharia A, Lupton M, Mallet D, Banks J, Bruza P, Siemens G (2016) The connected learning analytics toolkit. In: Proceedings of the sixth international conference on learning analytics & knowledge. ACM, pp 548–549
29. Kleesuwan S, Mitatha S, Yupapin PP, Piyatamrong B (2010) Business intelligence in Thailand's higher educational resources management. Procedia-Soc Behav Sci 2(1):84–87
30. Leony D, Pardo A, de la Fuente Valentín L, de Castro DS, Kloos CD (2012) GLASS: a learning analytics visualization tool. In: Proceedings of the 2nd international conference on learning analytics and knowledge. ACM, pp 162–163
31. Macfadyen LP, Dawson S (2012) Numbers are not enough. Why e-learning analytics failed to inform an institutional strategic plan. J Educ Technol Soc 15(3):149–163
32. Maniu I, Maniu GC (2015) Data analysis techniques for examining factors influencing student's enrollment decision. SEA-Pract Appl Sci 3(2):61–64
33. Minguillón J, Conesa J, Rodríguez ME, Santanach F (2018) Learning analytics in practice: providing E-learning researchers and practitioners with activity data. In: Frontiers of cyberlearning. Springer, Singapore, pp 145–167

34. Mitchell JJ, Ryder AJ (2013) Developing and using dashboard indicators in student affairs assessment. New Dir Stud Serv 2013(142):71–81
35. Moore LJ, Greenwood AG (1984) Decision support systems for academic administration. AIR Professional File
36. Mohd WMBW, Embong A, Zain JM (2008) A knowledge-based digital dashboard for higher learning institutions. In: Joint European conference on machine learning and knowledge discovery in databases. Springer, Berlin, Heidelberg, pp 684–689
37. Muntean M, Bologa A, Bologa R, Florea A (2011) Business intelligence systems in support of university strategy. Recent Res Educ Technol, 118–123
38. Park C (2003) In other (people's) words: Plagiarism by university students–literature and lessons. Assess Eval High Educ 28(5):471–488
39. Park Y, Jo IH (2015) Development of the learning analytics dashboard to support students' learning performance. J Univ Comput Sci 21(1):110
40. Piedade MB, Santos MY (2010) Business intelligence in higher education: enhancing the teaching-learning process with a SRM system. In: 5th Iberian conference on information systems and technologies. IEEE, pp 1–5
41. Rice A, Abshire D, Christakis M, Sherman G (2010) The assessment movement toward key performance indicators. In: Meeting of NASPA international assessment and retention conference, Baltimore, Md
42. Roberts LD, Howell JA, Seaman K (2017) Give me a customizable dashboard: personalized learning analytics dashboards in higher education. Technol Knowl Learn 22(3):317–333
43. Rodríguez ME, Baneres D, Ivanova M, Durcheva M (2017) Case study analysis on blended and online institutions by using a trustworthy system. In: International conference on technology enhanced assessment. Springer, Cham, pp 40–53
44. Schwendimann BA, Rodriguez-Triana MJ, Vozniuk A, Prieto LP, Boroujeni MS, Holzer A, Dillenbourg P (2017) Perceiving learning at a glance: a systematic literature review of learning dashboard research. IEEE Trans Learn Technol 10(1):30–41
45. Siemens G, Long P (2011) Penetrating the fog: analytics in learning and education. Educause Rev 46(5):30
46. Siemens G, d Baker RS (2012) Learning analytics and educational data mining: towards communication and collaboration. In: Proceedings of the 2nd international conference on learning analytics and knowledge. ACM, pp 252–254
47. Siemens G, Dawson S, Lynch G (2013) Improving the quality and productivity of the higher education sector. Policy and strategy for systems-level deployment of learning analytics. Society for Learning Analytics Research for the Australian Office for Learning and Teaching, Canberra, Australia
48. Stocker R (2012) The role of business intelligence dashboards in higher education. Credit Control 33(1):37–42
49. Turban E, Cameron Fisher J, Altman S (1988) Decision support systems in academic administration. J Educ Adm 26(1):97–113
50. Van Barneveld A, Arnold KE, Campbell JP (2012) Analytics in higher education: establishing a common language. EDUCAUSE Learn Initiative 1(1):l–ll
51. Verbert K, Duval E, Klerkx J, Govaerts S, Santos JL (2013) Learning analytics dashboard applications. Am Behav Sci 57(10):1500–1509
52. West DM (2012) Big data for education: Data mining, data analytics, and web dashboards. Gov Stud Brookings 4(1)

Blockchain: Opportunities and Challenges in the Educational Context

Victor Garcia-Font

Abstract A blockchain is a new technology that provides a distributed append-only ledger, which basically it means that it is almost unfeasible to modify or delete data once recorded. Furthermore, blockchains are maintained and governed by multiple nodes, which decide the state of the system following a consensus protocol. In this way, blockchains not only avoid single points of failure, but also create transparent systems that cannot be tampered or manipulated by a reduced group of entities with admin rights (unlike centralized systems). Attracted by these benefits, in the recent years many projects from many different sectors are proposing to use a blockchain to improve existing services or to deploy new business cases that require the consensus, the involvement and/or the cooperation of several actors. This chapter explores prominent proposals related to education and academia. In this context, trends are moving towards: financial applications, administrative efficiency, certification, immutable public registry, reputation systems, and identity systems and privacy. Moreover, this chapter also takes into account the problems associated with blockchains and discusses the main difficulties and challenges that proposals embracing this technology will have to address.

Keywords Blockchain · Cryptocurrencies · Decentralization · Education

V. Garcia-Font (✉)
Internet Interdisciplinary Institute (IN3), Universitat Oberta de Catalunya (UOC), CYBERCAT-Center for Cybersecurity Research of Catalonia, Barcelona, Spain
e-mail: vgarciafo@uoc.edu

D. Baneres et al. (eds.), *Engineering Data-Driven Adaptive Trust-based e-Assessment Systems*, Lecture Notes on Data Engineering and Communications Technologies 34,
https://doi.org/10.1007/978-3-030-29326-0_7

Acronyms

BOINC	Berkeley Open Infrastructure for Network Computing
CA	Certificate Authority Glossary: Certificate Authority
CV	Curriculum Vitae
DID	Decentralized Identifiers
GDPR	General Data Protection Regulation Glossary: General Data Protection Regulation
NIST	National Institute of Standards and Technology Glossary: National Institute of Standards and Technology
P2P	Peer-to-Peer Glossary: Peer-to-Peer
PKI	Public Key Infrastructure Glossary: Public Key Infrastructure
TSA	Timestamping Authority Glossary: Timestamping Authority
W3C	World Wide Web Consortium Glossary: World Wide Web Consortium

1 Introduction

In 2009, with the release of Bitcoin [30], not only the cryptocurrencies emerged as a new form of decentralized money, but also their underlying technology, the blockchain, became the basis of many other applications that enable the creation of collaborative business models that were not possible until then. The blockchain has been designed as an append-only ledger, where data can be easily read and appended, but where it is very hard to delete or modify any recorded information. These design principles make the blockchain a highly secure system containing traceable records that are considered to be immutable. Currently, these features and the different governance models of blockchain platforms are showing that it is possible to manage and control highly valuable digital assets, such as cryptocurrencies, in a collaborative manner involving multiple stakeholders with conflicting interests, removing middlemen and institutional silos.

Nevertheless, some projects have suffered from certain weaknesses of the blockchain (e.g. low throughput, high transaction fees) and the inconveniences of this collaborative way of managing a system (e.g. disagreements among the stakeholders involved in a project [21]). In order to mitigate the main drawbacks of blockchain technology and also to find alternative governance and consensus models that can be more suitable for different business cases, many variants from the original Bitcoin blockchain have been proposed. In fact, in the recent years, the blockchain ecosystem has evolved and has become highly complex. Many blockchain platforms have sprung up offering the promising features of this technology in many different flavours. At the same time, this complexity has mislead many projects to embrace blockchain technology in contexts where its benefits are dubious, or where other alternatives could have been more adequate to achieve similar results with less burdens.

Taking all the above into account, the purpose of this chapter is twofold. Firstly, this chapter is intended to be introductory to the blockchain. Thus, Sect. 2 explains the principles of this technology to non-technical readers. Secondly, this chapter focuses on the educational context. In this way, Sect. 3 points out current opportunities of blockchain in education and explores how this technology can enable new use cases and contribute to improve certain areas in this field. Then, Sect. 4 discusses the main difficulties and challenges in this scenario. Finally, Sect. 5 concludes the chapter.

2 Background

This section contains the necessary background for non-technical readers to be able to understand the basics of blockchain technology. First, an introduction to the blockchain is given in Sect. 2.1, and second, Sect. 2.2 describes the types of blockchain platforms.

2.1 Blockchain

In early 2009, the release of Bitcoin marked the beginning for a new form of money: cryptocurrencies. Although the main ideas behind this technology were already published some time ago, such as HashChash [3], B-money [11], smart contracts [35] or in [19], the publication of the Bitcoin software meant the first real implementation of a decentralized digital currency. Furthermore, this new technology goes far beyond a new payment system, and its core, the blockchain, enables the creation of decentralized applications that remove intermediaries, empower final users, and make possible new use cases and services that were not feasible until then.

The functioning of Bitcoin and the blockchain was first proposed in 2008 in the paper "Bitcoin: a peer-to-peer electronic cash system" [30] published by an anonymous author under the pseudonym Satoshi Nakamoto. Basically, Bitcoin, and in general any blockchain system, requires two main components: a software client installed in the user's equipment and a Peer-to-Peer (P2P) network that maintains the blockchain.

Regarding the software client, it is a computer program, known as wallet, that enables users to manage transactions that record data in the blockchain. In cryptocurrencies, these transactions are responsible of transferring funds. The basic mechanism enabling transactions is asymmetric cryptography. Briefly, a user creates a key-pair, which includes a public key, that can be disclosed, and a private key, that has to be kept secret. Using cryptographic protocols, the user can employ his or her private key to digitally sign a message. Then, anybody can use the public key to verify that the message was signed by the private key belonging to that key-pair. Similar to using a bank account number, in cryptocurrencies, users can store funds in an address, which is a value derived from the public key. Then, they can use the private key associated to that public key to digitally sign a transaction to move the funds to another address. In this way, a wallet is responsible of generating and storing the key-pairs, creating and digitally signing transactions, and, finally, interacting with the peer-to-peer network to deliver the transactions.

Regarding the peer-to-peer network, it is formed by nodes that have mainly the following functions: receiving and forwarding transactions and blocks, verifying that transactions and blocks are correct and follow a certain specification (e.g. in the Bitcoin protocol a transaction that is not properly signed is not valid, and a block over a certain size is not valid either), storing a copy of the blockchain, and generating new blocks. Nodes can have different responsibilities, being the nodes that generate new blocks a cornerstone of the network, because they are in charge of deciding the state of the system by recording the transactions received from the users. The following paragraphs give more details about how these nodes generate new blocks and maintain the blockchain.

In this type of system, transactions are from time to time packed in blocks, which are then appended to a data structure, forming in this way a sequential chain of blocks: the blockchain. Figure 1 includes a graphical representation of some blocks in a blockchain such as Bitcoin. The blockchain can be considered a kind of append-only ledger, where data can only be added and it can never be deleted or modified. Hence, in order to modify a balance of a variable, it is necessary to create a new transaction referring to that variable, instead of directly modifying its value as it would be done in traditional databases or in a spreadsheet. Hence, the primary role of the blockchain is to record all transactions that are considered as accepted in the history of a service.

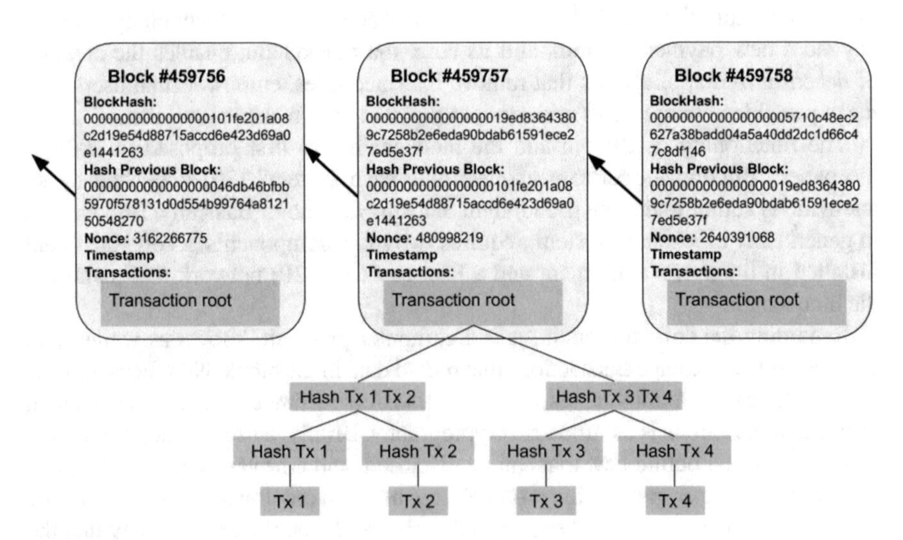

Fig. 1 Graphical representation of some blocks of the Bitcoin blockchain

The nodes in charge of packing the transactions in new blocks are called miners and the process of generating the new blocks is called mining. This process starts when the miners receive transactions of the users forwarded from other nodes of the network. Each miner might receive different transactions or in a different order depending on the connectivity among nodes. Then, the miners verify that the transactions follow the Bitcoin protocol and select a subset of the valid transactions to be included in a new block. These transactions are included in the block using a tree data structure, as Fig. 1 shows, known as Merkle tree [28]. This type of structure allows nodes to quickly find if a transaction is included or not in a block.

Once the tree is generated, the miner creates a block header including the root of the tree, a timestamp, a reference to the hash of the previous published block and a field called Nonce, explained in more detail below. As Fig. 1 shows, the links in the blockchain are the concatenation of block headers through the reference to the hash of the previous published block.

Regarding the Nonce, this field has no meaning by itself, and it can probably be ignored by anybody that only examines the content of the block, but it is not interested in verifying if it has been generated correctly. Actually, the Nonce is a tool that miners use to easily modify the block until the block follows a certain specification required by the Bitcoin protocol. In fact, finding an appropriate value for this field is what constitutes the primary effort of the miner during the mining process. In order to understand the importance of that field, first, it is important to highlight that a main goal of mining is to create a competition among the miners to select the node that publishes the next new block in the blockchain. This competition is carried out using a hash function, which is a non-invertible function that takes some data as input, and it generates a fixed-size short digest as output. With this type of function, the output cannot be predicted analyzing the input data without executing the function. In blockchain, a hash of the block header is computed as a kind of block identifier and, according to the Bitcoin protocol, this identifier has to be lower than a certain value. This means that when a miner computes the hash value of a block header, if this value is not lower than the threshold, then the miner has to somehow change the block header in order to recompute the hash. Instead of changing the transactions in the block or some other field that would require a long computation, the miners use the Nonce field to try different values until they find a block that fulfills the protocol requirements, like trying to randomly find the missing piece of a puzzle. In this way, miners compete to be the first one to create an appropriate block. Once a miner can create a new block, it disseminates it through the network and all the nodes can start working in generating a block that follows the recently published block. Due to the effort required in finding a proper Nonce, this process is known as Proof-of-Work (PoW). In fact, this task requires investing a large amount of energetic resources, and in order to incentivize participation and competition, a reward in bitcoins is given in every published block. On average, in the Bitcoin network, a new block gets published every 10 min.

Additionally, the PoW is used to securize the blockchain. As it can be seen in Fig. 1, the previous hash value is what links the blocks. In this way, once a block is published, any attempt to change any field included in the block would be easily spotted, because in order to go unnoticed, it would be also necessary to recompute the block hash, which would require a change of the pointer to this value in the next block. At the same time, this would change the next block hash, triggering changes like that in all the following blocks.

It is worth noting that with a data structure such as a blockchain, there is no principal node that decides which blocks are correct or which is the current state of the system. Conversely, each node participating on the network takes these decisions on its own. Thus, it is on the interest of all the participants to follow a common protocol and, for instance, it is on the interest of each miner to generate valid blocks that are not rejected by the other nodes. In fact, the blockchain was the first practical implementation on a large Internet-scale to achieve consensus on the state of a system, where participants do not need to know each other and, therefore, do not trust each other. Currently, other variants of this technology have been proposed that do not follow a sequential chain of blocks structure and that use alternative protocols to PoW. For these reasons, in a more general way, this type of protocols are called consensus protocols and this type of platforms are called distributed ledger technologies (DLT). Although not all DLT are a sequential chain of blocks, in this document, we use the terms blockchain and DLT indistinctly.

Furthermore, DLT can be used in many different use cases besides cryptocurrencies. In order to execute the program logic, DLT use what it is called smart contracts [35], which are a type of computer program that execute some actions defined beforehand when certain conditions are met. Combining smart contracts and blockchain enables parties that do not trust each other to automatically execute programs that record the result of the execution into a blockchain. Thus, without any intermediary, parties can automatically transfer cryptocurrency funds or digital assets, or modify the state of certain system variables according to the input parameters of the smart contracts.

In this way, cryptocurrencies can be considered payment systems enabled by asymmetric cryptography, smart contracts and a blockchain. As an example, a standard smart contract for these systems consists of transferring a certain amount of cryptocurrency to a recipient able to submit a digital signature performed with a private key associated to an address recorded in the smart contract.

Another type of smart contract is described in Hyperledger [22], where the system uses a blockchain to track seafood provenance. In this use case, transactions record every time that fish are sold or traded. Furthermore, besides recording the parties involved in the transactions, some supply chain mechanisms also propose to use Internet-of-Things (IoT) devices to leave evidence of important variables about the trade. For instance, in the seafood supply chain example, sensors can be used to ensure that temperature in fish containers has not gone beyond a threshold during transportation.

In summary, DLT can be considered a technology that integrates the following basic components:

- An append-only data model capturing the state of the system.
- Smart contracts as a programming tool to change the state of the system.
- A consensus protocol to agree on the accepted transactions and their order.

2.2 Types of Blockchain

The features mentioned up to this point in general refer to Bitcoin and, more generally, blockchain technology using PoW as consensus protocol. Nonetheless, the blockchain space is nowadays broad, and not only many other coins have been created, but also many projects propose to use blockchain in different environments besides cryptocurrencies. Depending on the context, certain blockchain features are more important than others and, therefore, researchers have designed some blockchain variants to overcome the drawbacks of Bitcoin and its consensus protocol. For example, Bitcoin aims to be public with traceable transactions. This means that the blockchain is accessible by anyone on the Internet, and the transactions show the address of the sender and the recipient. In this way, anyone can explore the blockchain and build a graph including the payment history of Bitcoin. On the other hand, other cryptocurrencies, such as Monero[1] and Zcash,[2] provide less transparent systems, implementing mechanisms to enhance privacy and to hide the senders and the recipients of the transactions.

In this way, several mechanisms can be used in different ways to adapt certain characteristics of the blockchain to the needs of the different projects. Some of the proposals have even designed new consensus protocols that change fundamental principles of the original blockchain in order to give the responsibility of generating and validating new blocks only to a reduced set of nodes, which is to detriment of decentralization. Nevertheless, by granting different permissions to the nodes that want to participate in a blockchain, the promoters of these platforms can keep the control of the blockchain and can achieve better performance.

Decentralization and the fact that anybody can participate in the consensus protocol are principal characteristics of most cryptocurrencies. However, some use cases have proven that closed systems with permissions can also be useful in many other contexts. Thus, in the blockchain space there is a major classification of blockchain platforms according to the permissions required to participate in the consensus protocol. In this way, these technologies are normally classified between permissionless and permissioned blockchains. More details on these two types are provided below.

[1]Monero. https://www.getmonero.org/.

[2]Zcash. https://z.cash/.

2.2.1 Permissionless Blockchains

Bitcoin blockchain is of the permissionless type. This means that special rights are not needed to participate in the mining process. Thus, any interested party can install a software compatible with the Bitcoin protocol, connect to the network and start competing with the other nodes to generate new blocks. Moreover, no permission is needed to record transactions in the blockchain. The only requirements are to follow the Bitcoin protocol and to control some bitcoins in order to pay the transaction fees. Additionally, anybody can download the Bitcoin blockchain, read it and trace the recorded transactions. Therefore, this is a public permissionless blockchain. In fact, public permissionless blockchains are common in cryptocurrencies, because cryptocurrencies aim at being fully transparent systems in order to show that users are treated equally and according to the predefined rules.

It is worth noting that in these blockchain systems it is necessary to give an incentive to the miners for their work. Mining is a resource intensive process and, therefore, some compensation is indispensable to attract miners that compete with each other and, in this way, create a fully decentralized system. For this reason, permissionless blockchains are normally associated with a cryptocurrency, because the cryptocurrency is used to incentivise miners. Generally, miners get two types of incentives when they generate a new block. Firstly, they receive a predefined reward for the block and, secondly, they can collect the fees paid by the users for each transaction.

In this way, public permissionless blockchains can be extremely decentralized systems that open up a lot of possibilities. Nevertheless, blockchains of this type have certain disadvantages compared to other systems, such as centralized databases, that make them often still not ready to be an actual game changer. Basically, the main problems of these blockchains are velocity and scalability.

Regarding velocity, as previously stated, the Bitcoin blockchain creates a new block every 10 min on average. Therefore, in order to see newly published information after a block has been published, users have to wait 10 min on average. Moreover, not all new transactions can always be included in the following block and, therefore, the waiting period can be longer. Additionally, for security reasons, transactions should not be considered fully valid until the block where they are published does not have 6 or more subsequent blocks. This means that, since each block is published every 10 min, then, in the best case scenario, users have to wait at least one hour. Although these facts are particular of the Bitcoin network, other permissionless blockchain technologies suffer similar problems. For instance, as it can be seen in Etherscan,[3] the block time in the Ethereum network generally stays below 20 s, but this is still some orders of magnitude higher than the almost instantaneous performance of conventional databases.

Furthermore, the block time has also consequences on scalability and throughput. It is common that blockchain protocols define a maximum block size. This, combined with the block time, considerably limits the amount of transactions that can get

[3]Etherscan. Ethereum Average Block Time Chart. https://etherscan.io/chart/blocktime.

published per second. Systems like Bitcoin or Ethereum cannot even handle 30 transactions per second, which is a really low amount compared to the ten thousand transactions per second that the Visa network can handle in peak times [34].

Another concern with permissionless blockchains is the great amount of resources consumed in the PoW. As explained above, miners compete to be the fastest creating valid new blocks in order to get the rewards. This has led participants to spend high amounts of energy using highly powerful hardware in mining activities. In fact, it is estimated that the Bitcoin network consumes an amount of power approximately equivalent to a country like Ireland [15]. Besides, it is common to use equipment specially designed for mining cryptocurrencies, which has to be replaced after few months because it becomes obsolete.

Finally, the facts explained in this section are common in many permissionless blockchains, mainly the ones based on PoW, such as Bitcoin and Ethereum. In order to address some of the disadvantages of PoW, for example to not require so much energy or to decrease block time, some alternatives have been proposed, among others: Proof-of-Stake (PoS), Delegated Proof-of-Stake (DPoS), Proof-of-Importance (PoI) or Proof-of-Burn (PoB). Nonetheless, these consensus protocols are still not ripe enough or they do not offer the same level of decentralization than PoW. More details about consensus protocols for permissionless blockchains can be found in Cachin and Vukolić [7].

2.2.2 Permissioned Blockchains

Permissioned blockchains have been proposed for contexts where openness and full transparency are not a requirement, or even undesirable, such as in enterprise environments. Actually, identity management is an important aspect in this type of blockchain. Only certain entities have special permissions to validate and generate new blocks, to take decisions regarding the protocol, and even to participate in transactions or view the recorded data and the state of the system.

Typically, permissioned blockchains are used in enterprise environments, where the information in the blockchain can be of interest and can even be recorded by many different entities related to a business, however the management of the platform can be left to few powerful stakeholders in a consortium (e.g. different governmental institutions, companies from the same holding, important stakeholders in a sector). For instance, Hyperledger [22] shows how to improve seafood provenance using a permissioned blockchain. In this example, every seafood transaction between different parties can be automatically recorded in the blockchain using smart contracts. In this way, with a permissioned blockchain, on the one hand, the different stakeholders regulating and controlling the seafood sector could be the only entities capable of validating and including transactions in the blocks, and also governing the protocol of the platform. These entities could also grant permissions to a more numerous group of people and companies in this sector (e.g. fishermen, restaurants) in order to participate in the system creating the transactions when fish get caught or traded. On the other hand, the information about seafood provenance gets immutably recorded and can be used in case of disputes or it can be made public to consumers.

Another outstanding use case of permissioned blockchains is in situations where transactions can be viewed or created by anybody, but just a reduced set of trusted entities can act as miners and have rights to take decisions concerning the protocol. This is the case of some cryptocurrencies like Ripple,[4] which offers a cryptocurrency that can be used by the general public in a platform managed by a few trusted stakeholders.

This type of blockchain is clearly less revolutionary than permissionless blockchains. Actually, permissioned blockchains are considered by many just a shared database [26]. Nevertheless, due to the more restrictive conditions to write new blocks, the consensus protocols in this type of blockchains can substantially differ from PoW and, therefore, overcome some of the limitations of permissionless blockchains. For instance, in permissioned blockchains it is not necessary to incentivize competition among miners with cryptocurrency rewards. Conversely, entities participating in the consensus protocol should be self-interested in running a full blockchain node in order to gain some control over the system, for example, to not have to totally entrust the platform to a single entity from the consortium. In this case, consensus protocols for this type of blockchains can accept thousands of transactions per second, can be faster and require less energy than PoW. These are clear advantages enabled by the fact that nodes do not have to compete among each other to generate the next block, but rather cooperate in validating that all the participating nodes are correctly following the protocol, and if not, excluding the misbehaving ones. Some popular consensus protocols used in permissioned blockchains are: Simplified Byzantine Fault Tolerance (SBFT), Redundant Byzantine Fault Tolerance (RBFT) and Proof-of-Elapsed-Time (PoET). More details about consensus protocols for permissioned blockchains can be found in Cachin and Vukolić [7].

In addition, privacy is another appealing feature of permissioned blockchains. In a permissionless blockchain, data are forwarded to all the nodes that connect to the P2P network. In many cases, such as in Bitcoin, data are not even encrypted and transactions are easily traceable by third parties. In other cases, such as Monero[5] or Zcash,[6] blockchains use cryptographic protocols to obfuscate transaction data and prevent extracting information from the transaction graph. Nonetheless, although data are encrypted, and therefore not directly accessible by the mining nodes, the fact that information are by design distributed to all the nodes already raises certain concerns. For instance, bugs in cryptographic protocols can eventually lead to information disclosure. Furthermore, law concerning data protection is usually written taking into account traditional centralized information systems and it is not clear what should be the data treatment in highly decentralized systems such as a permissionless blockchain.

[4]Ripple. https://ripple.com/.

[5]Monero. https://www.getmonero.org/.

[6]Zcash. https://z.cash/.

3 Applications of Blockchain in Educational Contexts

As seen so far, blockchain is a promising technology that can change the way some industries work and can enable new business cases. In an educational context, many projects propose to use blockchain tackling several different problems. This section reviews current proposals in this domain and classifies the projects in six different categories according to the specific purposes of each project. These categories are: financial applications, administrative efficiency, certification, immutable public registry, reputation systems, and identity systems and privacy. Sections below provide further details about these categories.

3.1 Finantial Applications

Cryptocurrencies have been the first application of blockchain and it is a typical use case in many scenarios. The usage of cryptocurrencies goes beyond simple payment systems. In fact, cryptocurrencies are a useful tool to easily provide rewards to users and create incentive-based systems.

In Swan [34], the author gathers several use cases of blockchain that can foster new economical models. Regarding learning, the author presents a use case based on smart literacy contracts. This type of smart contract is a kind of learning activity that has a reward attached after completion. The author argues that this can provide new forms of financial aid similar to microlending, but based on personal development instead of directly based on currency. In this way, donors can finance educational activities in bitcoins or in a specific coin created for this purpose (e.g. Learncoin).

Other authors propose to use cryptocurrencies to stimulate collaboration in academic environments and, at the same time, reduce educational costs for students with financial difficulties. For instance, in Devine [13] the author delineates how education institutions can create blockchain tokens transferable to the students that assist professors in academic activities. The institution then have to accept the tokens back as a discount to reduce the cost of certain campus services or tuition fees.

In Swan [34], the author also includes other usages of cryptocurrencies meant to improve research and publication procedures. The author presents the idea of creating blockchain based journals and to use a Journalcoin to reward any party involved in the publishing process, such as authors, peer reviewers, editors, service providers, etc. Besides, in this context cryptocurrencies can be used as tokens in order to create a reputation-based system (see Sect. 3.5 for further details about reputation systems). Using a blockchain system, scientific publishing can become more transparent and, unlike current review mechanisms, rewards can be sent to the different contributors according to their work. The author also outlines some ideas on how to use other tokens, such as ExperimentalResultscoin or Researchcoin, to stimulate reproducible research or to purchase paper reading rights avoiding the middleman.

At the same time, cryptocurrencies can be used to boost participation of the general public in research activities. For instance, Storm [33] is a platform designed to create microtasks with a gamification approach (i.e. applying game principles to increase engagement). After resolving a task, the resolver earns rewards in cryptocurrency. This can be indeed useful for many research projects, for example to find participants for surveys or to tag images to later test new machine learning models. Also, several cryptocurrency-based platforms have been proposed to trade computational resources, where users can lend their processing power [16] or their storage space [37] in exchange for coins. A software platform with similar goals, but directly focussed on science, is Berkeley Open Infrastructure for Network Computing (BOINC) [1]. Participants can install BOINC in their computers and contribute with their resources to scientific projects, such as SETI@Home[7] or Einstein@Home.[8] Gridcoin [18] has been created in order to give incentives to the participants using BOINC.

Finally, it is worth mentioning that although there are lots of proposals to create new coins to incentivize collaboration around specific use cases, there are not many empirical experiments or conclusive studies proving the feasibility of these proposals and, therefore, they should be treated with caution.

3.2 Administrative Efficiency

A common goal for many blockchain projects from different domains is to reduce paper-based workflows, decrease administrative costs and increase the efficiency in routinary procedures involving multiple parties.

The blockchain is a technology that enables creating a tamper-proof ledger shared by various stakeholders that can have competing interests and, therefore, that do not trust each other as the only source of information. In this way, the different stakeholders can use blockchain systems to create records in a secure and trustable manner. Furthermore, combining blockchain with smart contracts allows reliably automatizing many processes involving several of the stakeholders. In this way, data recorded by the smart contracts can be traced, making the parties accountable and creating a highly transparent system that can easily be verified by third parties. This can not only speedup administrative procedures, eliminate deduplicated data and information silos, but also help resolve possible conflicts among the parties. Obviously, privacy preserving techniques have to be considered to not reveal any sensitive information in these cases. Besides, since the information flowing in these systems generally concerns only a set of institutions and not the general public, then private and permissioned blockchains are commonly proposed in these situations to have a higher control over the governance of the systems and the data.

[7]SETI@Home. https://setiathome.berkeley.edu/.

[8]Einstein@Home. https://einsteinathome.org/.

Currently, good examples of this usage of blockchain can be seen in the supply-chain and logistics. A remarkable initiative is Everledger,[9] which claims to use rich forensics to identify diamonds and other gems and then, record in a blockchain any important detail about these assets in order to offer proof of origin for the precious stones. In the freight sector, Maersk and IBM [20] propose to use blockchain in order to offer a transparent way to know the location and other relevant features of cargo. The system interconnects numerous stakeholders involved in the transportation of goods, breaking some data silos and creating a common platform that has to eliminate some manual and paper-based procedures, saving in bureaucratic costs and accelerating administrative and operational actions. Although projects in this field are still very young and none has yet displaced completely existing systems, the need of common platforms with these characteristics are clear in logistics and, therefore, blockchain projects in this sector are advancing towards clear goals.

On the other hand, in the field of education, proposals in this regard are less clear and more immature. Nevertheless, the need to ease data exchange among academic institutions is strong and it is not a minor matter. Nowadays, mobility programs are in high demand, and society urges stronger interaction between industry and academia, which results in internships and training programs provided by third parties, but validated by high education institutions, which register them in the students transcripts. Mobility programs, internships and so on involve many different administrative procedures that are currently settled manually. Besides, these generally imply sharing large amounts of data among the different institutions involved in the agreements, such as academic achievements, transcripts or letters of commitment. Currently, these procedures are slow, cumbersome and entangle lots of paperwork. On this regard, Erasmus without papers [24] is an initiative by the European University Foundation partnering together with several universities. This initiative aims at creating an European network to electronically exchange student data, eliminating paper-based workflow and, in this way, reducing bureaucratic procedures and administrative costs. The initiative proposes to deploy a platform on top of the current information technology infrastructure in order to enhance interoperability and ease information exchange. Although this platform is still at its infancy, the project plans to deploy a decentralized P2P network, open to any trustworthy stakeholder involved in student mobility. Moreover, the project also aims at creating several data standards for all the steps in the procedures involved in students mobility. According to this project, it is not possible to identify a unified flow of data among institutions. Nonetheless, it is possible to identify information that is commonly required to be exchanged related to mobility, such as: personal data, study rights, course contents, learning agreements, learning agreement amendments, transcript of records, grade distribution, inter-institutional agreements, student nomination, and information on start and end date of mobility.

[9]Everledger. https://www.everledger.io/.

3.3 Certification

Currently, academic certificates and diplomas are normally issued on paper. Although these certificates generally have anti-counterfeiting features, it is still possible to forge them, specially when being shared as a copy or a scan of the original document. Furthermore, the built-in mechanisms that prevent forging the documents are expensive, which makes them not suitable for course certificates of minor importance. Additionally, verifying paper-based certificates involves many times contacting the issuing institution, which is time-consuming and, therefore, expensive. Sometimes the verification is even not possible if the institution that has issued the certificate no longer exists.

For these reasons, creating digital academic certificates that could easily be shared with potential employers and other third parties has been a hot topic for many years. Although technology to enable this has long been there, such as Public Key Infrastructure (PKI), Certificate Authority (CA) and Timestamping Authority (TSA), the existence of certain related problems have prevented replacing paper by digital diplomas. These technologies could be administered either by the same academic institutions that issue the certificates or by trusted third parties. In the first case, there would be security concerns on the way each institution deploys and administers the certificate tools or, at least, there would be asymmetries among institutions. In both cases, the validity of the certificates would be strongly bounded to the issuers or the third parties running the certificate and timestamping infrastructures. The demise of an institution would inevitably have negative consequences to the reliability of its certificates. It is worth noting that academic certificates may have a longer lifespan than their issuers or a PKI (taking this into account is especially important for titles issued by small schools). On the other hand, using a blockchain as an independent timestamping authority provides a reliable timestamping source (so, possible collusions among issuers and TSA are avoided), it prevents problems related to the demise of organizations running key services, and it also makes the system more resilient to cyber attacks, removing single points of failure. Furthermore, using a blockchain, the responsibility of managing the digital diplomas can be transferred to their holders, in this way releasing the academic institutions of the responsibility of storing personal data in centralized repositories, which can become targets of cyber attacks and ease massive data leaks.

One of the principal projects that enables publishing academic certificates on the blockchain is Blockcerts.[10] This is an open source project [6] started by the MIT's Media Lab and Learning Machine proposing a standard for creating, issuing, viewing and verifying blockchain-based certificates. Two main goals of this project are: enabling all participants to use self-sovereign identities (more details about decentralized identities in Sect. 3.6) and to control certificates themselves, and avoiding having to trust in any third party for any purposes. Therefore, Blockcerts has been designed to not require well known organizations to attest the identity neither of the issuer of the certificates nor the recipients. In this way, the identity of the partic-

[10]Blockcerts. https://www.blockcerts.org/.

ipants cannot be directly proven using Blockcerts. Participants are represented by their public keys, and it is in their interest (or not) to demonstrate the ownership of the keys. Furthermore, by enabling recipients to control their own certificates, then central repositories to store all these data become optional. Systems with these characteristics are specially adequate for people with difficulties to prove their identity with official documentation, such as asylum seekers. Also, the fact that they do not depend on official repositories makes this type of diploma censor resistant.

From a technical point of view, Blockcerts basically proposes that an issuer signs a digital file (containing the information to certify) using his/her private key. Then, the signature is appended to the certificate. Afterwards, a hash of the certificate (including the issuer's signature) is published in a blockchain record along with the date and the recipient's address. In this way, the certificate is protected against tampering and any interested party having the issuer's and the recipient's public keys can easily verify who issued the certificate and to whom. Finally, the recipient can store the certificate in his/her Blockcerts wallet. The wallet, similar to the cryptocurrencies wallet, is used to store private information related to certificates, such as the certificate document and the private key that is associated with the public key included in the certificate.

Lifelong learning passport [17] goes one step further than Blockcerts and proposes a machine verifiable certificate system based on Ethereum to administer digital diplomas managed using an identity hierarchy that can cope with basic needs of the academic certification system in a holistic way. On the top of the hierarchy, the authors propose to have accreditation authorities, which are responsible to authorize education institutions to issue diplomas. Then, in the second level of the hierarchy there are the certificate authorities. These are the ones in the name of whom the diplomas are being issued (i.e. universities, schools, etc.). Finally, in the third level there are the certifiers, who are the ones indeed certifying the diploma in the name of the certificate authority. For example, certifiers can be employees of the certificate authorities. This project is a first approach to create a blockchain-based platform to handle academic titles taking into account the needs of the academic system on a global perspective. Nevertheless, this type of systems have still a long way to go. The complexity of the academic system regarding certificates goes far beyond issuers and accreditation authorities. For example, including information about the academic personnel in the platform linked to the university certificates would enhance the value of the titles of prestigious institutions with highly qualified professors that hold accreditations issued by quality agencies.

At the same time, many of the initiatives to publish certificates on the blockchain have the goal of changing the way a typical Curriculum Vitae (CV) is being presented to employers. In this regard, this initiatives generally aim at making machine verifiable CVs that become more dynamic, where job seekers can give more importance to minor achievements, such as short summer courses. With paper-based certificates, including this type of achievement on the CV generally means that employers have to trust candidates on the veracity of the included information, because verifying such achievements is too costly. However, besides designing technological solutions for this use case, in order to have an efficient and effective mechanism to issue and share these achievements, it is important to agree on certain data structures to represent them and make them easily interoperable. In this way, Open Badges[11] is a project working on a specification in order to standardize the way people can create an ever-evolving set of badges crediting for small merits. In Tolbatov et al. [36] the authors discuss the sustainability of the current learning models centralized on brick-and-mortar higher education institutions. In the paper, the authors debate on the adequacy of blockchain technology to hold education data in a way that the students gain the control over their data in order to create a portfolio where they can show, in a verifiable manner, any type of accomplishment, including university degrees, contributions to projects, micro-accreditations, or any other result from a new distributed learning reality.

3.4 Immutable Public Registry

Transparency and immutability are two important properties of blockchain systems. In this way, any data recorded in a blockchain can be publicly accessible (in public blockchains), records include an approximate publication timestamp, and users have strong guarantees that the information has not been altered. All these can be used to create proof of existence. By recording the hash of a document in a blockchain, then it is possible to prove that the person that recorded this information had a copy of the document when the hash value was recorded. Furthermore, the content of the document can be kept private and it just has to be disclosed in dispute situation. In this case, the owner of the document can easily prove that the recorded value in the blockchain corresponds to the hash value of the document. In this way, intellectual property can be protected in a confidential manner.

In education, this type of immutable public registries can be used to enforce copyright and detect plagiarism. For example, Po.et[12] is a decentralized and permissionless protocol built on top of Bitcoin and IPFS [5] with which authors of academic content can have a prove that their work was published in a certain date. Many other applications of this kind have been presented in the blockchain space in order to eliminate middlemen like notaries and other kinds of public registries.[13,14,15]

3.5 Reputation Systems

Section 3.3 gathers some proposals to easily share achievements with which people can show their merits, and which are obvious ways to show the reputation of a person.

[11] Open Badges. https://openbadges.org/.

[12] Po.et. https://www.po.et/.

[13] Stampd. https://stampd.io/.

[14] Stampery. https://stampery.com/.

[15] ProofOfExistence. https://proofofexistence.com/.

Even though some of the proposals are focussed on sharing minor achievements, which enable dynamic portfolios, this is sometimes not enough because it cannot include non certifiable merits, such as the level of student satisfaction. Moreover, comparing certifications is sometimes difficult, subjective and not very machine-friendly. Therefore, some projects aim at going beyond certificates and propose to associate reputation scores to individuals. Actually, reputation systems (i.e. systems where users are associated with some type of metric or review about their behavior, involvement, etc.) are already being used in many online applications, like in vendor/buyer reviews in barter apps. Although these systems are practical to have a good user experience and avoid some scams, in general, they have the limitation of being strictly attached to a single application. This means that in every new system, users need to have a minimum interaction with the application in order to increase their reputation over the average. Furthermore, the way to compute the reputation measure and how reputation data are being administered normally depends on a single entity, which can raise some doubts on the rules to treat the reviews and the reputation data. For example, in issues regarding fraudulent reviews [25].

Currently, in academia there are already certain reputation measures linked to scientific productivity, such as H-index or number of citations. Although these are widely accepted, they have similar problems to the reputation systems mentioned above. For instance, there are different ways to compute these metrics, which results in different systems offering different values for the same metrics (e.g. Scopus,[16] ISI Web of Knowledge,[17] Google Scholar[18]). Besides, these are very specific scientific metrics and do not take into consideration other aspects of research or teaching, like student satisfaction. In order to avoid the problems related to having a single authority administering the reputation platform and also to open the system to different applications of the educational and academic community, some projects propose to build the reputation platforms on top of a blockchain. In this regard, the authors of [32] propose a blockchain reputation system for academia, where reputation becomes a transferable value. In order to boost their system, the authors propose to first give some reputation credit (named Kudos) to people and institutions according to certain classic metrics like university rankings, H-index and so on. Once an initial distribution of Kudos is done, these can be transferred among users. In this way, students could be rewarded with Kudos after completing tasks and passing exams. Also academic personnel could be rewarded in Kudos beyond scientific productivity. A system like this also allows to reward an author that has anonymously published interesting content online without the need of revealing the identity of the author. Another reputation system based on the blockchain is [12]. In this case the authors are not only focused on academia and propose a system with a general reputation score. Actually, the proposed system puts the responsibility to compute the reputation score on the clients. In this way, a client can be programmed to compute reputation in a different way than another one, giving different weights to different variables or taking into

[16]Scopus. https://www.scopus.com.

[17]ISI Web of Knowledge. https://www.webofknowledge.com/.

[18]Google Scholar. https://scholar.google.com.

account the type of transactions in the blockchain or how and by whom they have been registered.

Indeed these systems offer different capabilities than traditional ways of measuring reputation. They can give a general measure of reputation which is perfectly valid in certain situations. Besides, fine-grained systems are imaginable where reputation could be analyzed by area of expertise or where the measure would use dynamic mechanisms that could take into account different weights according the reputation of the awarding sources, the importance of the awarded situation, etc. However, some doubts arise by the fact that transferrable reputation opens the door to economically tradable reputation.

3.6 *Identity Systems and Privacy*

One of the most prominent research fields in blockchain is related to identity management. Nowadays, the most common authentication system still requires the usage of a user and a password. This has several drawbacks, for example user security depends on the proper implementation of the authentication mechanisms in each service; users tend to reuse passwords in different platforms; common user identifiers (or deducible identifiers) used in different services can be used to correlate information across sites; etc. Blockchain can help reducing some of these risks, enhancing security and privacy, decentralizing the storage of identifiers, and leaving aside traditional user/password login systems.

In this regard, the World Wide Web Consortium (W3C) is working on a new standard to use Decentralized Identifiers (DID) [9]. DID are globally unique identifiers that can be directly created by their owners, not depending in this way on a central service, as with cryptocurrency wallets. The owners can associate public/private keypairs to each DID that can be used for private communication and also to establish different identifiers not only for each different service, but also for different communications within the same application. The project Hyperledger Indy[19] is a well-known initiative in this regard. Furthermore, the usage of blockchain has relaunched the concept of self-sovereign identity, that aims to provide a system with which people are able to create and maintain themselves a digital identity, avoiding the need of a trusted entity issuing and storing this private information. A couple of relevant initiatives are Sovrin,[20] based on Hyperledger Indy, and uPort,[21] based on Ethereum. In the field of education, the characteristics of decentralized identity systems can be beneficial in many use cases. For example, StudyBits[22] is a project that proposes the usage of self-sovereign identities to issue educational certificates on the blockchain respecting privacy and anonymity. Also, as mentioned before, one of the goals of

[19] Hyperledger Indy. https://www.hyperledger.org/projects/hyperledger-indy.

[20] Sovrin. https://sovrin.org/.

[21] uPort. https://www.uport.me/.

[22] StudyBits. https://www.bcined.com/studybits.html.

Blockcerts is to enable all participants to use self-sovereign identities. In fact, this system does not provide any built-in identity mechanism. Thus, decentralized identity seems a natural system for these use cases. Furthermore, academic identity has to go beyond borders and be prepared to acknowledge and credit achievements to people and institutions from different backgrounds, which can include conflict areas. Hence, open and censor resistant mechanisms must be borne in mind.

At the same time, as it has been mentioned before, the blockchain can enable systems that help preserving privacy. Similarly to the aforementioned Open Badges, the W3C is working on verifiable claims and credentials.[23] According to [2], "a verifiable claim is a qualification, achievement, quality, or piece of information about an entity's background such as a name, government ID, payment provider, home address, or university degree. Such a claim describes a quality or qualities, property or properties of an entity which establish its existence and uniqueness". The purpose of the working group is to define a specification to provide a standard to manage and exchange the claims and credentials on the web in a way that can be automatically verified by a machine and ensuring that are cryptographically secure and that preserve privacy. This specification combined with cryptographic methods, such as zero-knowledge proofs, can enable ways to attest some characteristic about a person, revealing a minimum amount of information. This goal is aligned with a minimal disclosure approach required by regulations such as the General Data Protection Regulation (GDPR). A typical example for this is to generate a claim stating that a person is over 21 years old. This claim can then be used to acquire alcoholic beverages without revealing the date of birth nor the real age of the holder.

In the context of education, verifiable claims and credentials can be useful in several cases, for example: to attest academic achievements without having to disclose the complete transcript of records, to prove eligibility for a scholarship without having to reveal certain personal details (e.g. low income), to give proof of identity in online courses, to prove achievements without revealing the real identity (e.g. showing a verifiable claim in an anonymous forum attesting a certification in order to emphasize the holder's posts). Other interesting use cases for verifiable claims can be found in [2].

4 Discussion

First thing that should be taken into account before deciding the type of blockchain that suits best a certain project is to evaluate whether a blockchain is even necessary or there are other technological solutions that can fit the same purpose without the drawbacks that entail using a blockchain. In the literature there are many articles, such as [38] by National Institute of Standards and Technology (NIST) or [23] by Hyperledger giving guidelines to assess decision makers on whether a blockchain may be a convenient technology for their systems and if so, the most convenient

---

[23] Verifiable Claims Working Group. https://www.w3.org/2017/vc/WG/.

type of blockchain to use. From this type of papers, Koens and Poll [26] is especially relevant, because the authors make a comprehensive analysis of 30 existing decisional schemes proposed in other articles, and they propose a detailed decision flow diagram that not only assists on whether to use or not a blockchain, but also it points out the best possible technologies to use instead of blockchain considering the different situations specified in their flow diagram. These alternatives are: not using any database, a central database, a shared central database, a distributed database, a distributed ledger, currently no solutions available, or a blockchain.

Furthermore, as shown in Sect. 3.3, sometimes the blockchain is proposed to replace other mechanisms like PKI or TSA. As we have seen, Blockcerts[24] justifies this in order to not depend on third parties and to have independent timestamping services, taking into account that the certificates issued with Blockcerts may last longer than the issuing institutions.

At the same time, besides evaluating possible alternatives to a blockchain, educational projects that decide to finally include a blockchain in their architecture have to bear in mind that blockchain is still relatively new and immature and, therefore, they may have to face and overcome certain obstacles. Below, we list the main difficulties and challenges that have been encountered so far in the blockchain space that are also applicable in the educational context:

Transactions are not recorded immediately, specially in permissionless blockchains, which tend to be slow. Blockchains are very different to conventional databases in this sense. In a conventional database, programmers are used to commit large volumes of data almost instantaneously. In contrast, when a transaction is sent to the blockchain platform, it can take some time until a node in charge of committing the information can include the transaction in a block. Even if the transaction can get included in the following block, this can already take some time (e.g. around 10 min in average between blocks in Bitcoin[25] and around 15 s in Ethereum[26]).

The **blockchain is not suitable to store large volumes of data**. Blockchain data are shared among the nodes of the network and, therefore, it is important to economize the volume of information that is sent in the transactions. Actually, in permissionless blockchains, transaction fees are computed according to their size. Therefore, in order to record large amounts of data, normally, a hash value is computed from the data and then, the hash value is recorded in the blockchain and the actual data are recorded in other storage systems, such as IPFS [5].

The **low throughput** (i.e. maximum number of transactions per second) of permissionless blockchains can be a problem for systems that have to record many individual transactions. In Bitcoin, the throughput has been estimated to be around 7 transactions per second [10] (currently, this has slightly increased due to the deployment of SegWit [27], which proposed a new way to include the transactions in the block and establish their maximum size). Anyway, as mentioned above, this is far

[24]Blockcerts. https://www.blockcerts.org/.

[25]Bitcoin block time historical chart. https://bitinfocharts.com/comparison/bitcoin-confirmationtime.html.

[26]Ethereum average block time chart. https://etherscan.io/chart/blocktime.

from the ten thousand transactions per second that can be handled by the Visa network [34]. Although Bitcoin has one of the lowest throughputs of the blockchain space, other permissionless platforms have similar numbers. In order to overcome this problem, some projects create a Merkle tree of hashes from data that require immutability and then, only the root of the tree is recorded in the blockchain, reducing the number of transactions and also saving in transaction fees.

Scalability is a well known problem by the blockchain and cryptocurrencies community [10]. The scalability issues arise when the usage of a system gets increased massively in a certain dimension. For example, exponentially increasing the number of users of the system or the number of transactions. Bitcoin has proven to be a secure and useful system to digitally transfer value. Nevertheless, in its current state, Bitcoin is able to handle only a reasonably low number of transactions. As mentioned above, it is far from being able to support amounts of transactions similar to a typical credit card system. Hence, currently permissionless blockchains cannot cope with high loads of concurrent users and transactions.

Experiencing **long delays in transaction commitment** in high activity periods is a consequence of the low throughput and the scalability problems of permissionless blockchains. When there are many more received transactions than the amount that the blocks can handle, then the transactions have to wait to be included in the blocks. For instance, on the 12 of November 2017 the median time required to include new transactions in a mined block in Bitcoin was 27 min.[27]

High transaction fees have to be paid in high demand periods. When many more transactions are received than can be included in the blocks, then miners can be picky and select the transactions with the highest fees. On the 22th of December 2017 bitcoin transaction fees reached an average of 55 USD.[28] In any case, projects using permissionless blockchains need to acquire cryptocurrencies in order to create transactions and, therefore, they should have a strategy contemplating any possible situation as a consequence of the volatility in cryptocurrency prices.

Privacy requirements can become an actual challenge in highly transparent and traceable systems, especially when these systems are immutable and, therefore, rollbacks to erase mistakes leading to a privacy leak are not possible. A general rule is to not record any personal information in the blockchain, even if the data are encrypted. Blockchain data may be publicly available for decades and it should be taken into account that flaws may be eventually discovered in cryptographic algorithms or the key used to encrypt may be not long enough to be considered secure in the future. Thus, when personal data are involved in a transaction, normally a hash value of the data is recorded in the blockchain and the personal data are stored outside of the blockchain. Nonetheless, it should be considered what the consequences are of somebody being able to establish a link between the recorded hash value and the real information. This is particularly important in those initiatives transferring the

[27]Median confirmation time. https://blockchain.info/charts/median-confirmation-time.

[28]Bitcoin Avg. Transaction Fee historical chart. https://bitinfocharts.com/comparison/bitcoin-transactionfees.html.

responsibility of managing the personal data to the users, where these may not be aware of the consequences of sharing their personal information with third parties.

Governance disagreements. Blockchains are platforms administered by many different parties. This implies that changes on the protocol, and even software updates to repair bugs, have to be accepted by a great majority of the stakeholders. Conflicting interests among these can result in lockout situations. For instance, in 2017 in the Bitcoin community, various groups of interest had different positions regarding how to address scalability problems in this cryptocurrency. These differences created a conflict that lasted several months and Bitcoin ended up divided in two different cryptocurrencies (Bitcoin and Bitcoin Cash[29]) applying two different solutions.

Unexpected flaws in blockchain platforms can jeopardize basic features of systems built on top. For example, the authors of [29] have empirically proven that it is possible to link several transactions in Monero,[30] a cryptocurrency where anonymity and non-linkability are its main characteristics. This is just an example that blockchain technology is still immature and no project should blindly rely on it. Naturally, programing flaws and bugs are common in all the systems. However, immutability, transparency and governance disagreements can have a higher impact in these cases than in centralized systems. Besides, it should be considered that possible future findings can become a threat to some element in the blockchain. For instance, quantum computing could break some cryptographic mechanisms [31].

Attacks against blockchains. Although blockchains are considered highly secure, there are several studies reporting vulnerabilities of these systems [8]. In this sense, due to their open nature, public permissionless blockchains are being more studied than private permissioned blockchains. Hence, in the future, unexpected discovered vulnerabilities should not surprise users relying on the permissioned and close nature of the latter.

Usability is nowadays a principal concern among blockchain programmers. Currently, cryptocurrencies are the main application of the blockchain. Although using these new payment systems is not difficult, it represents indeed an insurmountable hurdle for some people. Educational platforms using blockchain technology should learn from this and create application interfaces specially designed taking into account the needs and the capabilities of their users.

Interoperability between blockchain platforms is currently a challenge and, therefore, interacting with digital assets recorded in different blockchains, or transferring assets from one blockchain to another one is still difficult.

Regulation and law have to be taken into account by any project using blockchain technology. The blockchain is a disruptive technology that represents new ways of interacting with the digital world and enables new business cases that may still not be clear in regulatory documents. In other cases, users may not be aware of the legal implications of certain actions. For example, it is specially important to follow GDPR when using personal data; or to take into account the monetary value of cryptocurrencies, pay the necessary taxes and follow anti money laundry regulations.

[29]Bitcoin Cash. https://www.bitcoincash.org.

[30]Monero. https://getmonero.org/.

Obviously, most of the problems and drawbacks mentioned in this section do not only affect educational projects, but in general any project adding a blockchain in its architecture. Furthermore, Sect. 3 has shown that many different types of applications can be built in this context, which makes it difficult to limit the list of drawbacks to just common issues for all educational initiatives. Nevertheless, the educational community has not been the first one to join the blockchain hype and, therefore, it can learn from the mistakes previously done by researchers and practitioners from other disciplines. For example, one of the first use cases of blockchain was in payments. The first applications were not design with user-friendly interfaces, which caused many people to not go on board and it even caused economic losses due to mistakes when transferring or storing cryptocurrencies [14]. Sometime later, not only users and developers of payment systems, but also of other types of applications realized that public blockchains had scalability problems. A single game called CryptoKitties was responsible for causing important congestion problems to the Ethereum network [4]. Moreover, as previously seen, transaction fees can increase a lot during high usage periods, which can disrupt many business cases based on public blockchains. These are just some examples to show to the educational community, where blockchain adoption is still low, that besides the potential benefits of this technology, the blockchain means a change of paradigm that can also cause many inconveniences and unexpected problems. Therefore, before including a blockchain in any education project, it is necessary to analyze that the blockchain is the best alternative in each case, and there are no more conventional and more mature technologies to achieve the same results.

5 Conclusions

Blockchain is a new technology considered by many as a game changer, because it provides a highly secure append-only ledger, which is distributed and governed by multiple parties.

In this chapter, we have seen that, in the context of education, blockchain has many applications that are just springing up these recent years. Basically, the proposals cover six different areas: financial applications, administrative efficiency, certification, immutable public registry, reputation systems, and identity systems and privacy.

The proposals use blockchain with different goals. The financial applications and the reputation systems use cryptocurrencies to have secure means to transfer economic value or status tokens. Applications aiming to improve administrative efficiency propose to use a blockchain in order to have a system where commonly share certain information, which could reduce paperwork and, in this way, cutdown costs. Projects regarding certification and immutable public registries use the blockchain as boards where information can be made public as well as untamperable. Finally, the blockchain enables the creation of self-sovereign identities, which can return the control of the data to their owners and help deploying censorship-resistant systems where merits and claims can be shared in a private manner.

Nonetheless, blockchains are still new and immature. Therefore, they still have certain technical problems that can become a challenge or represent an insuperable difficulty for many projects. For instance, permissionless blockchains have a low throughput and they are still not scalable. These can result in long delays in transaction commitment and high fees in certain periods of congestion. Compared to conventional databases, blockchains are slow and they are not meant to store large volumes of data. Besides, systems built on top of a blockchain have to be aware of the new attack vectors introduced by this technology. Finally, it is also important to take into account other issues regarding usability, interoperability, the governance models, the implications on privacy of such transparent and traceable systems, and other legal and regulatory matters.

Acknowledgements This work was supported by the Spanish Government, in part under Grant RTI2018-095094-B-C22 "CONSENT", and in part under Grant TIN2014-57364-C2-2-R "SMART-GLACIS."

References

1. Anderson DP (2004) BOINC: a system for public-resource computing and storage. In: IEEE/ACM international workshop on grid computing, pp 4–10
2. Andrieu J, Lee S, Otto N (2017) Verifiable claims use cases. In: World Wide Web Consortium. https://www.w3.org/TR/verifiable-claims-use-cases/
3. Back A (2002) Hashcash—a denial of service counter-measure. ftp://sunsite.icm.edu.pl/site/replay.old/programs/hashcash/hashcash.pdf
4. BBC (2017) CryptoKitties craze slows down transactions on Ethereum. https://www.bbc.com/news/technology-42237162
5. Benet J (2014) IPFS-content addressed, versioned, p2p file system. https://github.com/ipfs/papers/raw/master/ipfs-cap2pfs/ipfs-p2p-file-system.pdf
6. Blockcerts (2019) Blockchain certificates: Blockcerts. https://github.com/blockchain-certificates
7. Cachin C, Vukolić M (2017) Blockchain consensus protocols in the wild
8. Conti M, Lal C, Ruj S (2017) A survey on security and privacy issues of bitcoin
9. Credentials Community Group (2018) A primer for decentralized identifiers. https://w3c-ccg.github.io/did-primer/
10. Croman K, Decker C, Eyal I, Gencer A, Juels A, Kosba A, Miller A, Saxena P, Shi E, Sirer E (2016) On scaling decentralized blockchains. In: International conference on financial cryptography and data security, pp 106–125
11. Dai W (1998) B-money. http://www.weidai.com/bmoney.txt
12. Dennis R, Owen G (2016) Rep on the block: a next generation reputation system based on the blockchain. In: International conference for internet technology and secured transactions (ICITST). Infonomics Society, pp 131–138
13. Devine P (2015) Blockchain learning: can crypto-currency methods be appropriated to enhance online learning?
14. Frauenfelder M (2017) 'I forgot my PIN': an epic tale of losing $30,000 in bitcoin. https://www.wired.com/story/i-forgot-my-pin-an-epic-tale-of-losing-dollar30000-in-bitcoin/
15. G.F. (2018) Why bitcoin uses so much energy. https://www.economist.com/the-economist-explains/2018/07/09/why-bitcoin-uses-so-much-energy

16. Golem.network (2017) Golem: the Golem project. https://golem.network/crowdfunding/Golemwhitepaper.pdf
17. Gräther W, Schütte J, Kolvenbach S (2018) Blockchain for education: lifelong learning passport. In: Proceedings of the 1st ERCIM blockchain workshop 2018. Reports of the European Society for Socially Embedded Technologies', pp 1–8
18. Gridcoin Foundation (2018) Gridcoin white paper. https://gridcoin.us/assets/img/whitepaper.pdf
19. Haber S, Stornetta WS (1990) How to time-stamp a digital document. In: Conference on the theory and application of cryptography. Springer, Berlin, pp 437–455
20. Haswell H, Storgaard M (2017) Maersk and IBM unveil first industry-wide cross-border supply chain solution on blockchain. https://www-03.ibm.com/press/us/en/pressrelease/51712.wss
21. Hertig A (2018) A fight is breaking out over bitcoin cash—and it just might split the code. https://www.coindesk.com/a-fight-is-breaking-out-over-bitcoin-cash-and-it-just-might-split-the-code
22. Hyperledger (2017) Hyperledger: seafood supply chain traceability. https://sawtooth.hyperledger.org/examples/seafood.html
23. Hyperledger (2018) Hyperledger: lessons learned from hyperledger fabric PoC projects. https://www.hyperledger.org/blog/2018/04/19/lessons-learned-from-hyperledger-fabric-poc-projects
24. Jahnke S (2017) Erasmus without paper. https://www.erasmuswithoutpaper.eu/sites/default/files/pages/EWP%20desk%20research%20final%20version.pdf
25. Kinstler L (2018) How TripAdvisor changed travel. https://www.theguardian.com/news/2018/aug/17/how-tripadvisor-changed-travel
26. Koens T, Poll E (2018) What blockchain alternative do you need? In: Data privacy management, cryptocurrencies and blockchain technology. Springer, Berlin, pp 113–129
27. Lombrozo E, Lau J, Wuille P (2018) Segregated witness (consensus layer). https://github.com/bitcoin/bips/blob/master/bip-0141.mediawiki
28. Merkle RC (1980) Protocols for public key cryptosystems. In: Symposium on security and privacy. IEEE Computer Society, pp 122–134
29. Miller A, Möser M, Lee K, Narayanan A (2017) An empirical analysis of linkability in the Monero blockchain
30. Nakamoto S (2008) Bitcoin: a peer-to-peer electronic cash system. https://bitcoin.org/bitcoin.pdf
31. Proos J, Zalka C (2003) Shor's discrete logarithm quantum algorithm for elliptic curves. Quantum Inf Comput 3(4):317–344
32. Sharples M, Domingue J (2016) The blockchain and kudos: a distributed system for educational record, reputation and reward. In: European conference on technology enhanced learning, vol 9891. Springer, Berlin, pp 490–496
33. StormX.io (2017) StormX: storm token. Gamified micro-task platform. https://s3.amazonaws.com/cakecodes/pdf/storm_web/STORM_Token_White_Paper_Market_Research_Network_Development_vFINAL_.pdf
34. Swan M (2015) Blockchain: blueprint for a new economy. O'Reilly Media
35. Szabo N (1997) Formalizing and securing relationships on public networks. First Monday 2(9)
36. Tolbatov A, Ahadzhanova S, Viunenko A, Tolbatov V (2018) Using blockchain technology for e-learning. Measuring Comput Dev Technol Process 110–113
37. Vorick D, Champine L (2014) Sia: simple decentralized storage. https://sia.tech/sia.pdf
38. Yaga D, Mell P, Roby N, Scarfone K (2018) NISTIR 8202 (DRAFT), blockchain technology overview. https://csrc.nist.gov/publications/detail/nistir/8202/draft

Design and Execution of Large-Scale Pilots

Roumiana Peytcheva-Forsyth and Harvey Mellar

Abstract This chapter presents the piloting process across the various stages of the TeSLA system development. The chapter discusses the contexts of the pilot institutions and the way these affected the planning, execution and results of the pilots at project and institutional levels, the development of the communication protocols, the development of contingency plans based on risk assessments, and the reporting procedures. The chapter also presents an outline of the success of the pilots in capturing data for use by the instrument developers as well as for the development of pedagogic approaches to the integration of TeSLA instruments in course assessment and the use of TeSLA with SEND students. Advice for future TeSLA users is presented, as well as lessons learned from the piloting process.

Keywords Assessment activities · TeSLA instruments · TeSLA integration · Case studies

Acronyms

AU	Anadolu University
CPD	Continuous Professional Development
JYU	University of Jyväskylä
OUNL	Open University of Netherlands
OUUK	The Open University of United Kingdom
PDB	Project Demonstration Board
PL	Pilot Leaders
PMB	Project Management Board

R. Peytcheva-Forsyth (✉)
Sofia University, 15 Tzar Osvoboditel blvd, Sofia, Bulgaria
e-mail: r.peytcheva@fp.uni-sofia.bg

H. Mellar
UCL Institute of Education, London, UK
e-mail: h.mellar@ucl.ac.uk

D. Baneres et al. (eds.), *Engineering Data-Driven Adaptive Trust-based e-Assessment Systems*, Lecture Notes on Data Engineering and Communications Technologies 34,
https://doi.org/10.1007/978-3-030-29326-0_8

PM	Pilot Manger
SEND students	Students with Special Educational Needs and Disabilities
SU	Sofia University
TL	Technical Leader
TUS	Technical University of Sofia
VLE	Virtual Learning Environment
UOC	Universitat Oberta de Catalunya

1 Background—Large Scale Pilots in an EU Innovation Project

The requirement for piloting of innovation actions was an integral part of the requirements in the Horizon 2020 call to which TeSLA responded [3]:

Innovation actions: *Support to large scale pilots (in real settings) that develop and integrate innovative digital educational tools, solutions and services for learning and teaching, and supporting engagement of teachers, learners and parents. They should aim at reducing the current restrictions of time and physical space in learning and teaching. They should foster greater connection between formal, non-formal and informal learning and remove obstacles for ubiquitous learning. The pilots should link all relevant stakeholders in educational technology. As part of piloting scenarios, a specific target group to address is children and adults with mental or physical disabilities who undergo general education, lifelong learning or vocational training.*

For TeSLA, therefore, the purposes of the pilots were both technical (to refine and test the individual technologies and their integration within the TeSLA system; and to refine and test TeSLA integration with institutional technological platforms) and pedagogical [to refine and test e-assessment activities that are able to make effective use of the TeSLA system and instruments; to refine integration within institutional online learning and assessment cultures and academic integrity regimes; and to address the needs of students with special educational needs and disabilities (SEND)]. These objectives could only be achieved by large scale piloting across a range of European Higher Education institutions with significant numbers of teachers and courses and large numbers of students, including significant numbers of students with special educational needs and disabilities. This piloting would need to be developed over a series of stages, with parallel development of technological and pedagogical strands of work going hand in hand.

This chapter is organized as follows: Sect. 2 provides an overview of the TeSLA piloting process, Sect. 3 provides some background information on the institutions which took part in the pilots, Sect. 4 sets out in detail the process of planning the pilots and their execution, Sect. 5 reports on the level of participation in the pilots and outlines the results of the pilots, Sect. 6 provides a discussion of the recommendations made by the participants in the pilot for those who will deploy TeSLA in their institutions in the future and Sect. 7 sets out the lessons learned from the piloting process.

2 TeSLA Pilots: An Overview

A plan for the pilots was set out in the proposal incorporated in the Grant Agreement (GA)—that is the agreement between the TeSLA Consortium and the Europe Commission as the funder of the project 'An Adaptive Trust-based e-assessment System for Learning—TeSLA'. This initial plan is described in Rodríguez et al. [21] and the plan was closely followed during the execution of the project.

The pilots tested the TeSLA system extensively in stages as it was developed. They were carried out in seven universities situated in Bulgaria, Finland, Netherlands, UK, Spain and Turkey, over a three-year period. The outline organization of the pilots is shown in Fig. 1.

At all stages, the project sought to ensure a reasonable distribution of participation in the pilots across partner institutions, across the various TeSLA instruments (face recognition, voice recognition, keystroke dynamics, forensic analysis, plagiarism detection), across types of students (including SEND students), and across types of educational contexts, courses and assessments, including assessment response types and formats. It was important to ensure that the students gave informed consent, firstly for their participation in the pilot trials in order to ensure compliance with research ethical guidelines designed to protect participants from any negative consequences of participation in the pilots, and secondly for the collection and processing of personal data (including biometric data) by the TeSLA system in order to ensure compliance with data protection legislation. Draft consent forms were developed by Université de Namur (who were partners in the TeSLA project and concerned principally with

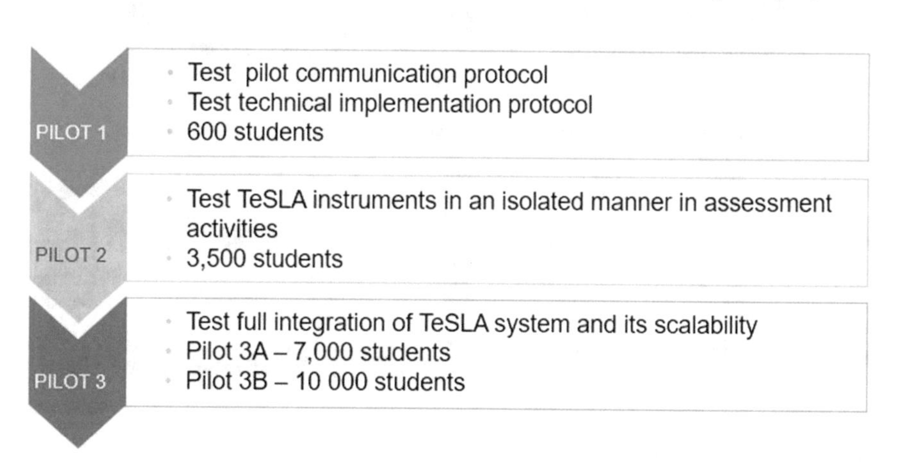

Fig. 1 The TeSLA pilots

ethical and legal aspects) and then adapted to local requirements by each institution. Institutions also obtained ethical clearance for the pilot studies via the approved channels in the institution. Students were required to complete the consent form before they were able to access the TeSLA system, and were able to withdraw from the study at any point and request the deletion of their data.

The collection of data during the piloting process is summarized in Fig. 2.

After completing the consent form, students were asked to use the BOS online survey tool [5] to complete a pre-pilot questionnaire, and then, after the pilot, to complete a post-pilot questionnaire. The data from the student questionnaires, along with teacher questionnaires and online and face-to-face focus groups was used to inform the pilot evaluation. Students' and teachers' involvement in the evaluation activities was voluntary, though teachers encouraged all students to take part.

During the pilot, students carried out enrolment activities for the TeSLA instruments that they were going to use in the assessment activities in their course (except for the test of the plagiarism detection instrument which did not require an enrolment), and then carried out one or more assessment activities using the TeSLA instruments during the course.

In order to improve the delivery of the pilots at each iteration, each pilot stage captured a variety of forms of feedback, namely:

- Feedback on the piloting process itself
- Detailed feedback on instrument use and technological issues
- Feedback on pedagogic integration
- Feedback on use of TeSLA by SEND students.

The following sub-sections provide descriptions of each of the four stages of the piloting process.

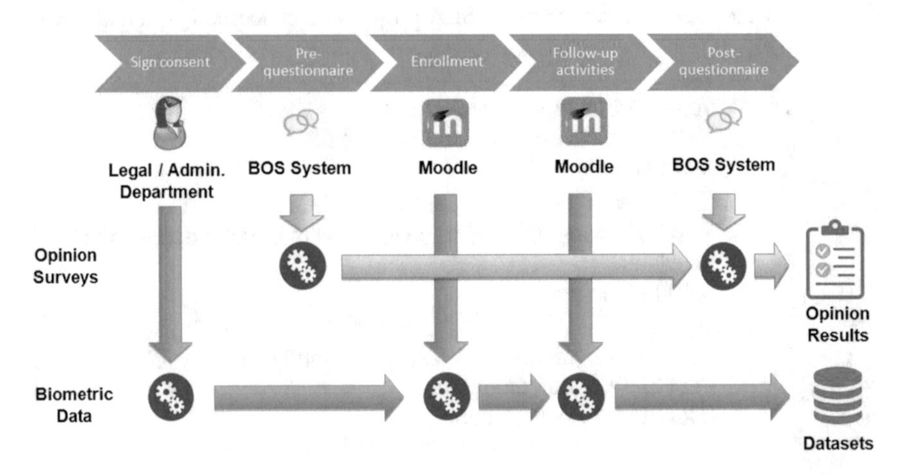

Fig. 2 Data flow in the pilots

2.1 Pilot 1

By month 8 of the project the individual TeSLA instruments were available and Pilot 1 (Small Educational Pilots) was carried out during months 9–13. The main goals of Pilot 1 were to:

- Check and improve the coordination between the Pilot Manager and other participants (TeSLA project team, teaching staff, technical staff, administrators (including quality assurance personnel) and students).
- Identify the legal and ethical issues at the institutional level.
- Identify the critical risks at the institutional level.
- Trial the pilot data collection instruments and analysis process.
- Test the technical implementation protocol at partner level.
- Determine most suitable activities for the e-assessment process at subject level.
- Integrate TeSLA instruments within assessment activities to ensure students' authentication and authorship in variety of educational contexts.
- Identify how the needs of SEND students might be met.
- Collect evaluation data about expectations, opinions and personal experiences of e-assessment and the integration of the TeSLA authentication and authorship tools in assessment.

2.2 Pilot 2

By month 12, the alpha version of the TeSLA system was available and Pilot 2 (Medium Test-bed Pilots) was carried out during months 14–20. The main goals of Pilot 2 were to:

- Test TeSLA biometric instruments (face recognition, voice recognition, and keystroke dynamics) and document analysis instruments (forensic analysis and plagiarism) in an isolated manner in assessment activities.
- Test security and integrity aspects of the system, such as the encrypted channels, the digital signature and timestamp of learners' deliverables.
- Refine learning activities for e-assessment, and integrate TeSLA instruments within assessment activities to ensure students' authentication and authorship in a variety of educational contexts.
- Collect evaluation data about expectations, opinions and personal experiences of e-assessment and the integration of the TeSLA authentication and authorship tools in assessment.
- Arrange for collection of materials and information for quality assurance purposes.

Additionally, a special study was integrated in this pilot to test processes for identification of SEND students, and initial testing of their use of the TeSLA instruments.

2.3 Pilot 3A

By month 18, the beta version of the TeSLA system was available and Pilot 3A (Large Scale Pilots) was carried out during months 14–20. The main goals of Pilot 3A were to:

- Test full integration of the TeSLA system and its scalability.
- Refine the modular development of the TeSLA system and all its components.
- Verify the reliability of the authentication and authorship.
- Refine learning activities for e-assessment, and integrate TeSLA instruments within assessment activities to ensure students' authentication and authorship in a variety of educational contexts.
- Collect evaluation data about expectations, opinions and personal experiences of e-assessment and the integration of the TeSLA authentication and authorship tools in assessment.

Additionally, a special study was integrated in this pilot to determine the authentication accuracy when TeSLA uses a combination of authentication instruments.

2.4 Pilot 3B

By month 24, the release candidate of the TeSLA system and portal was available and Pilot 3B (Large Scale Pilots) was carried out during months 26–30. The main goals of Pilot 3B were the same as those for Pilot 3A, and additionally three special studies were integrated in this pilot:

- Capture of detailed descriptions of TeSLA integration in assessment activities in courses.
- Construction of detailed case studies of SEND students working with TeSLA.
- Carrying out cheating tests on the instruments.

3 The Institutional Context: The Pilot Institutions

The Grant Agreement set out a plan for the pilots that anticipated equal distribution of the targets across the seven pilot institutions, but it was found that the institutional context of the pilot universities might have a strong effect on their ability to integrate TeSLA and its instruments into real educational contexts and therefore to achieve the targets. This necessitated flexible planning and frequent changes in the preliminary plans at both project and institutional levels. To understand the effect of the institutional context on the pilot planning, execution and results it is necessary to present the context of the seven pilot institutions, which is the aim of this section.

An important feature of the institutional context is the form of learning delivery. The TeSLA system was tested in both online and blended learning contexts. The pilot was carried out a range of contexts in order to test its appropriateness in each context, as well as to take advantage of the special features of each to support specific aspects of the piloting. Two universities—Sofia University (SU) and Technical University of Sofia (TUS)—are campus-based universities that also have a number of blended courses. Two others—Anadolu University (AU) and University of Jyväskylä (JYU)—are universities that have both traditional on-campus provision and also open learning systems that employ online teaching. Three universities – Open Universiteit Nederland (OUNL), the Open University of UK (OUUK), Universitat Oberta de Catalunya (UOC)—are open online universities without traditions in on-campus education. The institutional context of each of these universities is set out below.

3.1 Institutions that Are Campus Based and also Offer a Range of Blended Courses

3.1.1 Sofia University

SU has 25,000 students. It was founded in 1888 and is a campus-based university that is in the process of transforming some master degree programs from face-to-face to online mode, and which was recently accredited for distance education provision. It has faculties of: History; Philosophy; Classical and Modern Philology; Slavic Studies; Law; Education; Educational Studies and the Arts; Journalism and Mass Communication; Theology; Economics and Business Administration; Mathematics and Informatics; Physics; Chemistry and Pharmacy; Biology; Geology and Geography; and Medicine.

Courses are designed by a professor (as course leader) who is responsible for the teaching methodology, authoring the course (including any online elements), assignments and assessment. In online courses (which are hosted in Moodle) and sessions in a virtual classroom, an assistant professor acts as a facilitator supporting students' learning, responding to individual queries, guiding the online discussions, and providing assessment feedback. The most common assessment formats are a combination of written assignments or mid-term tests during the course (usually online) and a final face-to-face examination. Bulgarian law requires that final exams be conducted in the university building under teacher supervision.

3.1.2 Technical University of Sofia

TUS has 10,000 students. It has its origins in the Higher Technical School which opened in Sofia in 1945 and is an on-campus university which has developed a range of blended learning courses in recent years. It has faculties of: Automatics; Electrical Engineering; Power Engineering and Power Machines; Industrial Technology; Mechanical Engineering; Electronic Engineering and Technologies; Telecommunications; Computer Systems and Control; Transport; Management; and Applied Mathematics and Informatics.

The lead teacher of a course team designs the course and assessment. The design and approval cycle takes about three months. The course team presents most of the material for the course, designs the assessment (which can be varied each time the course is run), marks the assessments and provides feedback to the students. For blended courses the Moodle virtual learning environment (VLE) is used. Assessments during the course include laboratory tests (in labs), multiple-choice tests, coursework submitted online, and oral presentations. Bulgarian law requires that final exams be conducted in the university building under teacher supervision.

3.2 *Institutions that Offer Both on-Campus and Online Courses*

3.2.1 Anadolu University—On-Campus Courses

AU has 22,000 on-campus students. It was founded in 1982, and has its origins in the Eskişehir Academy of Economic and Commercial Sciences founded in 1958. Taking the on-campus and Open Education System students together Anadolu University is the fourth largest university in the world by student enrollment.[1] AU is developing a range of blended learning courses within its on-campus provision. The university also includes an Open Education System which is described below. It has nine faculties: Pharmacy; Humanities; Education; Fine Arts; Law; Economics and Administrative Sciences; Communication Sciences; Health Sciences; and Tourism.

Course design and presentation is the responsibility of the teacher. In blended courses the Blackboard VLE is used. Assessment is determined by the faculty and the teacher, who also marks the assessments and provides feedback to the students. Assessment may be by: face-to-face or online examinations, written assignments, multiple-choice questions or other e-assessment activities. The assessment methods used in a course can be varied by the teacher each time the course is run.

3.2.2 Anadolu University—Open Education System

The Open Distance Education System of AU has 1,100,000 students. It was established in 1982, and has three faculties: Open Education; Business Administration; and Economics.

[1] https://en.wikipedia.org/wiki/List_of_largest_universities_and_university_networks_by_enrollment#.

Courses and assessments are designed by a group of course designers and academics in the related field. Students have access to textbooks, TV programs, e-learning materials (using the Blackboard VLE), face-to-face classes and a variety of student support services: asynchronous support, face-to-face classes, face-to-face lab sessions and teacher moderated video-conferences. Examinations and other assessment activities are prepared by the Assessment Department. There are face-to-face mid-term and final examinations, largely consisting of multiple-choice questions.

3.2.3 University of Jyväskylä—On-Campus courses

JYU has 15,000 on-campus students. It was founded in 1966; and has its origins in the Jyväskylä Teacher Seminary founded in 1863. It has six faculties: Education and Psychology; Humanities and Social Sciences; Information Technology; Mathematics and Science; Sport and Health Sciences; and the School of Business and Economics.

Course curricula setting out the main teaching materials and assessment models are developed by faculties and are intended to last for three years. Teachers are then free to design courses and assessment to meet the curriculum. Teachers provide the structure and the content of the course, give guidelines and orientations, tutor the learning processes, engage in the dialogue with the students, and assess the outcomes. Online learning uses Koppa, an in house VLE that can be used only for written texts, and, more recently, Moodle. Assessments include face-to-face exams, online exams in a VLE, individual written assignments, multiple-choice questions, group exams or assignments, and video presentations.

3.2.4 University of Jyväskylä—Open University

JYU Open University has 13,000 students. It was established in 1984 and is an independent institution within the University of Jyväskylä whose aims is to promote educational equality and lifelong learning. It offers around 40 bachelor-level subjects from the faculties of the University of Jyväskylä, as well as a wide range of general studies, including language and communication courses. The Open University does not give degrees, but students can transfer credits to degree courses within the on-campus university.

Teaching is mainly distance education, with web-based learning environments playing a significant role, but otherwise course design, teaching and assessment are organized in a very similar way to the procedures in the on-campus university.

3.3 Institutions that Offers Courses Wholly or Principally Online

3.3.1 Open Universiteit Nederland

OUNL has 15,000 students. It is a public university founded in 1984 with the specific objective to develop, innovate and offer open distance higher education. It has four faculties: Humanities and Law; Science & Technology; Psychology and Educational Sciences; and the Open Universiteit Graduate School.

Courses are developed by a small team of teachers over a period of six months to a year. The lead teacher distributes tutoring and assessment tasks among co-teachers. Learning is online, using activating and interactive teaching formats, and a mix of self-study, virtual and face-to-face meetings. The in-house online learning environment (yOUlearn) was built and implemented in order to enhance facilitation of online learning. It is optimized for activating course design, aimed at cohorts and collaborative learning. Much of the assessment is done in study centers where regular written exams are carried out as well as computer-based exams and oral exams. The assessment in some courses is via 'special assignments' (an essay, literature study, assignment, piece of software, game, presentation, poster, attendance at a conference etc.) which are produced off-line and submitted in the VLE, or carried out in a virtual classroom. Most, but not all courses, combine these kinds of assessments with some form of supervised assessment.

3.3.2 Open University (UK)

OUUK has 174,000 students. It was established in 1969 with a mission to be open to people, places, methods and ideas. It has five faculties: Arts & Social Sciences; Open University Business School; Open University Law School; Science, Technology, Engineering and Mathematics; and Wellbeing, Education and Language Studies.

Modules are designed and produced by module teams comprising academics, artists, designers, editors and support staff usually over a period of two years but sometimes in as little as three months. Students are taught mainly through the OUUK VLE (Moodle), though materials are often available through other platforms for greater study flexibility. Some modules have elements of face-to-face tuition and undergraduate tutors usually have groups of around 15 students each. Assessment is carried out through the VLE and other assessment tools with some face-to-face exams. There is continuous assessment throughout the module which is marked by tutors electronically—this is sometimes used purely formatively but often it also counts towards the final grade usually making up 50% of the overall grade. The end of module assessment consists of a piece of coursework or a face-to face examination—usually making up 50% of the overall grade.

3.3.3 Universitat Oberta de Catalunya

UOC has 70,000 students. It was founded in 1994 with an educational model based on personalization and accompanying students using e-learning. It has eight faculties: Arts and Humanities; Economics and Business; Health Sciences; Information and Communication Sciences; Computer Science; Multimedia and Telecommunications; Law and Political Sciences; and a Center for Modern Languages.

Programs of study are designed by a committee consisting of a small set of teachers and administrative staff and this generally takes between 12 and 18 months. These programs include a list of the related courses, which are then delegated to faculty staff who design the courses (competences to be achieved, syllabus, learning resources, assessment model, etc.). The preparation of a new course normally takes from 6 to 12 months. A professor (a full-time member of staff) is responsible for the program and designs the program and learning plans for all courses and takes care of quality assurance. Course instructors—part-time staff—monitor learning, and conduct activities and assessment. Tutors—part-time staff— give support and advice during the whole academic life of the student. Course instructors, under the supervision of the responsible teacher, design the assessment activities and the planning of course refinements. The UOC educational model is based on the learner and in the activities she/he performs with the most appropriate learning tools and resources through the UOC Virtual Campus, being always guided by online teachers. The assessment model is mainly based on continuous and formative assessment; which may be complemented by final face-to-face examinations at the end of the semester. Assessment activities are diverse, and may include: debates, oral presentations, written essays, Moodle quizzes, product creation (e.g. programming), problem resolution etc. and may be individual or collaborative assessments.

4 Planning, Execution and Reporting of the Pilots

The research papers, focused on pilot studies' design and implementation in the field of e-assessment (see [1, 2, 8, 10]) discuss some common stages of pilots, though these discussions are not always structured in an explicit way by the authors.

The first stage could be defined as a preparatory stage and relates to the pilot planning and design as well as to the preparation of the key stakeholders' (teachers, students, administrators) involvement in the pilots. Introducing the potential participants to the pilot and motivating them to take part in it is considered to be a very important activity at this stage. Another key aspect at this stage, when technology is involved, is to ensure that the participants are acquainted with the technology and to make sure it runs smoothly.

The reviewed papers describe similar information and motivational strategies for reaching the potential participants in the pilots. The most important one is the introduction of the potential participants to the key aspects of the pilot: its aims and the conditions for their participation (voluntary/compulsory); the technology and how it works; the time schedule; the way in which feedback will be provided; informing them about how the pilot study will affect their usual learning environment, and whether their achievements during the pilot study will count towards their final grade, etc.

Nilsson et al. (2014) highlight the importance of the pilot manager for the smooth running of the pilot at all its stages. The authors also draw attention to the ethical issues of pilot studies in accordance with the requirement "all study procedures to follow the ethical recommendations for human subjects' research, including informed consent, privacy, and confidentiality of data and avoidance of dependency".

What comes as a conclusion from the literature review is that all pilot stages should be planned very carefully and need to be explicit and clear for the key participants in the pilot study. In the case of large-scale pilots, having in mind the large number of participants and the variables that affect the pilot's execution, detailed and careful planning is even more important for the successful achievement of the objectives.

The remainder of this section presents the key stages of the large-scale pilots performed in the frame of TeSLA project: planning—at both project and institutional level—execution and reporting the pilots' results.

4.1 *Planning at Project Level*

A plan for all pilots was set out in the proposal incorporated in the Grant Agreement (which is described in detail in Rodríguez et al. [21]. This plan was closely followed during the execution of the project but nevertheless flexible planning, including anticipation of risks at the project level, was necessary to allow flexible responses in case of deviations from the plans and the implementation of contingency measures.

The pilot planning process at the project level was based on several factors: the specific objectives of the particular pilot (described in Sect. 2); operationalization of the objectives in terms of achievable and measurable success metrics; the specific institutional context of the individual pilot institutions; the lessons learned from the execution of previous pilots within the overall plan; the specific requirements of the technical team related to data required for the specific phase of TeSLA development in each pilot.

To facilitate the planning, the pilot management team elaborated an exhaustive list of activities to guide and monitor the progress of the pilots. The list of activities included: identification of the courses and target numbers of students to be involved in each pilot; identification of the SEND students who could participate; determining the timing of enrollment and assessment activities using TeSLA instruments; incorporating enrollment and assessment activities into the pilot courses; preparation of information for students and staff about the project (in the appropriate language); development and administration of consent forms; development of instructions for online students on carrying out enrolment and assessment activities using the TeSLA instruments; and development of contingency plans describing strategies to mitigate or respond to any deviations in implementation of the planned trials.

A project level risk management plan was elaborated building on the contingency plans developed by the partner institutions—the details of the institutional contingency plans are given below in the section on-risk management.

Other than for Pilot 1, specific success metrics were defined in order to support the planning, monitoring and evaluation of the pilots, and to ensure that the appropriate data about the use of individual TeSLA instruments was collected for the instrument developers. These metrics included the total numbers of students (and SEND students) per institution, and their distribution in terms of the use of the TeSLA instruments and the variables described in Table 1.

In Pilots 3A and 3B it was important to have deep and sustained interactions with SEND students on an individual basis in order to learn about any problems arising from their use of TeSLA instruments, and to see how TeSLA might be best adapted to support their needs. So specific success metrics were adopted for this aspect of the pilots: 300 students, across as many of the pilot institutions as possible, and representing each of the following categories of disabilities: blind or partially sighted; deaf or hearing loss; restricted mobility or motor disability; specific learning difficulty (e.g. dyslexia); chronic illness; psycho-social problems.[2]

Whilst project level plans started from the position of sharing all targets equally across the seven partners, this was then adapted to suit the specific institutional contexts—for example some institutions made much greater use of final text-based

Table 1 Data categories for pilots

Category	Sub-categories
Students	• Gender (male/female) • Age (≤21, 22–30, 31–40, 41–50, ≥51) • Students with special educational needs and disabilities (SEND students)—types of disabilities
Educational context	• Online • Blended
Courses	• Field of study (Arts and Humanities, Engineering and Architecture, Health Sciences, Sciences, Social and Legal Sciences) • Course level (continuous professional development (CPD), undergraduate, post-graduate) • Language of study (Bulgarian, Catalan and Spanish, Dutch, English, Finnish, Turkish)
Assessment	• Purpose (diagnostic, summative, formative) • Where the assessment took place (at home, at university) • Whether the assessment was supervised (supervised/unsupervised) • Whether the activity was individual or collaborative
Assessment response	• Select answer • Create answer or product (format: Audio, Video, Text (natural language or code), Artifact (e.g. a painting, a meal, sheet music) • Perform/enact/demonstrate

[2]The classification of disabilities was drawn up in consultation with academics in the institutions with specific expertise in teaching SEND students.

assessments than others, and some made greater use of online multiple-choice tests—this impacted on the choice of instruments which it would be easier to pilot in a particular institution. Some institutions had access to much larger numbers of students than others and so were able to provide contingency arrangements in case the targets elsewhere were not met. Some institutions had a majority of female students, some a majority of male students, some were equally balanced. Some institutions had a preponderance of young students, and some had large numbers of more mature students. Some institutions had quite inflexible assessment regulations that meant that forms of assessment could not be changed to meet the project's needs, whereas others had more flexible regulations enabling changes to be made at short notice. Some institutions had easy access to SEND students, whereas some did not. Across the seven pilot institutions the project was able to create individual institutional targets that enabled the overall goals to be achieved.

4.2 Planning at Institutional Level

Given the overall plan and the participation targets at project level, each institution drew up its own plan taking into consideration their local context. Some of the relevant factors considered were: modes of delivery, course sizes, forms of assessments, organization of course design and student support, fields of study, organization of the academic year, and timing of assessments. It was found that dropout rates varied greatly between institutions (and in particular there was a much higher rate of dropout in online contexts than in blended contexts) and so each institution needed to accordingly adapt its initial plans for inviting students to take part in order to ensure that targets for participation were met.

The planning process was supported by the use of shared databases (one for each institution) in order to provide an overview of the number of participants and assessment activities and their distribution within the categories set out in Table 1.

Using this data, the Pilot Manager could identify gaps (or numbers exceeding goals) with respect to targets—especially in relation to the distribution of testing for individual instruments—and then negotiate changes in institutional plans to enable overall project targets to be achieved. Besides supporting the planning stage, the databases also played an important role in tracking pilot progress by identifying problems as they arose and triggering corrective actions firstly at the institutional level and, where this was not possible or failed, then at the project level. The databases also supported and simplified final reporting.

4.3 Communication Protocol

The success of the pilots depended to a great extent on the establishment of efficient and effective communication protocols among all actors involved. Communication protocols were drawn up setting out the roles and responsibilities of the various actors and the processes of coordination between project partners, and between the members of each institution—learners, teachers, technicians, administrators and managers. The communication protocols and associated metrics were revised for each pilot in response to the stages of development of the TeSLA system and to feedback from previous pilots as to how well the communication protocol was seen to work.

The communication between institutions and key actors in the pilots was implemented in three 'communication circles':

- **Project communication circle**—between the Project Management Board (PMB), the Pilot Manager (PM) representing the Project Demonstration Board (PDB) and other key personnel: Technical Coordinator (TC), and Educational Manager (EM), Technical instrument developers, and designated leaders for data privacy and ethics, quality assurance in online higher education and pilot evaluation (see Fig. 3). These key personnel liaised with the Project Demonstration Board via the Pilot Manger and also attended PDB meetings in order to take part in discussions related to their areas of responsibility.
- **Pilot management communication circle**—between the Pilot Manager and Pilot Leaders (PL) representing the pilot institutions. This group constituted the Project Demonstration Board (See Fig. 3).
- **Institutional communication circle**—within the pilot institutions (see Fig. 4).

As the Pilot Manager played a very important role in assuring the coordination between partners in the pilots, the functions and responsibilities of this role were set out in detail in the communication protocol.

The progress of the pilots at institutional level was the responsibility of the institutional Pilot Leader (PL). Their responsibilities were to: monitor the work and validate

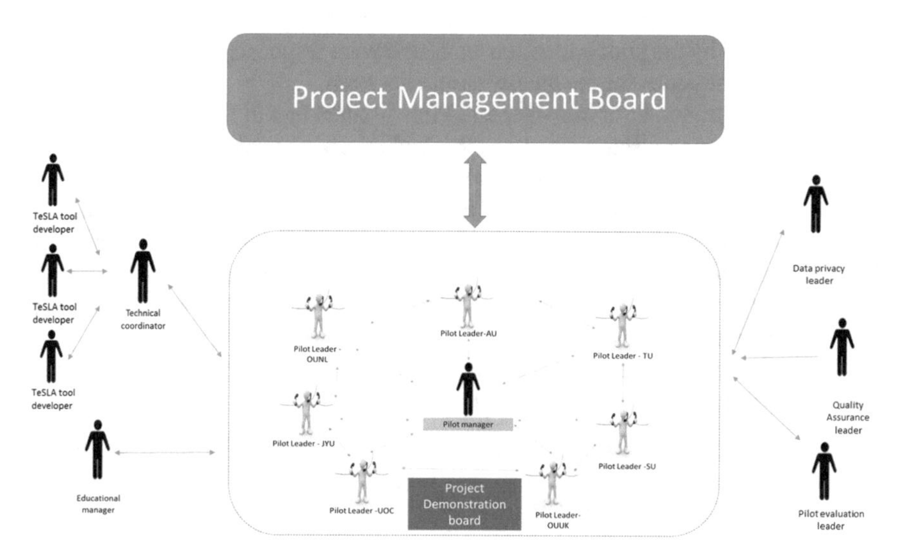

Fig. 3 Project communication circle

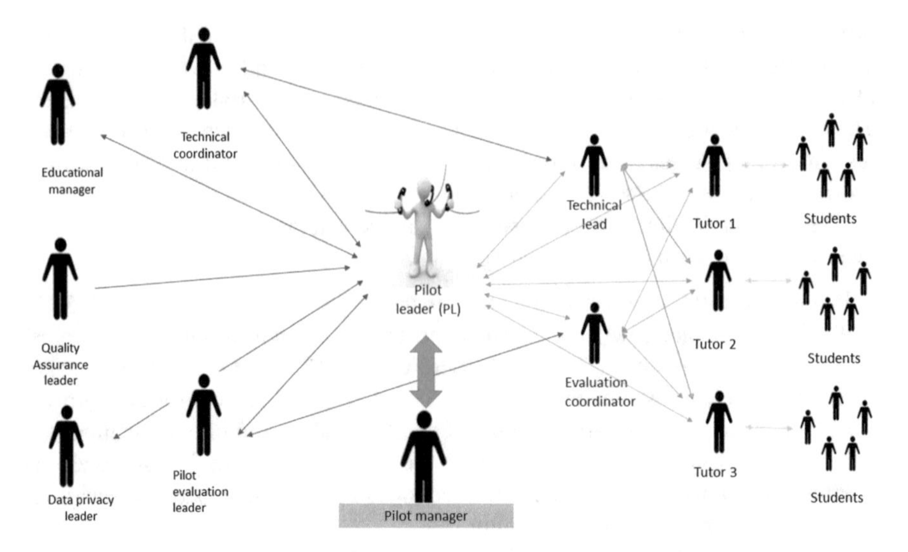

Fig. 4 Institutional communication cycle

the results within their pilots; control the progress of the work plan and development process; control and ensure the integrated e-assessment system built is a scalable and sustainable solution; monitor user acceptance and provide feedback as to technical viability; provide support to all the parties involved in their pilot, with special attention to the end users (teachers and learners); and to attend the PDB meetings to report on pilot progress and share experience gained during the development and validation of the pilot. The Pilot Leaders were also responsible for ensuring that students and teachers participating in the project completed the evaluation questionnaires provided by the pilot evaluation team, and were responsible for organizing the quality review sessions for the quality assurance team.

Twice a month the Pilot Manager and Pilot Leaders met in the PDB to report on pilot progress and to discuss the issues that had arisen, possible actions to take, and implementation of contingency plans where required.

Central to the successful testing of the TeSLA system and the instruments was the coordination between those conducting the pilots and the technical teams. The Technical Coordinator was responsible for bringing together information from the various TeSLA instrument developers, and for liaising with the Technical Leaders at the pilot institutions.

The communication protocol defines the processes to follow when technical issues arose during the pilots, which are as follows:

- The institutional Pilot Leaders maintain communication with the teachers and students in their institution, and so they receive information about any problems that arise in their use of TeSLA.

- When a technical issue arises, the Pilot Leaders communicate with the Technical Leader of the institution, who must then decide if this is an issue with the institutional VLE or with TeSLA.
- When the issue is related to TeSLA, the institutional Technical Leader posts the issue in a designated GitLab repository and monitors progress on the issue there.
- The TeSLA project technical team uses GitLab to manage the TeSLA software development. All the technical staff have access to the repositories, which allows the team to track, classify and prioritize open and closed issues. The Pilot Leaders also have access to GitLab and are able to track the progress on issues.
- More general discussion about technical issues takes place in a designated section of the management platform used by the project (Basecamp[3]).

In order to keep the pilot team up to date with developments, the Technical Coordinator posts a monthly update on the technical issues on Basecamp that have been raised in Basecamp and GitLab and how they have been addressed, as well as updates on TeSLA development and upcoming changes. The Pilot Manger then convenes a meeting with the Technical Coordinator and the Technical Leaders from the pilot institutions in which this report is discussed.

These communication protocols were generally found to work well for this project and, based on the Pilot Leaders' feedback and the Pilot Manager's' experiences and reflections, they were modified and improved in minor aspects for each pilot.

4.4 Risk Management

Risk management formed an integral part of the planning process. Risk assessments were carried out for each of the pilots and contingency plans were elaborated to address the identified risks. The risk assessments in later pilots were able to build on the experience of earlier pilots (particularly Pilot 1) to better identify possible risks, and as new potential risks arose as the TeSLA system was developed and new features needed to be tested. These plans were refined during the progress of the project. Four major sources of risk were identified:

Pilot execution: problems related to the recruitment of students, the information to be delivered to the main actors involved in the pilot, and coordination amongst partners. The specific risks identified included: insufficient number of students; insufficient number of SEND students; excessive information; coordination problems; complexity in analyzing the results; and scalability in the management of the pilot.
Academic: risks related to the need for teachers to integrate the TeSLA instruments into assessment activities possibly requiring changes in the design of learning activities and assessments which might impact on students' learning and achievements. The specific risks identified included: activity adaptation; activity adaptation for SEND learners; and impact of the adapted activity on students learning and achievements.

[3]https://basecamp.com/.

Technological: risks associated with the integration of TeSLA instruments in real assessment scenarios and the potential problems related to software bugs, hardware problems and implementation delays. The specific risks identified included: compatibility with browsers; VLE integration problems; technological background of learners and teachers; unauthorized access to data; technical problems when performing the activities; TeSLA development risks (VLE, TeSLA e-assessment Portal and instruments); quality of feedback to the instrument developers; and software readiness.

Legal: risks related to legal issues arising from delays in the legal documentation signature, the process of accepting the informed consent and other restrictions due to institutions legal framework. The specific risks identified included: informed consent acceptance; data protection framework; participation of new institutions data identification of student; management of private data; and rectification or erasure.

During the two stages of Pilot 3—Pilots 3A and 3B—the pilot institutions also developed their own institutional risk assessments and contingency plans to reflect the specific risks that might arise in their execution of the pilots, based on the experience of earlier pilots and the specific context of these institutions.

Risks were reviewed during the PDB meetings and strategies to address the risks that arose were discussed. It was important to identify these risks in a timely manner, and in the event that an institution was not able to address the risk with its own resources and efforts, then the PDB made decisions related to the implementation of contingency measures at project level. An example of implementation of contingency measures at project level was where there was a shortage of students able or willing to participate in the pilot in one pilot institution then other partners were asked to increase the number of participants. Another example was where a particular institution was not able to integrate all of the TeSLA instruments into students' assessments, and so other partners were asked to increase the number of testing activities for the instruments that were not being used.

4.5 *Reporting*

Each pilot institution produced reports for each of the pilots. The statistical data related to participation and assessment activities was entered by the Pilot Leaders in the database described earlier. Each Pilot Leader also produced a narrative report about the execution of the pilot according to a template developed by the Project Manager. The structure and the content of these narrative reports varied depending on the particular pilot objectives. In the reports for Pilots 1 and 2, the pilot institutions were asked to reflect and report on successes and the failures in the piloting process as they saw them, the strategies they adopted for addressing technological, educational, organizational and administrative issues, and lessons learned which might facilitate the execution of the later (larger) pilots. In the reports for Pilots 3A and 3B (the Large Scale Pilots), they were asked to focus on aspects of TeSLA integration which could be important and useful for its future utilization, such as descriptions of the pedagogical integration of TeSLA instruments in a variety of learning and assessment contexts, at both assessment activity and course level, and analysis of the SEND students use of TeSLA instruments.

The institutional reports were compiled by the Project Manager into project level reports for each pilot and the relevant feedback sent to the technical teams, evaluation team, quality assurance team, Educational Manager and the Project Manager. Specific recommendations were also incorporated into recommendation regarded the future use of TeSLA.

5 Participation and Results

5.1 Statistical Reports on Participation and Activities

All pilots achieved their main quantitative goals for participation and distribution of activities.

5.1.1 Pilot 1 (Small Educational Pilots)

The total number of students participating in Pilot 1 was 637. Most of the pilot institutions met or exceeded their institutional target (75 students). The face recognition, voice recognition and keystroke dynamics instruments were tested in both enrolment and follow-up assessment activities in five out of the seven pilot institutions. In this pilot most of the targeted courses were in the field of Engineering and Architecture with 37% of participants studying in computing and related areas. This was probably because these students felt most comfortable (and perhaps were most interested) in experimenting with new technologies still under development.

Most institutions had difficulties in identifying and involving SEND students in the pilot. However, 24 SEND students did participate, which enabled the testing of some usability aspects of the instruments for this group of students. The SEND students involved in Pilot 1 indicated the following disabilities: visually impairment, blindness, physical and mental disabilities. Attempting to involve SEND students in this pilot helped to identify the problems that institutions had with recruiting SEND students to the TeSLA pilots and so let to the development of a range of improved approaches to recruitment for the following pilots.

All institutions found a significant drop out—close to 50% overall—between initial recruitment and signing the consent form on the one hand, and participating in the assessment activities and completing evaluation questionnaires on the other, and so they built this into their planning for future pilots both in terms of initial over recruitment and strategies to encourage continuation in the pilot.

5.1.2 Pilot 2 (Medium Test-bed Pilots)

The total number of students participating in Pilot 2 was 4931 including 204 SEND students, which comfortably exceeded the target set (3500 students), though some institutions had difficulties in meeting their individual targets.

A reasonable distribution of tested instruments was achieved, exceeding the target of 700 students testing each instrument. A large number of students used the face recognition instrument alone as a result of the incorporation of this instrument within some large online courses (with 1443 students using this instrument in Anadolu University alone, and another 1191 in the other institutions). Pilot Leaders commented on the difficulty of testing the voice recognition instrument because of the difficulties of the enrollment process, though 893 students did successfully test the instrument. Similarly, Pilot Leaders commented that the requirements for a large amount of typing for the keystroke dynamics instrument caused difficulties, though 1095 students did successfully test the instrument. The least used instrument was the forensic analysis instrument (865 students) which required a lot of typing. This discouraged some of the teachers from integrating this instrument in assessment activities and it was also unpopular with some students.

It became clear in this pilot that the integration of TeSLA instruments into real assessment activities depended on a number of factors. Technical issues with some instruments made their trialing more difficult, resulting in some teachers switching to using instruments with fewer technical problems. The teachers' awareness of the instruments and of the ways in which they could be integrated into assessment activities was also limited. These two issues would be well addressed in the later pilots. Two other factors identified related to the institutional context and would continue to have an influence in future pilots.

One factor related to the requirements for enrollment and use of the instruments in terms of the number of required data recordings, their duration and timing, requirements which were easier to accommodate in some contexts rather than others (this mainly impacted on the voice recognition, keystroke dynamics and to a lesser extent the forensic analysis instruments).

The second factor related to the predominant forms of assessment in particular institutions (and their flexibility) and the extent to which a TeSLA instrument could be appropriately integrated in these forms of assessment. In universities where assessment was more flexible, the teachers had greater freedom in selecting and changing assessment methods, and so were able to more easily integrate the TeSLA instruments. In universities where the assessment was more strictly determined, the choice of possible TeSLA instruments was very restricted—for example in courses assessed purely by written assignments only the plagiarism detection and forensic analysis instruments could be readily integrated.

In this pilot, students in a number of courses tested more than one instrument, and others tested an instrument in more than one assessment activity. A significant number of tests were carried out in assessment activities that formed part of the actual assessment for the course, though many other tests were additional activities incorporated simply for the purposes of testing the TeSLA system. This use of multiple instruments and increased incorporation into real assessment activities indicated that the project was on track for the full integration into authentic learning and assessment activities that would be the goal of Pilots 3A and 3B (Large Scale Pilots).

In Pilot 2 most of the students who participated were studying in the field of Social and Legal Sciences (29% in Education and Psychology, 28% in Social Science and 9% in Business and Economics), indicating a move away from the technology-oriented courses that dominated Pilot 1. Students in blended education courses constituted 48% of the participants, with 52% in online or distance education courses.

The TeSLA instruments were used in formative, summative, and diagnostic assessment activities, though because of the technical issues with the use of the TeSLA instruments at this stage of their development, teachers were reluctant to use them for summative assessment purposes.

TeSLA instruments were usually integrated in individual assessment activities (93%) and only in a few cases were they used in collaborative activities. Pilot Leaders commented that they found it difficult to integrate TeSLA instruments into collaborative activities as, whilst TeSLA was integrated into the Assignment and Test activities in Moodle, it was not integrated into the Forums, Wikis or Workshop activities. The Pilot Leaders requested further development of the TeSLA instruments to support collaborative assessment activities.

5.1.3 Pilots 3A and 3B (Large Scale Pilots)

The quantitative targets for Pilots 3A and 3B were met at project level in six of the seven institutions. The TeSLA system was tested in terms of integration, reliability and scalability by 17,373 students who completed enrolment and at least one assessment activity, and a further test of the reliability of TeSLA authentication and authorship mechanisms was carried out in a special authentication accuracy study by close to 600 students who completed assessment activities which incorporated three, or all four, of the authentication instruments.

TeSLA was integrated and tested in 310 courses by 392 teachers in undergraduate, continuous professional development (CPD), and post-graduate courses. Whilst individual institutions had different balances of students by gender and age, the pilots as a whole achieved an almost equal balance for gender and a reasonable balance across age ranges. The project had encouraged Pilot Leaders to involve a wider range of courses in Pilots 3A and 3B in order to test the integration of the instruments in a wide variety of educational contexts. In Pilot 3B, the majority of students were in courses in the field of Social and Legal Sciences (59%), but there was also a good representation of Arts and Humanities courses—16%, Science—9%, Health Sciences—10%, Engineering and Architecture—7%. Whilst the majority of students were in undergraduate courses (75%), there were also significant numbers of students in post-graduate and continuing professional development courses. The percentage of students in online (as distinct from blended) courses rose from 51% in Pilot 3A to 69% in Pilot 3B, which most likely reflects an increasing trust in TeSLA by partner institutions and teachers (because of greater technological stability and increased familiarity with TeSLA on the part of teachers), leading to greater willingness to integrate the instruments into activities that are carried out entirely from a distance, thus providing evidence that TeSLA was achieving acceptance in its role of supporting student assessment in the context of e-learning.

Courses were delivered in seven languages. TeSLA instruments were integrated in a variety of cultural contexts defined by the specificity of the particular language of study and some of the comments of the students and teachers on TeSLA utilization reflect on the specifics of the language when TeSLA is in use.

Altogether 27,524 tests with the TeSLA instruments (in enrolment and follow-up activities) were performed during Pilots 3A and 3B—mostly in authentic educational contexts. The selection of the particular instruments to be used was found to be dependent on the institutional context and though there was a somewhat unequal distribution of the number of tests per instrument, the TeSLA developers received a sufficient amount of data to test both the accuracy of each of the individual instruments and the scalability of the whole TeSLA system.

Table 2 provides a description of the assessment activities using the categories given in Table 1. It should be noted that because of the large numbers of trials even

Table 2 Descriptive data for activities in Pilots 3A and 3B

Query	Response	% of student-activities N = 27,524 (%)
How was the assessment used?	Diagnostic	1
	Formative	22
	Formative and summative	58
	Summative	11
	N/A	8
Where did the assessment activity take place?	At home	88
	At university	12
Was the assessment supervised?	Supervised	12
	Unsupervised	88
Was the assessment individual or collaborative?	Collaborative	2
	Individual	94
	Individual and collaborative	4
What was the type of response required in the assessment activity?	Select answer	17
	Create answer or product	73
	Perform/enact/demonstrate	11
What was the response format for response type 'Create answer or product'?	Audio	11
	Video	12
	Text (natural language or code)	76
	Artifact (e.g. a painting, a meal, sheet music)	1

small percentages reflect quite large numbers of student activities. The predominant type of assessment activity (73%) was 'Create answer or product (short answer; essay; poster; report; thesis; software system ...)' which probably reflects the predominance of students from courses in the social sciences where the assessment is commonly based on the production of reports.

5.2 *Results*

5.2.1 Pilot 1 (Small Educational Pilots)

The project achieved its main goals for Pilot 1, in that it was able to:

- Check how the logistics aspects affect the pilot—the place, the time, the technology, and the support.
- Check and improve the coordination between the Pilot Manager and other participants.
- Identify the legal and ethical issues at the institutional level.
- Identify the critical risks at the project and institutional level.
- Make a first trial of the data collection and analysis process for monitoring purposes.
- Test the technical implementation protocol at partner level.
- Begin to identify the most suitable activities for the e-assessment process at subject level and so to integrate tools to ensure students' authentication and authorship in variety of educational contexts.
- Collect and archive data captured by the TeSLA instruments for analysis by technical partners and instrument developers.
- Begin to identify how to meet the needs of SEND students.
- Study the opinions and attitudes of pilot participants towards the integration of the authentication and authorship tools in assessment through the collection of evaluation data.

The technological aspects of the project dominated Pilot 1 preparation and execution, and through the process of piloting those instruments that had initial versions ready for testing we met and overcame a wide range of problems, providing a very good basis for the larger scale trials in Pilot 2, and captured some initial data for the instrument developers to refine their instruments, and the TeSLA technical team to refine the TeSLA system.

The communication protocol was found to work well. Regular monitoring of the pilot progress through the Project Demonstration Board, and the use of the management platform used by the project—Basecamp—for discussing technical issues were both found to be effective. Where difficulties arose—for example the under recruitment of participants at two institutions—then the contingency plans were applied effectively.

An important element of the approach to Pilot 1 was flexibility in the strategies used in both planning and implementation, allowing us to explore options and determine suitable approaches for future pilots.

5.2.2 Pilot 2 (Medium Test-bed Pilots)

The project achieved its main goals for Pilot 2, in that it was able to:

- Test the TeSLA biometric instruments (face recognition, voice recognition, and keystroke dynamics) and document analysis instruments (forensic analysis and plagiarism) in an isolated manner in assessment activities.
- Enable the technical team to test security and integrity aspects of the system, such as the encrypted channels, the digital signature and timestamp of learners' deliverables.
- Refine learning activities for e-assessment and integrate TeSLA instruments within assessment activities in courses to ensure students' authentication and authorship in variety of educational contexts.
- Test processes for identification of SEND students, and initial testing of their use of the TeSLA instruments.
- Collect evaluation data about teachers' and students' expectations, opinions and personal experiences on e-assessment and TeSLA instruments.
- Collect materials and information for quality assurance purposes.

During Pilot 2, the partner institutions began to work together more effectively as a team. They gained valuable experience in this pilot which was then applied in the following pilots where the challenges would be greater in that much larger numbers of students, courses and teachers would be involved and a higher level of integration of the TeSLA instruments into the normal assessment processes of the partner universities would be required.

Pilot Leaders provided feedback on the progress of Pilot 2 and recommendations related to technical, pedagogical, and legal aspects of the project were collected and summarized in the report on the pilot. These recommendations were then taken into consideration in the planning and execution of the Large Scale Pilots.

Many of the participants (students and teachers) provided feedback on their experiences of the TeSLA authentication and authorship instruments as well as sharing their opinions about cheating and academic dishonesty in general by taking part in the survey, individual interviews and focus groups after each of the TeSLA pilots. This information forms the basis of reports by the project teams concerned with quality assurance and evaluation. Institutional TeSLA teams either individually or in teams with colleagues from other partner institutes analyzed the collected data and reported the results at a number of scientific forums and published papers in research journals [4, 6, 7, 9, 11–21].

5.2.3 Pilots 3A and 3B (Large Scale Pilots)

The project achieved its goals for Pilots 3A and 3B. The TeSLA system was tested in terms of integration, reliability and scalability by 17,373 students who completed an enrolment and at least one assessment activity, and a further test of the reliability of TeSLA authentication and authorship mechanisms was carried out in a special authentication accuracy study by close to 600 students who completed assessment activities incorporating three or all four of the authentication instruments.

In these pilots most pilot institutions achieved a good level of pedagogical integration of the TeSLA instruments, and developed interesting ideas and models that will be useful for others adopting TeSLA later. Several pilot institutions also worked closely with SEND students and produced specific reports on this work. These two aspects of the work of Pilots 3A and 3B are described in more detail below. The collected data from the study of the SEND students' experiences with TeSLA were analyzed and the results were reported to a wide audience at three conferences in Europe and published in the conference proceedings (see [11, 15, 18]).

5.2.4 TeSLA Integration in Assessment Activities

One of the main objectives of Pilots 3A and 3B was to test the pedagogical integration of TeSLA in a variety of educational contexts. In general, most pilot institutions achieved a good level of integration, and interesting ideas and models have been developed that will be useful for others adopting TeSLA later. The following is a short overview of the results of this work, which is set out in detail in Chap. 9: *Integration of TeSLA in assessment activities* in this volume.

TeSLA instruments were integrated into pilot courses in four ways: an involvement in low stakes and formative assessments; an involvement in coursework constituting less than a half of the overall mark supplemented by face-to-face supervised examinations; use with final written assessments that carry the whole of the course mark; and integration into the majority or all of a series of assessment activities that carry the whole of the course mark (continuous assessment). Institutions coming new to TeSLA can initially select from these approaches one that best suits their present context and then work to expand the usage.

The analysis of the case studies describing the assessment activities using TeSLA instruments in the pilot identified seven common categories of TeSLA instrument use, which varied according to the design of the assessment activity and the teacher's purpose in seeking to authenticate student identity and/or check the authorship of assignments. These descriptions provide a good basis for teachers new to TeSLA to select an appropriate approach that suits specific course needs.

The course and assessment scenarios described are based on actual successful practice by the pilot institutions. Other approaches can readily be imagined and will no doubt develop as TeSLA is used in the future, but these scenarios provide an assured starting point, illustrating approaches that have been demonstrated to be practicable and to make effective use of the TeSLA instruments.

5.2.5 SEND Students

A special study on SEND students' use of TeSLA was carried out during Pilots 3A and 3B. Some 72 students from four pilot institutions with a wide variety of disabilities and from a range of age groups took part in individual tests and interviews, which allowed us to identify issues with the accessibility and usability of the TeSLA instruments for this group of students and to provide feedback to the technical team for improvement of the TeSLA instruments. Some issues with the accessibility and usability for students with specific disabilities that cannot be overcome were also identified (see [11, 15, 18]).

All the SEND students involved were highly appreciative of the availability of the TeSLA, which they saw as: allowing for more equal opportunities for all students; reducing the pressures on students with severe physical disabilities to drop out from the university; saving time and money for travel; improving productivity; increasing flexibility; and, increasing the reliability of assessment in comparison with face-to-face examinations.

Most of the SEND students appreciated the opportunities that TeSLA provides for conducting exams from home on certain occasions, but they did not see online assessment as alternative replacement for all face-to-face assessment because this would limit the opportunities for their socialization.

The students' experiences and opinions in relation to accessibility and usability of the TeSLA system and instruments varied a great deal according to the type and degree of their disability. The analysis of the data suggests that, due to the heterogeneity of SEND students and the specificity of different disability groups, it is not possible for a system such as TeSLA to satisfy equally the needs of all of such a diverse group of learners in terms of accessibility and usability. The groups of SEND students that are most vulnerable in this respect are blind and partially sighted students and deaf students. They are the groups most affected by the identified limitations of the TeSLA system. This means that the optimization of the user interface, flexibility and adaptability of the individual instruments and feedback from the system should take into account the specific needs of these particular groups of students. The interface should be easy to navigate and usable with assistive technologies such as screen readers, and the system should be accessible through a full range of browsers as some SEND students relied on assistive technologies, such as screen readers in their native language, that were only available in some browsers. In order to increase the usability of the system, the enrollment procedure needs to be as simple and clear as possible and the real assessment activities should be tailored to the student's individual abilities.

In terms of the incorporation of TeSLA instruments into assessment activities, it was concluded that tutors should design the assessment activities with alternative options for the use of TeSLA instruments, which might be a combination of instruments, allowing a student to choose which of them is most accessible and least disturbing for her in terms of her disability.

6 Discussion

The Pilot Leaders were asked to propose recommendations to future users of the TeSLA system. Based on their comments and on feedback from the Pilot Manager we have drawn up the following recommendations for consideration by potential TeSLA users that we hope will be helpful.

Any future deployment of TeSLA in a university will, of course, be dependent on the arrangements made by the TeSLA Consortium for the licensing and support of the TeSLA system. Any university that intends to use TeSLA will need to make arrangements with the TeSLA Consortium relating to licensing and support and be responsible for the associated costs. It is important that the implementation and ongoing use of TeSLA within an institution be guided by a local team consisting of academic, administrative and IT staff who will work with the TeSLA Consortium support personnel.

In preparing for the implementation of TeSLA, the team should:

- Explore how TeSLA would fit into the management of academic integrity in the institution. What are the opinions and attitudes of university management, academic staff, and students to cheating and plagiarism? Is there a sufficient concern about academic dishonesty to merit the introduction of the TeSLA system?
- Ensure that the value of using TeSLA is clearly presented to all stakeholders, setting out the advantages, and also discussing potential problems.
- Ensure that issues around privacy, use of (sensitive) personal data, confidentiality, ethics, and informed consent are clearly explained including how the biometric data of students will be used and protected, and how the feedback from the system will be managed, as well as possible issues that could arise.
- Decide upon the rules and procedures for using TeSLA and for using the feedback from the instruments at both institutional and course level within the overall context of the regulation of academic integrity within the institution. Ensure that these procedures are clearly communicated to all stakeholders.

In most institutions, the teachers will play a pivotal role in designing the use of TeSLA, though in some institutions some aspects of this work will be dealt with by other staff, for example by course design teams or by assessment staff who are not directly involved in teaching, so the following suggestions need to be interpreted in terms of the organization in a particular university. The important issues here are that the team should:

- Provide training materials for teachers and students on basic use of the TeSLA system.
- Develop and provide guidelines for teachers on technological aspects of TeSLA integration in the assessment activities depending on the specifics and the functionality of the institutional VLE.
- Encourage alignment of learning objectives, teaching and learning methods, and assessment in all course designs.

- Explore the typical assessment models in the institution and create a teacher manual that presents a variety of TeSLA integration scenarios in order to help teachers develop effective use of the instruments. This manual should provide examples of alternative assessments scenarios to provoke creativity in TeSLA integration in a variety of assessment activities.
- Support the teachers in designing assessment activities with a combination of instruments for SEND students to allow them choose which instruments are most accessible and least disturbing for them.
- During the day to day provision of courses using TeSLA, it is important for the team to ensure that tutorials for use of TeSLA are readily available; technical support is available for detecting and solving problems with TeSLA use; TeSLA is used in a way that provides some flexibility to students, by allowing at least some parts of the assessment activities to be carried out unsupervised and outside the university; teachers and administrators understand the meaning of the feedback from the TeSLA system; and the results of authentication and authorship checking provided by TeSLA are used appropriately and in accordance with the institutional regulations related to academic integrity.

7 Conclusions

Most of the lessons learned from this piloting are well known but are worth reiterating:

The use of multiple rounds of piloting with gradually increasing numbers of trials was very effective in supporting the development of the piloting process and ensuring that procedures were well developed by the time of the final Large Scale Pilots. The development of a communication protocol with clear definitions of roles and responsibilities in the team was very helpful. The pilots were too large for everyone to always have an overview of what was going on, and so it was important for everyone to be clear as to what their own responsibilities were.

In these pilots, a range of educational institutions with a large number of teachers were working with technologies developed by a number of technological partners. It was extremely important to set up clear processes for liaison between the technical and pedagogical teams to ensure rapid responses to issues. This was difficult to get right, and the project adjusted the way in which this was done in each pilot, learning from the issues which arose in previous pilots. It was important to monitor progress continually and adjust plans accordingly. To have detailed contingency plans developed at the start of the pilots was very important, as problems frequently arose, both with the technology and in terms of students' and teachers' involvement, and the use of existing contingency plans meant that responses could be made quickly.

Another lesson, which should have been obvious, but nevertheless threw up surprises again and again, was that individual universities have very different cultures and ways of doing things, this is true even for universities in the same country, but across a range of countries these differences are even greater. It was all too easy to assume that things in another university worked in a similar way to those in one's own and so to misunderstand the nature of problems faced, and the opportunities available in that other context. Pedagogic ideas and vocabulary also varied and this could cause difficulties in communication if these were not made explicit. However, this variety of contexts was also a strength for the project in that it enabled some tests to be done in some contexts which could not have been carried out in another, and helped to develop TeSLA in a way that meant that it could be adapted for use in the even wider range of university contexts in which we hope it will be used in the future.

Acknowledgements We want to thank to all Pilot Leaders for their hard work and contribution to the running of the institutional pilots, and the collection and analysis of the data: Serpil Kocdar (AU), Tarja Ladonlahti (JYU), Francis Brouns (OUNL), Wayne Holmes, Chris Edwards (OUUK), Lyubka Aleksieva (SU), Mariana Durcheva, Ana Rozeva (TUS), M. Elena Rodríguez (UOC).

References

1. Chatzigavriil A (2014–2015a) GV100 e-assessment pilot study. Retrieved 20 Feb 2016 from http://eprints.lse.ac.uk/64329/1/Government-e-assessment-Pilot-study.pdf
2. Chatzigavriil A (2014–2015b) Law e-assessment pilot study. Retrieved 20 Feb 2016 from http://blogs.lse.ac.uk/lti/files/2015/10/Law-e-assessment-Pilot-study.pdf
3. European Commission (2015) Call ICT-20-2015. Retrieved 10 Apr 2019 from https://ec.europa.eu/info/funding-tenders/opportunities/portal/screen/opportunities/topic-details/ict-20-2015
4. Edwards C, Holmes W, Whitelock D, Okada A (2018) Student trust in e-authentication. In: L@S '18: proceedings of the fifth annual ACM conference on learning at scale. ACM, New York, Article no. 42
5. JISC (n.d.). Online surveys (formerly BOS) (2019). Retrieved 12 Apr 2019 from https://www.onlinesurveys.ac.uk/
6. Karadeniz A, Peytcheva-Forsyth R, Kocdar S, Okur MR (2018) Approaches to prevent cheating and plagiarism in e-assessment: higher education students' views. In: Proceedings from EDULEARN18: 10th annual international conference on education and new learning technologies. IATED Academy, Spain. https://doi.org/10.21125/edulearn
7. Kocdar S, Karadeniz A, Peytcheva-Forsyth R, Stoeva V (2018) Cheating and plagiarism in e-assessment: students' perspectives. Open Praxis 10(3):221–235
8. Litherland K, Carmichael P, Martínez-García A (2013) Ontology-based e-assessment for accounting: outcomes of a pilot study and future prospects. J Acc Educ 31(2013):162–176
9. Mellar H, Peytcheva-Forsyth R, Kocdar S, Karadeniz A, Yovkova B (2018) Addressing cheating in e-assessment using student authentication and authorship checking systems: teachers' perspectives. Int J Educ Integrity. Retrieved 12 Feb 2018 from https://link.springer.com/article/10.1007/s40979-018-0025-x
10. Nilsson A, Andrén M, Engström M (2014) E-assessment of prior learning: a pilot study of interactive assessment of staff with no formal education who are working in Swedish elderly care. BMC Geriatr 14(1). http://www.biomedcentral.com/1471-2318/14/52
11. Noguera I, Guerrero-Roldán A, Peytcheva-Forsyth R, Yovkova B (2018) Perceptions of students with special educational needs and disabilities towards the use of e-assessment in online and blended education: barrier or aid? In: Proceeding from 12th international technology, education and development conference, Valencia, Spain, 5–7 Mar 2018

12. Noguera I, Guerrero-Roldán AE, Rodríguez ME (2017) Assuring authorship and authentication across the e-assessment process. In: Joosten-ten Brinke D, Laanpere M (eds) Technology enhanced assessment. Communications in computer and information science. Springer, Cham, pp 86–92
13. Okada A, Whitelock D, Holmes W, Edwards C (2019) e-Authentication for online assessment: a mixed-method study. Br J Edu Technol 50(2):861–875
14. Okada A, Noguera I, Aleksieva L, Rozeva A, Kocdar S, Brouns F et al (2019) Pedagogical approaches for e-assessment with authentication and authorship verification in Higher Education. Br J Edu Technol 50(2):861–875
15. Peytcheva-Forsyth R, Aleksieva L (2019) Students' authentication and authorship checking system as a factor affecting students' trust in online assessment. In Proceedings from INTED2019: 13th annual international technology, education and development conference, Valencia, Spain, 11–13 Mar 2019
16. Peytcheva-Forsyth R, Aleksieva L, Yovkova B (2018a) The impact of technology on cheating and plagiarism in the assessment—the teachers' and students' perspectives. In: AIP conference proceedings, vol 2048, p 020037. https://doi.org/10.1063/1.5082055
17. Peytcheva-Forsyth R, Aleksieva L, Yovkova B (2018b) The impact of prior experience of e-learning and e-assessment on students' and teachers' approaches to the use of a student authentication and authorship checking system. In Proceedings from 10th annual international conference on education and new learning technologies, Palma de Mallorca, Spain, 2–4 July 2018
18. Peytcheva-Forsyth R, Yovkova B, Ladonlahti T (2017) The potential of the TeSLA authentication system to support access to e-assessment for students with special educational needs and disabilities (Sofia university experience). In: Proceedings from ICERI2017, Sevilla, Spain, 16–18 Nov 2017
19. Peytcheva-Forsyth R, Yovkova B, Aleksieva L (2018c) Factors affecting students' attitudes towards online learning—the case of Sofia University. In: AIP conference proceedings, vol 2048, p 020025 (2018). https://doi.org/10.1063/1.5082043
20. Peytcheva-Forsyth R, Yovkova B, Aleksieva L (2019) The disabled and non-disabled students' views and attitudes towards e-authentication in e-assessment. In: Proceedings from INTED2019: 13th annual international technology, education and development conference, Valencia, Spain, 11–13 Mar 2019
21. Rodríguez ME, Baneres D, Ivanova M, Durcheva M (2018) Case study analysis on blended and online institutions by using a trustworthy system. In: Ras E, Guerrero Roldán A (eds) Technology enhanced assessment. TEA 2017. Communications in computer and information science, vol 829, pp 40–53. Springer, Cham

Integration of TeSLA in Assessment Activities

Harvey Mellar and Roumiana Peytcheva-Forsyth

Abstract During Large Scale Pilots of the TeSLA system, teachers in six European universities constructed case studies of the use of the TeSLA instruments in assessment activities in authentic courses. A thematic analysis of these case studies identified seven common categories of TeSLA instrument use, which varied according to the design of the assessment activity and the teacher's purpose in seeking to authenticate student identity and/or check the authorship of assignments. These seven categories are described in detail in this chapter. These descriptions are intended to provide a starting point for teachers new to the use of student authentication and authorship checking instruments to identify appropriate ways to integrate the instruments in assessment activities in their courses. The chapter goes on to identify some of the issues arising from the context of the collected data and implications of the study for the future implementation of TeSLA and similar systems incorporating instruments for student authentication and authorship checking.

Keywords Assessment activities · TeSLA instruments · TeSLA integration · Case studies

Acronyms

ICT	Information and Communication Technology
VLE	Virtual Learning Environment

H. Mellar (✉)
UCL Institute of Education, University College London, London, UK
e-mail: h.mellar@ucl.ac.uk

R. Peytcheva-Forsyth
Sofia University, Sofia, Bulgaria
e-mail: r.peytcheva@fp.uni-sofia.bg

D. Baneres et al. (eds.), *Engineering Data-Driven Adaptive Trust-based e-Assessment Systems*, Lecture Notes on Data Engineering and Communications Technologies 34,
https://doi.org/10.1007/978-3-030-29326-0_9

TeSLA Instruments

FA	Forensic Analysis
FR	Face Recognition
KD	Keystroke Dynamics
PD	Plagiarism Detection
VR	Voice Recognition

Pilot Institutions

AU	Anadolu University
JYU	University of Jyväskylä
OUNL	Open University of Netherlands
OUUK	The Open University of United Kingdom
SU	Sofia University
TUS	Technical University of Sofia
UOC	Universitat Oberta de Catalunya

1 Introduction

E-learning and e-assessment are increasingly used in European universities [8] as, indeed, is the case in many other regions and institutions. However, the openness of e-assessment to cheating has been a cause of much concern for universities [14, 24, 32], and plagiarism and 'contract cheating' [19] similarly present threats to both on campus assessment and e-assessment. Besides their direct impact on the integrity of assessment, these threats also have the potential to constrain the forms of assessment that universities are willing to adopt, with a possible retreat away from more innovative and flexible forms of assessment towards supervised face-to-face examinations where cheating is easier to control. The European Commission funded the TeSLA project to develop a system for student authentication and authorship checking that can be used with e-assessment in order to enable universities to have a greater degree of trust in e-assessment, so that online students can conduct their assessments at a distance alongside learning at a distance [20, 22, 23, 28].

Over a period of three years (2016–2018), the TeSLA project developed a modular system incorporating five instruments. Four of these instruments involve the creation of a learner model during an enrollment phase: Face Recognition (FR)—which compares images of faces with a learner model; Voice Recognition (VR)—which compares voice structures with a learner model; Keystroke Dynamics (KD)—which compares rhythm and speed of typing with a learner model; Forensic Analysis (FA)—

which compares writing style to a learner model. The fifth instrument—Plagiarism Detection (PD)—detects textual similarities between documents created by students.

These instruments were incorporated into the Virtual Learning Environments (VLEs) of the seven higher education institutions participating in the project in order to carry out a series of pilot studies, with a gradually increasing number of students, rising from 600 in the first pilot to over 11,000 in the second phase of the third pilot (for a detailed account of these pilots see Chap. 8 *Design and execution of TeSLA pilots* in this volume). The pilots were carried out in online and blended learning contexts in order to test TeSLA's appropriateness in both contexts. Two universities—Sofia University (SU) and Technical University of Sofia (TUS)—were campus-based universities that also have a number of blended courses. Two others—Anadolu University (AU) and University of Jyväskylä (JYU)—were universities that have both traditional on-campus provision and open learning systems that employ online teaching. Three universities—Open Universiteit Nederland (OUNL), the Open University (OUUK), and Universitat Oberta de Catalunya (UOC)—were open online universities without traditions in on-campus education.

In this chapter, we aim to describe the ways in which TeSLA was integrated within e-assessment activities in the partner institutions during the final year of the pilots in order to present a picture of the degree of integration that was actually achieved in the pilots, which it is hoped will act as a guide to those looking to integrate TeSLA, or similar systems, into their own courses in the future.

The remainder of this chapter is organized as follows: Sect. 2 provides some background on assessment and e-assessment in higher education; Sect. 3 gives an overall picture of the use of TeSLA instruments in the final year of the pilots; Sect. 4 describes the collection of detailed descriptions of scenarios of TeSLA use in real assessment activities; Sect. 5 describes seven categories of the use of TeSLA instruments identified in the analysis of these scenarios; Sect. 6 illustrates these categories with a description of how these categories come together across whole courses; Sect. 7 discusses the relationship of our analysis to previous work on the analysis of the use of the TeSLA instruments; and, Sect. 8 sets out some conclusions and recommendations for future integration of TeSLA and similar systems in assessment activities.

2 Assessment in Higher Education

The *Report to the European Commission on Improving the quality of teaching and learning in Europe's higher education institutions* by the High Level Group on the Modernisation of Higher Education [21] notes that whilst there had been considerable progress in adopting some aspects of the qualification framework of the European Higher Education Area, that there is still a need to move beyond a concentration on knowledge and understanding alone, to encompass a wider range of competences such as: applying knowledge and understanding; making judgments; communication skills; and learning skills. To achieve this, they argue that assessment procedures need to be changed:

> … they must no longer simply check taught facts and knowledge, but rather measure the competences the student obtained as a result of a process of learning. In some cases this may require new formats, for example role plays or simulated situations that anticipate what the graduate might encounter later in the labour market. Institutions need to define overarching standards not only for teaching requirements, but also regarding these innovative forms of assessment. (p. 42)

However, reform in assessment lags behind reform in other areas of Higher Education, in part due to the lack of assessment expertise [5]. There is also an identified need for improvements in the constructive alignment [1] of assessment and e-assessment to course learning objectives [3, 9, 13].

The extensive academic literature on assessment and e-assessment in higher education (see for example: [15, 16, 26, 27, 33]) provides a wide range of perspectives on assessment including forms of assessment that go beyond teacher assessment to include self and peer assessment, and also identifies many possible formats for assessment activities. Guàrdia et al. [12] include the following in a list of formats for assessment activities: multiple-choice questions, short answer, judged mathematical expression (i.e. short answer question allowing mathematical expressions which may be evaluated as to their equivalence to the correct answer), free text responses and essays, e-portfolio assessment, blogs, online discussion, concept maps, online role-plays and scenario-based activities.

The use of e-assessment has raised concerns about cheating, plagiarism and academic integrity more generally. Studies of the implementation of academic integrity in EU universities have shown considerable variability between countries, and whilst they have identified some examples of good practice, they commonly found a lack of awareness and inadequate institutional response [7, 10, 11].

Technological support for academic integrity has so far mainly centered on plagiarism detection, and a number of studies have examined this use [29, 34]. The use of linguistic analysis tools to identify authorship is a relatively new approach that has only recently begun to be explored in educational contexts [17, 30]. On-line proctoring systems have not been widely use in Europe to date though some experimentation has taken place [6]. However, before the TeSLA project, there was a lack of studies describing the use of e-assessment with instruments for e-authentication and authorship verification other than plagiarism detection and so this chapter is exploring a relatively new area, building on earlier work from the TeSLA project [23].

3 The Use of TeSLA Instruments

This section provides a summary of basic descriptive data for the full set of assessment activities that were used in the pilots in the final year of the study, and then in Sects. 5

and 6 we will provide detailed descriptions of the use of TeSLA instruments in a number of specially selected scenarios. The data presented in these sections relates only to teachers' descriptions of their assessment activities and courses, as data on students' and teachers' opinions about TeSLA and their reactions to its use has been reported elsewhere (Okada et al. [23], and see Chap. 11 about the evaluation study in this volume).

The pilots were carried out in seven institutions, and the pilot was coordinated by a designated Pilot Leader in each institution who was responsible for organizing and monitoring the pilot. In the final year of the project we collected data from the seven institutions in relation to 532 assessment activities in 310 courses, involving 19,599 students and 34,909 'student-activities' (i.e. individual student assessments). Table 1 provides an overview of the basic descriptive data provided by the Pilot Leaders in relation to these activities, and an explanation and rationale for the categories used is presented below.

Firstly, we wanted to know the purpose of the assessment activity as this was likely to impact on the importance given to student authentication and authorship checking. Was the purpose diagnostic (to identify students' prior knowledge), for-

Table 1 Basic descriptive data for assessment activities

Query	Response	% of student-activities
How was the assessment used?	Diagnostic	1
	Formative	22
	Formative and summative	58
	Summative	11
	N/A	8
Where did the assessment activity take place?	At home	88
	At university	12
Was the assessment supervised?	Supervised	12
	Unsupervised	88
Was the assessment individual or collaborative?	Collaborative	2
	Individual	94
	Individual and collaborative	4
What was the type of response required in the assessment activity?	Select answer	17
	Create answer or product	73
	Perform/enact/demonstrate	11
What was the response format for response type 'Create answer or product'?	Audio	11
	Video	12
	Text (natural language or code)	76
	Artifact (e.g. a painting, a meal, sheet music)	1

mative[1] (to collect information about student learning and evaluate it in relation to prior achievement and attainment of learning outcomes so as to allow the teacher or the student to adjust the learning trajectory), or summative (to contribute towards the overall course mark), or was it being used for both formative and summative purposes?

Secondly, we wanted to describe the context in which the assessment activity was taking place as this was likely to impact on the importance given to student authentication and authorship checking, so we collected data on where the assessment activity took place (at home or university), and whether the assessment was supervised or not.

Thirdly, we wanted to know if the assessment activity was purely individual or whether it was in whole, or in part, a collaborative assessment activity. We were conscious of the increasing use of a range of collaborative forms of assessment in higher education, and of the possible challenges that this might present to the use of TeSLA and so wished to look specifically at assessment activities where this had been attempted.

Fourthly, we wanted to characterize the assessment activity in terms of what was done in the activity, because we wished to examine the suitability of TeSLA across a wide range of assessment activities. This was done using the categories for response type and response format that are described in Okada et al. [23]. The type of response required in the assessment activity was classified as (a) select answer, (b) create answer or product, or (c) perform/enact/demonstrate, and where the type of response was 'create answer or product', we also classified the response format as (a) audio, (b) video, (c) text (natural language or code), or (d) an artifact (e.g. a painting, a meal, sheet music).

From Table 1 it can be seen that most of the assessment activities were individual assessments (rather than collaborative), most were used for summative assessment, and most involved the creation of text. There is a significant amount of use for summative assessment because the project had put emphasis in the final pilots on the use of TeSLA instruments for summative assessment as we wished to demonstrate the ability of TeSLA to meet the need to for student authentication and authorship checking in high stakes summative assessments. Because the project needed to involve very large numbers of students in the final pilots in order to test the ability of the TeSLA system to handle such large student numbers, the project involved many courses using individual text based assessments as these were easier to scale, and as a consequence this form of assessment is somewhat over represented in the data. However, Table 1 does indicate that there was also a significant use of more varied forms of assessment, and descriptions of some of these are provided later in this chapter. The low level of use of TeSLA with collaborative assessment was something identified as a concern in the initial TeSLA pilots and the project had worked to improve on

[1]The formative use of assessment is perhaps best defined in this quotation from Black and Wiliam [2]: "An assessment functions formatively to the extent that evidence about student achievement is elicited, interpreted, and used by teachers, learners, or their peers, to make decisions about future instruction that are likely to be better, or better founded, than the decisions that would have been taken in the absence of that evidence".

this by, for example, incorporating the TeSLA instruments into forum activities, as is exemplified in the descriptions of assessment activities provided later in this chapter.

During the pilots, we noted four general approaches to experimentation with, and integration of, the TeSLA instruments in activities and courses. Firstly, there were initial tentative steps in which TeSLA instruments were used only in low stakes and formative assessments, where the failure of the TeSLA instruments, or difficulties of use, would not dramatically affect examination results. Secondly, there was an expansion of TeSLA use into more of the course assessment activities, but where still less the half the assessment was carried out at a distance with TeSLA instruments, and the remainder of the assessment consisted of face-to-face supervised examinations. This reflected a growing confidence in the use of the instruments, whilst retaining a reliance on face-to-face examinations in case of problems with the TeSLA system. Thirdly, there was a use of some TeSLA instruments with one or more final written assessments that carried the whole course mark. This kind of use tended to be easy to implement as it had little impact on existing assessment arrangements because plagiarism detection instruments were already commonly used in these kinds of assessment. Fourthly, there was a full integration of TeSLA in which TeSLA instruments were used within the majority, or all, of a series of assessment activities carrying the majority, or all, of the course marks. One pilot institution (OUUK) was not able to integrate the TeSLA pilot into authentic assessment activities because of university regulations related to the use of pilot software in ongoing courses, and so was only able to use TeSLA in specially constructed activities outside normal courses.

An analysis of the descriptions of assessment activities using TeSLA instruments provided by the Pilot Leaders identified seven common categories of scenarios of use, which varied according to the design of the assessment activity and the teacher's purpose in seeking to authenticate student identity and/or check the authorship of assignments. These seven categories are described in detail in Sect. 4, and illustrations of the activities in these categories are provided in Sects. 5 and 6. It is hoped that the descriptions of these categories of TeSLA integration will provide a useful basis for institutions new to TeSLA to identify appropriate scenarios that match their assessment activities and their needs for student identity authentication and authorship checking.

The course and assessment descriptions given in this chapter are based on actual successful practice by the pilot institutions. Additional methods using different kinds of assessment activities can certainly be imagined and will no doubt develop as TeSLA is used in the future, but these scenarios provide an assured starting point, providing examples of what has been demonstrated to be practicable and to make effective use of the TeSLA instruments.

4 Developing Detailed Descriptions of TeSLA Integrations

In order to obtain a detailed picture of how TeSLA was being integrated in the assessment activities in pilot institutions, Pilot Leaders provided detailed case studies of

Table 2 Use of TeSLA instruments

TeSLA instrument	Frequency
Face Recognition (FR)	35
Voice Recognition (VR)	19
Keystroke Dynamics (KD)	23
Forensic Analysis (FA)	22
Plagiarism Detection (PD)	24

the assessment activities and courses in which they had integrated TeSLA. They were asked to use a pre-prepared template to provide descriptions of three courses in which TeSLA instruments had been integrated into some (or all) of the assessment activities, and to describe in detail these assessment activities. In the course descriptions, the Pilot Leaders were asked to describe the full set of assessment activities used for the course, whether using TeSLA or not, and to indicate the percentage contribution of each assessment to the course mark. In the assessment activity descriptions they were asked to describe the learning objectives, the assessment activity, how the assessment results were being used (diagnostically, formatively, summatively etc.), and how and why the chosen TeSLA instruments were being used. Our intention in this study was to collect examples of the most successful practice (in the sense of practical implementation) and we did not ask for information on difficulties and problems as this was reported separately as part of the project evaluation. In total, we obtained detailed descriptions of 72 assessment scenarios in 18 courses from six pilot institutions.

In order to provide some background about the use of instruments in these 72 assessment cases, the frequency of use of each instrument in shown in Table 2 and the use in combination in individual activities is shown in Table 3. These tables show that the 72 case studies selected provide a reasonable distribution across the instruments and of the possible combinations of instruments.

It should be noted at this point, as it will form part of the descriptions in the next section, that the FR and VR instruments may be used either in their basic form or used together with 'attack detection'. The purpose of attack detection is to detect 'presentation attacks'—that is when person A claims to be another person B, and presents to the verification instrument a pre-recorded copy (print, or digital) of the face of B (in the case of FR), or a pre-recorded version of B's speech, or synthetically generated speech (in the case of VR). These instruments can be used with attack detection when students are working online, but the use of attack detection is not relevant when dealing with uploaded materials.

Table 3 Use of TeSLA instruments alone and in combination

TeSLA instruments	Frequency
FR	10
VR	5
KD	7
FA	3
PD	6
FR + VR	9
FR + KD	6
FR + FA	3
FR + PD	3
VR + KD	4
KD + FA	1
FA + PD	8
FR + FA + PD	2
KD + FA + PD	3
FR + KD + FA + PD	1
FR + VR + KD + FA + PD	1

5 Integrating TeSLA with the Assessment Activities

The descriptions provided by the Pilot Leaders were analyzed thematically making use of the six-phase approach to thematic analysis outline by Braun and Clarke [4]. Any thematic analysis is influenced by the intentions of the analyst. Our purpose was to develop accounts of TeSLA use that would provide starting points for teachers new to TeSLA to help them integrate TeSLA (or similar systems) into their courses. To achieve this purpose, the case studies we examined were not typical examples of TeSLA use in the project, but, rather, examples selected by the Pilot Leaders as being amongst the best available in their institutions. Our starting point in the analysis was teachers' expressed purposes in using TeSLA instruments in a specific assessment activity—often expressed most simply in terms of identity authentication or assuring authorship, or some mixture of the two, but connected to actual perceived academic integrity issues in that assessment activity in that course. Starting with the teachers' purposes was seen by the authors as a good way of connecting the activity descriptions to the teachers' interests in using TeSLA, and hence in making the descriptions useful to other teachers.

Teachers looking at using our examples to support their own implementation of TeSLA integration should bear in mind that in the pilots all the students received training on how to use the instruments before using them for the first time, and that all the assessment activities were carried out in, or submitted via, the institutional

VLE in which the TeSLA instruments were integrated—which means that the use of TeSLA was well integrated into the students' work flow.

In this section, we describe how the TeSLA instruments were integrated into individual assessment activities, and in Sect. 6, we will provide two examples of courses that combine a series of such assessment activities into an assessment regime for the whole course. We identified seven categories of ways of using TeSLA instruments in assessment activities in the data we examined. These categories are briefly described in Table 4, and then fuller descriptions and examples of these categories will be set out below. Some previous work on mapping the use of TeSLA instruments to their use with specific kinds of learning and assessment activities is reported by Okada et al. [23], and we will consider the relationship of that work to our own work in Sect. 7.

In some assessment activities, assessment of more than one kind of material (multiple-choice response/written/spoken/multimedia) was taking place, and this meant that the activity could be classified into more than one category—so an activity might be classified as being both in Category 2 and in Category 3 when part of the assessment is based on text and part on audio.

Further, in some assessment activities, the teacher wished both to authenticate the student's identity and to check the authorship of submitted material. So, the use of TeSLA instruments for an assessment activity using written materials might be classified as belonging to Category 2 (for identity authentication) and to Category 7 (for authorship checking). Similarly, an assessment activity might be classified in Category 6 and in Category 7 when there was both an identity check for online submission (Category 6) and authorship checking of uploaded materials (Category 7).

The remainder of this section provides descriptions of specific assessment activities drawn from the case studies illustrating each of these categories and combinations.

5.1 Category 1—Identity Authentication in Multiple-Choice Tests

Identity authentication for: multiple-choice and short responses in tests in the VLE (including tests in which the responses are read aloud by the student).

TeSLA tools for identity authentication can be straightforwardly incorporated into online quizzes and tests, and there were many examples of this in the TeSLA project. Two examples are given below.

The course 'Social Research' (SU) is an undergraduate level blended course covering research methods, design and statistical analysis. The final assessment task is on online test, which is an individual activity carried out unsupervised at home, and is used for formative and summative assessment carrying 50% of the course marks. This online quiz contains multiple-choice questions and one short response item. The quiz is open for only a short period and the instrument FR with attack detec-

Table 4 Classification of assessment case studies

Category	Description	TeSLA instruments (one of more of these may be used)
1	Identity authentication for: – *multiple-choice* and *short response items* in tests in the VLE (including tests in which the responses are read aloud by the student)	FR with attack detection VR with attack detection
2	Identity authentication for: – tests in the VLE that include open questions with *written* responses (other forms of questions may also be included in the test) – discussion forums in the VLE – written assignments prepared in the VLE	FR with attack detection VR with attack detection KD FA
3	Identity authentication for: – tests in the VLE which include open questions with *spoken* responses including reading student written responses aloud (other forms of questions may also be included in the test) – assessments in the VLE with spoken elements	FR with attack detection VR with attack detection
4	Identity authentication for: – the preparation or delivery of *multimedia* materials in the VLE, including: presentations, multimedia reports, videos, working with *technology* and *software* (including working with mathematical software)	FR with attack detection VR with attack detection KD
5	Identity authentication for: – uploaded completed *multimedia* materials consisting in part of video or audio recordings of the student, prepared outside the VLE including: presentations, multimedia reports, and videos	FR VR
6	Identity authentication for: – uploading and *submission* of material prepared outside the VLE	FR with attack detection VR with attack detection
7	Authorship checking for: – *written* assignments (including written responses to open questions in a test) prepared in the VLE, which are checked either during the activity or after being uploaded at the end of the activity – *written* assignments prepared outside the VLE	FA to check that the submission matches with earlier verified work by the same student PD to check that there is no overlap with submitted material from other students

tion is used for identity authentication during the assessment activity, capturing the student's image while answering the questions.

The course 'English' (AU) is an undergraduate level online course in A1 level English, it is a compulsory course for all distance education students and so has a very large number of students. The assessment activity 'Introducing yourself' is an individual activity carried out unsupervised at home and is used for formative assessment, providing feedback in preparation for the midterm exams. The assessment activity is an online quiz in which the students are required to speak aloud their choice of response as well as select the answers, something that is felt to be appropriate as this is a language course. The VR instrument with attack detection is used. This use of TeSLA in a formative assessment activity helps to familiarize students with the use of TeSLA before it is used in later summative assessments.

5.2 *Category 2—Identity Authentication for Written Responses*

Identity authentication for: tests in the VLE that include open questions with written responses; discussion forums in the VLE; and written assignments prepared in the VLE.

The majority of assessment activities in the case studies incorporated some form of text input, sometimes combined with other forms of input. A variety (or a combination) of TeSLA instruments may be used in this context—in the first of the examples below FR is used and in the second KD is used.

The course 'Creating Educational Multimedia' (SU) is a postgraduate level blended course. The assessment activity 'Forum—creating and evaluating multimedia' is an individual and collaborative activity carried out unsupervised at home, which is used for formative and summative assessment, carrying 20% of the overall mark. In the activity, each student develops a multimedia collage on the topic 'Multimedia learning' using the Pixlr tool (Pixlr n.d.). The students share their collages and present their ideas in a forum in the VLE. Each student then evaluates and justifies their assessment of the work of two other students. Students also receive detailed feedback from the teacher and they are then given the opportunity to improve their work, on which they receive their final grade. The instrument FR with attack detection is used online whilst the student is writing in the forum, providing identity authentication, and, at the same time, posting in the forum inhibits copying because the artifacts created by the students are visible to all the other students, and so any copying would be easily noticed.

The course 'Online Teaching Planning' (UOC) is a postgraduate level online course. The assessment activity 'Case Study Analysis' is an individual and collaborative activity carried out unsupervised at home, which is used for formative and summative assessment, carrying 30% of the overall mark. Students conduct a critical review of a case study of online teaching planning. A set of elements for analysis are developed to review the case and detect training needs and solutions. This is done in two phases. Firstly, there is a group phase where students analyze the given case and provide improvement proposals in an online forum. The FR instrument is used online whilst the students write their descriptions in the forum. Students develop these documents in a Google Doc file, which allows the teacher to check the contribution of each student to the document. Secondly, there is an individual phase where students self-evaluate and co-evaluate the group work and the work of their teammates, and provide a written evaluation (up to 1000 characters). KD is used online while students perform the individual part of the activity in order to authenticate the students' identity. The final outputs of the assessment activity are the template created to analyze the case and the document with the case analysis and the improvement proposal (e.g. a new teaching plan, or the design of an e-assessment activity).

Some assessment activities can be placed both in Category 2 as well as in Category 3 when speech is used in the activity as well as text, and examples of this combination are given below in the description of Category 3. It quite commonly happens that an activity can be placed both in Category 2 and in Category 7, when authorship checking is carried out as well as student authentication, and examples of this combination are given in the description of Category 7 below.

5.3 Category 3—Identity Authentication for Spoken Responses

Identity authentication for: open questions with spoken responses including reading student written responses aloud (other forms of questions may also be included in the test); and assessments in the VLE with spoken elements.

Most of the case studies of assessment activities that were provided by the Pilot Leaders that were classified in Category 3 also involved written work and were therefore also classified in Category 2. The activity described below is an example of an activity of this type that was classified in Categories 2 and 3.

The course 'Applied Computer Graphics' (TUS) is an undergraduate level blended course. The assessment activity 'Self-evaluation' is an individual activity carried out under supervision at the university, which is used for formative and summative assessment, carrying 50% of the overall mark. In the activity, the student comments orally on their course work, describing what they did and the difficulties they encountered. The VR and KD instruments are used during the activity, with VR being used for the spoken elements of the self-reflection, and KD being used for the text elements of the activity.

5.4 Category 4—Identity Authentication for Multimedia

Identity authentication for: the preparation or delivery of multimedia materials in the VLE, including presentations, multimedia reports, videos, working with technology and software (including working with mathematical software).

A common form of assessment activity in the case studies involved students preparing or delivering a multimedia presentation, or carrying out a task working with technology or software. Activities where TeSLA was used to provide student authentication during the process of carrying out the activity are placed in Category 4; activities where the resultant multimedia materials are submitted for authentication after completion are placed in Category 5 and are described in Sect. 5.5.

The course 'Differential Equations and Applications' (SU) is an undergraduate level blended course. The activity 'Course project' is an individual activity carried out unsupervised at home and is used for summative assessment, carrying 20% of the overall mark. Students are required to solve a set of equations. There are three parts to the project: (1) a theoretical part in which the students present the analytical calculations, preliminary results and solutions necessary for the visualization, and the answers to the set questions; (2) writing MATLAB code; (3) production of the graphics that are required in the problem solution (including animations, if any). The students develop the project online in the VLE within a pre-defined period. The TeSLA instrument FR with attack detection is used online during the assessment activity.

5.5 Category 5—Identity Authentication for Uploaded Multimedia

Identity authentication for: uploaded completed multimedia materials consisting in part of video or audio recordings of the student, prepared outside the VLE including presentations, multimedia reports, and videos.

Another common form of assessment activity in the case studies involved students preparing or delivering a multimedia presentation that included video recording or audio recording of the student presenting some, or all, of the material. Category 5 contains those activities where the resultant product is uploaded for analysis rather than tested during the development process. Assessment activities placed in this category were quite varied, and were perhaps the most interesting activities in terms of the way in which TeSLA instruments were used that are provided in the case studies.

The course 'Experiment in biology education' (SU) is an undergraduate level blended course. The assessment activity 'Conducting a biological experiment' is carried out at the end of the course, it is an individual and collaborative activity carried out unsupervised at home, which is used for summative and formative assessment, carrying 40% of the overall mark. The student is required to produce a video recording of themselves performing a biological experiment, which could be used as a learning resource in secondary school biology lessons. It is easier for the students to carry out this work and to video it at home (possibly making several attempts) rather than to do this at the university. Students record their videos and then submit their video file to the VLE where the FR instrument is used to confirm that the student is the person in the video.

The course 'Labor Law' (UOC) is an undergraduate level online course. The activity 'Labor Contract' is an individual activity carried out unsupervised at home, which is used for formative and summative assessment, carrying 12% of the overall mark. The assessment activity uses a case-based methodology in which several cases are presented to students to immerse them in professional situations. The students are required to solve the cases supported by answering some guiding questions. As part of the assessment activity, students must select one case and deliver a short (two minutes) video synthesizing the main arguments and presenting them to a potential stakeholder (e.g. the customer or the judge). The FR and VR instruments are used offline[2] once the video has been uploaded to the VLE.

5.6 *Category 6—Identity Authentication for Submission of Pre-prepared Material*

Identity authentication for: uploading and submission of material prepared outside the VLE.

The cases in this category differ from those in Category 5 in that the identity authentication is carried out at the point of submission and uploading of the materials. Only a small number of the case studies belong exclusively to Category 6, and an example of this kind is given below. When the uploaded product consists, at least in part, of textual material then it is often also passed to the FA and PD instruments for authorship checking, and an example of this combination is given below in the examples for Category 7.

The course 'Information and Communication Technology (ICT) in Education and work in Digital Environment (Physical Education)' (SU) is an undergraduate level blended course. The activity 'Creating a multimedia learning resource' is an individual activity carried out unsupervised at home, which is used for formative and summative assessment, carrying 20% of the overall mark. The students are required to create a multimedia resource to be integrated in a Physical Education lesson. Students choose the school grade and the multimedia applications to be used, and they can produce a video, PowerPoint presentation, multimedia poster or use any

[2]The description of the FR and VR instruments as being used 'offline' refers to the process of uploading a video and/or audio recording of the activity to the VLE after the completion of the activity in order for this to be processed by the FR and VR instruments. This is in contrast to the use of the FR and VR instruments during the activity itself, which we refer to as 'online' use.

other multimedia app. In the VLE, the student briefly explains orally their ideas for implementing multimedia in Physical Education lessons with VR being used for student authentication, and then the student uploads the multimedia product. After teacher feedback, students submit their improved work, and the VR instrument is used again in the same way as before for student authentication during the submission process.

5.7 Category 7—Authorship Checking for Written Materials

Authorship checking for: written assignments (including written responses to open questions in a test) prepared in the VLE, which are checked either during the activity or after being uploaded at the end of the activity; and written assignments, prepared outside the VLE.

The submission of textual material for authorship checking using the FA and/or the PD instruments is another kind of assessment activity frequently found in the case studies. Below, we present first an example where the activity fits within Category 7, and then present an example where the activity fits within both Categories 2 and 7, and finally one where it fits within both Categories 6 and 7.

The course 'Procedure Law' (UOC) is an undergraduate level online course. The activity 'Civil Process Development—Case Study' is an individual activity carried out unsupervised at home, which is used for formative and summative assessment, carrying 13% of the overall mark. In the assessment activity, several cases are presented to students to immerse them in professional situations. They are, firstly, required to solve the cases answering some guiding questions, producing a text file of 2000–3000 words, and then, secondly, they must select one case and write a short answer (around 300 words) synthesizing the main arguments for one of the potential stakeholders (e.g. the customer or the judge). FA and PD are used for authorship checking of the text file that the student delivers in the first part of the activity, and KD is used for student authentication while the student is completing the second part of the assignment online.

The course 'Applied Computer Graphics' (TUS) is an undergraduate level blended course. The activity 'Research report' is an individual activity carried out unsupervised at home, and then submitted in a supervised context at the university, which is used for formative and summative assessment carrying 15% of the overall mark. The activity involves summarizing and analyzing material on a given topic. FR with attack detection is used to check the student identity during the submission process and PD is used to check the authorship of the submitted report.

6 Examples of TeSLA Integration Across Courses

Incorporating a variety of forms of student authentication and authorship checking across several assessment activities within a course provides the teacher with a higher degree of confidence in student identity and authorship than simply incorporating TeSLA into one or two assessments. In this section, we present detailed descriptions of two courses to illustrate the way in which TeSLA can be integrated across a whole course.

Course: ICT in Education and work in Digital Environment (Physical Education) (SU)

This is a blended undergraduate course. Assessments accounting for 70% of the overall course mark use TeSLA instruments (FR, VR, FA, KD, PD) for student authentication and/or authorship checking. The remaining assessment, contributing 30% of the overall course mark, is carried out as a face-to-face oral presentation.

Assessment activity 1—'Short academic essay'—contributes 10% of the overall mark. The activity uses the KD instrument for identity authentication of written responses (Category 2). Students are required to research the use of ICT for Physical Education and its terminology, and then to write a short essay describing a term of their choosing associated with ICT in Education—explaining their choice and why the term is important for Physical Education. The KD instrument is integrated in the activity and the writing is carried out online in class using the VLE.

Assessment activity 2—'Creating learning activities plan'—contributes 20% of the overall mark. The activity uses the FR, FA, PD instruments for identity authentication for written responses (Category 2) and authorship checking of written materials (Category 7). Students are required to suggest and plan a lesson activity that integrates ICT in a school Physical Education lesson. They publish their ideas in a VLE forum, where other students and the teacher can read and comment on their ideas. The students may revise their ideas based on this feedback, and the students then assess their colleagues' work. The teacher assigns a final mark based on the student's work and their assessments. The activity is conducted online during both the initial and later (peer assessment) part of the assessment and FR is used in order to provide student authentication. The use of FA and PD on the completed text provides assurance that the student has not created the text for assessment before the activity and then copied and pasted it into the forum.

Assessment activity 3—'Creating lesson plan'—contributes 20% of the overall mark. The activity uses the FA and PD instruments for identity authentication for submission of pre-prepared material (Category 6) and authorship checking for written materials (Category 7). Students are required to create a lesson plan that integrates ICT for a Physical Education lesson. They choose the school grade and the instrument to be integrated, and write a lesson plan (about 1000 words) in a pre-defined format. The assessment activity is conducted offline and then the prepared document is submitted in a VLE assignment. FA is used to identify and confirm the student's authorship and PD is used to establish whether a student copies materials from another student.

Assessment activity 4—'Creating multimedia learning resource'—contributes 20% of the overall mark. The activity uses the VR instrument for identity authentication for submission of pre-prepared material (Category 6). Students are required to create a multimedia resource to be integrated in a Physical Education lesson. Students choose the school grade and the multimedia applications to be used (they can produce a video, PowerPoint presentation, multimedia poster or use another multimedia app). The assessment activity is prepared outside the VLE and then the multimedia resource is submitted in a VLE assignment. At the point of submission, the student provides a brief oral explanation of their ideas for implementing multimedia in Physical Education lessons, and this is authenticated by the VR instrument. After teacher feedback, students submit their improved work, authenticating their voice again.

Assessment activity 5—'Final exam'—contributes 30% of the overall mark. This activity did not use any TeSLA instruments. The final exam consists of a presentation of all the artifacts created in the previous four assessment activities and is conducted in face-to-face mode.

Course: Online Teaching Planning (UOC)

This is an online postgraduate course. Assessments accounting for 90% of the overall course mark use TeSLA instruments (FR, VR, FA, KD, PD) for student authentication and/or authorship checking. The remaining assessment, contributing 10% of the overall mark, is a collaborative activity for which no presently available TeSLA instrument is suitable.

Assessment activity 1—'Virtual debate'—contributes 20% of the overall mark. The activity uses the FR, and VR instruments for identity authentication for uploaded multimedia (Category 5). The students are required to prepare a video presentation of the conclusions of an on-line debate. The students are required to analyze the main factors that affect the planning of the teaching-learning process in virtual and blended learning environments. The students participate in a debate, which takes place in a forum moderated by the teacher. The teacher proposes questions, organized in discussion threads, and students respond based on suggested readings, contrasting their opinions with those of other students. It is mandatory to participate regularly, citing information sources and their classmates. The final part of the activity is devoted to drawing conclusions, and the students submit a video where they present the main conclusions from the debate. This video constitutes the output of the assessment activity. FR and VR are used offline once the activity has been uploaded to the VLE for student authentication purposes.

Assessment activity 2—'Drawing up workgroup agreements'—contributes 10% of the overall mark. This activity uses no TeSLA instruments. This assessment activity is a collaborative activity for which no presently available TeSLA instrument is suitable.

Assessment activity 3—'Case Study Analysis'—contributes 30% of the overall mark. The activity uses the KD instrument for identity authentication for written responses (Category 2). Students conduct a critical review of a case study of planning online teaching. A set of elements for analysis are used to review the case and detect training needs and solutions. This activity consists of two parts. Firstly, a group phase where students analyze the given case and provide improvement proposals. Secondly, an individual stage where students self-evaluate and co-evaluate the group work and their teammates' work. Students are required to develop a written response (up to 1000 characters) in the VLE where they evaluate themselves and their teammates. The outputs for this assessment activity are the template created to analyze the case and the document with the case analysis and the improvement proposal (e.g. a new teaching plan or the design of an e-assessment activity). KD is used for authentication purposes to verify the identity of students in the individual part of the activity. As for the collaborative part, students develop the documents in a Google Drive file, which allows the teacher to check the contribution of each student to the document. The FR instrument is used online for student authentication, capturing the student's image whilst they write their descriptions in the forum.

Assessment activity 4—'Designing an online activity/subject/platform'—contributes 40% of the overall mark. The activity uses the KD, FA, PD instruments for identity authentication for written responses (Category 2) and authorship checking for written materials (Category 7). Building on the work of the previous three activities, students are required to design a teaching action in a virtual environment (e.g. plan a course, design an e-assessment activity or design a learning platform). This activity comprises two parts. Firstly, there is an individual phase where the student develops his or her project. To start this process students are required to provide an account in the VLE of their initial ideas for their individual project (i.e. title, type of design, approach). Once the proposal is accepted by the teacher, the student develops the project. Secondly, there is a group phase where students share with their teammates their projects, receive group feedback and improve the learning product. The output for this activity is the document containing the planning or design of an online teaching action. KD, FA and PD are used online for student authentication and authorship checking while students perform the individual part of the activity.

7 Discussion

Okada et al. [23] present an account of the choice of appropriate TeSLA instruments for assessment activities arising from work within the TeSLA project in the second round of pilots. This paper mapped the TeSLA classification of response types to the revised version of Bloom's taxonomy [18] which describes six categories (or levels) of skills: remember, understand, apply, analyze, evaluate and create. This was then mapped to a classification of online activities and hence to an indication of the appropriate TeSLA instruments to use in conjunction with these activities. This mapping was carried out, in the main, from a theoretical perspective rather than being based on an analysis of empirical data about usage.

Whilst this mapping provides pointers as to appropriate instruments to consider in planning the use of TeSLA, it suffers from a number of shortcomings as a descriptive framework for the case studies that were collected by the Pilot Leaders.

Firstly, it starts from looking at the instruments that were thought to be appropriate for specific online activities that could then be connected to the assessment of the levels in Bloom's taxonomy, rather than starting from teachers' actual purposes in using the TeSLA system in a specific context.

Secondly, it describes the use of single types of online activities, whereas the descriptions provided by the Pilot Leaders showed that learning and assessment activities often combined a variety of types of online activities and, indeed, often combined them in a number of ways with offline activities.

Thirdly, it assumes that specific online activities can be related to specific levels of Bloom's taxonomy, which was not found to be the case in the case study descriptions. Whilst it is clear that some online activities might be more commonly associated with some levels of Bloom's taxonomy, we found no consistent mapping. Indeed, whilst it is clear that multiple-choice questions can be used to test for factual recall (as argued in [23]), it is also the case that with careful design they can test understanding, application and evaluation of knowledge (see for example [31]). Similarly, essays may be ideally suited for testing analysis and evaluation but one can readily image the use of essays to test recall of facts, understanding and application of knowledge. There may well be strong connections for some online activities to levels of Bloom's taxonomy, for example, simulations and game playing may be particular well suited to the assessment of the application of knowledge—but as we found no examples of game playing or simulation in the case studies we are not able to provide any evidence for that.

The categories arising from our analysis of the case studies do show an association between the declared purpose for the use of the instruments (i.e. reasons for requiring authentication, authorship checking, or both, in a particular context) and also the format of the assessment (that is, what is described in the column 'Description' in Table 4). The analysis, however, did not find clear associations of the learning objectives or competences to be assessed (which might be mapped to levels in Bloom's taxonomy) either with assessment formats or with the ways in which the TeSLA instruments were used. This is not to say, of course, that such a mapping might not exist, or that such a mapping could not be encouraged by way of targeted teacher training related to this issue. However, a discussion of this would take us beyond our present goal of generating descriptions of the existing data related to successful practice in order to provide useful summaries intended to support teachers in incorporating TeSLA instruments in assessment activities.

8 Conclusions

The descriptions of the ways in which TeSLA instruments were integrated into assessment activities in the TeSLA project demonstrate well-tried models that should provide useful starting points for those looking to use TeSLA for the first time as well as provide a good basis from which to experiment and explore ways of integrating TeSLA with other kinds of assessment activities.

There are a number of issues that require further development, which are, perhaps, to be found in those things that are *not* discussed in the category descriptions, rather than those things that are discussed.

Firstly, the range of assessment activities found was relatively restricted; there are many forms of assessment—including e-portfolios, simulations, and games—that are not covered. This most probably indicates that these forms of assessment were not widely used in the universities in the study.

Secondly, there is no description of the ways in which the results of students' authentication and authorship checking were used in discussions of academic integrity. This might include issues related to the accuracy of the instruments, and to ways of responding to indications of cheating—both at the student level in terms of identifying cheating, and at the course and institutional level in terms of making suggestions for modifying forms of assessment. This data was not available.

Thirdly, the analysis found that the relationship between learning objectives, teaching and learning activities and forms of assessment (in other words 'constructive alignment') was weakly described in the case studies and so no conclusions could be drawn about this issue. Some of this may be due to weak course design, but it seems likely that one of the reasons for this is that assessment activities using TeSLA instrument were retroactively fitted into existing courses. None of the courses in which TeSLA was used was developed right from the initial course design stage with the use of TeSLA in mind.

These three issues indicate a need for a considerably longer planning phase in the use of student authentication and authorship checking instruments than was available in this project. Instruments need to be in a stable state before teachers begin to plan and design courses that might incorporate them. Many universities have planning schedules that require a full design many months before the commencement of a course, and the design process itself can take several months. Alongside this, the fact that assessment design is often a weak point in course design as teachers get little explicit training in assessment design [5] means that teachers may need support in this issue at the design stage. These considerations should be taken into account in future developments of similar systems.

Looking to the future, the widespread use of mobile technologies will have important implications for the future of e-assessment, as will emerging innovative pedagogies and assessment methodologies. Changes in higher education teaching and learning mean that the distinction between learning and assessment may well be less clearly defined in the future, with assessment being seen as part of an ongoing learning process. Assessment is likely to become more student-centered (including greater use of collaborative assessments), there may be an increase of use of 'stealth assessment' in which assessment is incorporated into existing learning tools such as games, simulations and tutoring systems, and there will be a greater emphasis on formative assessment practices (including self-assessment, peer feedback, learning diaries, and e-portfolios). These changes will present new challenges to the integration of systems of student authentication and authorship checking like TeSLA, and generate requirements for the development of new instruments and methods of integrating them with assessment activities.

Acknowledgements The case studies of the assessment scenarios were provided by Serpil Kocdar (AU), Tarja Ladonlahti (JYU), Francis Brouns (OUNL), Lyubka Aleksieva, Blagovesna Yovkova, Stoyan Suev (SU), Malinka Ivanova (TUS), M. Elena Rodríguez, and Ingrid Noguera (UOC).

References

1. Biggs J, Tang C (2011) Teaching for quality learning at university: what the student does, 4th edn. Open University Press, Maidenhead
2. Black P, Wiliam D (2009) Developing the theory of formative assessment. Educ Assess Eval Accountability 21(1):5–31
3. Bocconi S, Trentin G (2014) Modelling blended solutions for higher education: teaching, learning, and assessment in the network and mobile technology era. Educ Res Eval 20(7–8):516–535
4. Braun V, Clarke V (2006) Using thematic analysis in psychology. Qual Res Psychol 3(2):77–101
5. Coates H (2015) Assessment of Learning Outcomes. In: Curaj A, Matei L, Pricopie R, Salmi J, Scott P (eds) The European higher education area: between critical reflections and future policies. Springer International Publishing, New York, pp 399–413
6. Draaijer S, Jefferies A, Somers G (2017) Online proctoring for remote examination: a state of play in higher education in the EU. In: Ras E, Guerrero Roldán A (eds) Technology enhanced assessment. TEA 2017. Communications in computer and information science, vol 829. Springer, Cham, pp 96–108
7. Foltýnek T, Dlabolová D, Linkeschová D, Calhoun B, Glendinning I, Lancaster T et al (2018) South-East European project on policies for academic integrity. Council of Europe, Strasbourg
8. Gaebel M, Kupriyanova V, Morais R, Colucci E (2014) E-Learning in European higher education institutions: results of a mapping survey conducted in October–December 2013. European University Association
9. Gil-Jaurena I, Softic KS (2016) Aligning learning outcomes and assessment methods: a web tool for e-learning courses. Int J Educ Technol High Educ 13:17
10. Glendinning I (2014) Responses to student plagiarism in higher education across Europe. Int J Educ Integrity 10(1):4–20
11. Glendinning I (2016) European perspectives of academic integrity. In: Bretag T (ed) Handbook of academic integrity. Springer, Singapore, pp 55–74
12. Guàrdia L, Crisp G, Alsina I (2016) Trends and challenges of e-assessment to enhance student learning in Higher Education. In: Cano E, Georgeta I (eds) Innovative practices for higher education assessment and measurement. IGI Global, Hershey, Pennsylvania, pp 36–56
13. Guerrero-Roldán A-E, Noguera I (2018) A model for aligning assessment with competences and learning activities in online courses. Internet High Educ 38:36–46
14. Harmon OR, Lambrinos J (2008) Are online exams an invitation to cheat? J Econ Educ 39(2):116–125
15. Jackel B, Pearce J, Radloff A, Edwards D (2017) Assessment and feedback in higher education: a review of literature for the higher education academy. Higher Education Academy, York, UK
16. JISC (2010) Effective assessment in a digital age. Joint Information Systems Committee, London
17. Juola P (2017) Detecting contract cheating via stylometric methods. In: Plagiarism across Europe and beyond 2017—conference proceedings. Brno, Czech Republic, pp 187–198
18. Krathwohl DR (2002) A revision of Bloom's taxonomy: an overview. Theory Pract 41(4):212–218

19. Lancaster T, Clarke R (2016) Contract cheating: the outsourcing of assessed student work. In: Bretag T (ed) Handbook of academic integrity. Springer, Singapore, pp 978–981
20. Mellar H, Peytcheva-Forsyth R, Kocdar S, Karadeniz A, Yovkova B (2018) Addressing cheating in e-assessment using student authentication and authorship checking systems: teachers' perspectives. Int J Educ Integrity 14:2. https://doi.org/10.1007/s40979-018-0025-x
21. McAleese M, Bladh A, Berger V, Bode C, Muehlfeit J, Petrin T, Schiesaro A, Tsoukalis L (2013) Report to the European Commission on improving the quality of teaching and learning in Europe's higher education institutions. Publication Office of the European Union, Luxembourg
22. Noguera I, Guerrero-Roldán AE, Rodríguez ME (2017) Assuring authorship and authentication across the e-assessment process. In: Joosten-ten Brinke D, Laanpere M (eds) Technology enhanced assessment. Communications in computer and information science. Springer International Publishing, Cham, pp 86–92
23. Okada A, Noguera I, Aleksieva L, Rozeva A, Kocdar S, Brouns F, et al (2019a) Pedagogical approaches for e-assessment with authentication and authorship verification in higher education
24. Okada A, Whitelock D, Holmes W, Edwards C (2019) e-Authentication for online assessment: a mixed-method study. Br J Educ Technol 50(2):861–875. https://doi.org/10.1111/bjet.12733
25. Pixlr (n.d.) https://pixlr.com. Retrieved 17 Apr 2019
26. Ras E, Whitelock D, Kalz M (2015) The promise and potential of e-assessment for learning. In: Reimann P, Bull S, Kickmeier-Rust M, Vatrapu R, Wasson B (eds) Measuring and visualizing learning in the information-rich classroom. Routledge, Abingdon, pp 21–40
27. Redecker C (2013) The use of ICT for the assessment of key competences. Joint Research Centre of the European Commission Scientific and Policy Report. Publication Office of the European Union, Luxembourg
28. Rodríguez ME, Baneres D, Ivanova M, Durcheva M (2018) Case study analysis on blended and online institutions by using a trustworthy system. In Ras E, Guerrero Roldán A (eds) Technology enhanced assessment. TEA 2017. Communications in computer and information science, vol 829. Springer, Cham, pp 40–53
29. Selwyn N (2008) 'Not necessarily a bad thing …': a study of online plagiarism amongst undergraduate students. Assess Eval High Educ 33(5):465–479
30. Sousa-Silva R (2017) Detecting plagiarism and contract cheating: new academic integrity challenges. In: Plagiarism across Europe and beyond 2017 conference proceedings. Brno, Czech Republic, pp 12
31. Tractenberg RE, Gushta MM, Mulroney SE, Weissinger PA (2013) Multiple choice questions can be designed or revised to challenge learners' critical thinking. Adv Health Sci Educ 18(5):945–961
32. Underwood J, Szabo A (2003) Academic offences and e-learning: individual propensities in cheating. Br J Edu Technol 34(4):467–477
33. Webb M, Gibson D, Forkosh-Baruch A (2013) Challenges for information technology supporting educational assessment. J Comput Assist Learn 29(5):451–462
34. Weber-Wulff D (2016) Plagiarism detection software: promises, pitfalls, and practices. In: Bretag T (ed) Handbook of academic integrity. Springer, Singapore, pp 625–638

Ensuring Diverse User Experiences and Accessibility While Developing the TeSLA e-Assessment System

Tarja Ladonlahti, Merja Laamanen and Sanna Uotinen

Abstract The TeSLA project, with its new, innovative approaches for e-assessment, offers a great possibility for increasing the educational equality and making higher education studies available for all. It has been estimated that 10–15% of students in higher education institutions have some disabilities or special educational needs. At online universities or in online programmes, the number is even higher. These numbers emphasise the importance of the universal design for learning as a leading principle while developing the digital learning environments and e-assessment procedures. In this chapter, we describe the key elements of ensuring the accessibility of the TeSLA e-assessment system during the TeSLA project. In the cooperation among seven universities participating in TeSLA pilots, different national or institutional rules and ways of meeting the students' individual needs have been recognised. The main goal of the project, in terms of accessibility, has been developing an instrument that is accessible and easy to use for all types of students. We also discuss technical and pedagogical solutions that support use of the TeSLA e-assessment system by diverse students.

Keywords Accessibility · Usability · Disability · Special educational needs · Student with special educational needs and disabilities (SEND student) · User experience

Acronyms

ADHD Attention Deficit Hyperactivity Disorder

T. Ladonlahti (✉) · M. Laamanen · S. Uotinen
University of Jyväskylä, P.O. Box 35, 40014 Jyväskylä, Finland
e-mail: tarja.ladonlahti@jyu.fi

M. Laamanen
e-mail: merja.h.laamanen@jyu.fi

S. Uotinen
e-mail: sanna.uotinen@jyu.fi

D. Baneres et al. (eds.), *Engineering Data-Driven Adaptive Trust-based e-Assessment Systems*, Lecture Notes on Data Engineering and Communications Technologies 34,
https://doi.org/10.1007/978-3-030-29326-0_10

AU	Anadolu University
ICT	Information and Communications Technologies
ISO	International Organization for Standardization
JYU	University of Jyväskylä
LTI	Learning Tools Interoperability
OUNL	Open University of the Netherlands
OUUK	Open University United Kingdom
SEND	Special Educational Needs or Disabilities
SU	Sofia University
TUS	Technical University of Sofia
UDL	Universal Design for Learning
UOC	Open University of Catalonia
VLE	Virtual Learning Environment
WAI	Web Accessibility Initiative
WCAG	Web Content Accessibility Guidelines

1 Introduction

Higher education programmes supported by systems like TeSLA will offer new opportunities for all students to study in online environments and increase educational equality and make higher education studies available for all. Hopefully, it will open new possibilities for students with special educational needs or disabilities (SEND students) to participate in education. The TeSLA project has a strong commitment to considering the accessibility issue, meaning that SEND students are included as potential users of the TeSLA system. This commitment follows the EU action to promote inclusive education and lifelong learning for students with disabilities [9].

Accessibility means that 'people with disabilities have access, on an equal basis with others, to the physical environment, transportation, information and communications technologies and systems (ICT) and other facilities and services' [9]. In the TeSLA context, accessibility is seen in relation to e-learning. It means that learners are not prevented from accessing technologies, content or experiences offered by technologies on the grounds of their disability (see [29]).

The accessibility issue is highly topical at present. In 2016, the European Parliament approved the directive on making the websites and mobile apps of public sector bodies (including public universities and libraries) more accessible, ensuring that people with disabilities would have better access to them. It has been recognised that EU member states have had different approaches to and legislation on accessibility and disability issues: Some have underlined anti-discrimination laws, while others have focussed on public procurement or detailed technical requirements [10]. Following the directive, new national laws and regulations related to new directives should have come into effect in September 2018 (Directive [EU] 2016/2102). This means a big change for public universities, especially if the strict national legislation related to accessibility has been missing. Higher education institutions can no longer ignore the accessibility issue related to online education.

When the TeSLA project started, there were no common EU legislations for accessibility, and the national laws varied greatly. However, website accessibility has long been an EU policy priority. It can be seen as an obvious part of the growth of 'e-government services'. While evaluating the development of e-government schemes, Easton [7] states, 'The ability to harness technology's potential to enhance the relationship between the individual and the State can, without a policy focus on inclusion, strengthen existing socio-economic divides and exclude already marginalized groups'. Applying this idea to the educational framework, this could mean that, in strengthening the role of the technology in education, without inclusive practices and accessibility guarantees, in contrast to the original goal, we may promote the marginalisation of SEND students.

Accessible e-learning is not an independent pedagogical or technological issue. Even in the educational context, it has a strong reliance on many social phenomena. According to Seale [28], the development of accessible e-learning is a practice that can and should be mediated. A contextualised model of accessible e-learning practice in higher education considers the following factors:

- All the stakeholders of accessibility issues in a higher education institution (students, lecturers, learning technologists, support workers, staff developers, managers);
- The context in which to operate: drivers (legislation, guidelines and standards) and mediators (stakeholders' views of, e.g. disability, accessibility, integration and segregation, responsibility, community and autonomy); and
- How the relationship between the stakeholders and context affect their responses and the accessible e-learning practices that develop.

In the beginning of the project, it seemed obvious that, to be able to establish the best practices, the universities participating in the TeSLA pilots needed to find a common core and build a variety of local, good practices around it. Still, the main goal or outcome of the project, in terms of accessibility, was developing an instrument that is accessible and easy to use for all types of students. By adapting the idea of Seale's [28] original 'contextualized model of Accessible E-learning Practice in Higher education', we describe the complexity of elements encountered at the institutional level during the pilots (see Fig. 1). Figure 1 visualises the complexity of the factors recognised during the TeSLA pilots while piloting the TeSLA e-assessment system and its accessibility.

The common European legislation and national legislation of higher education institutions in partner countries offer the basic guidelines for the accessible e-learning practices. In the following sections, we discuss the dimensions included in Fig. 1. First, we describe the effects of e-learning on SEND students and how we ensured a wide variety of user experiences and participation of SEND students while piloting the TeSLA e-assessment system. We also discuss how to recognise students' diversity and organise support for them based on the literature and pilot experiences. Accessibility of the TeSLA e-assessment system is strongly related to the online course

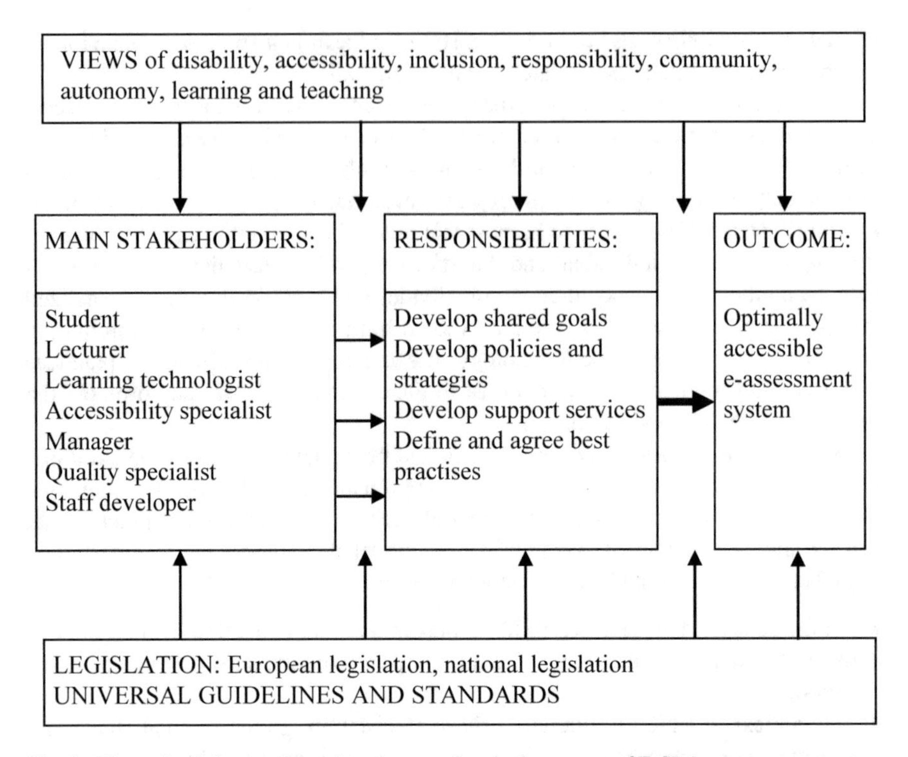

Fig. 1 Dimensions of accessible e-learning practices in the context of TeSLA e-assessment system

design and learning environment; therefore, the main elements of accessible online education are also defined. Finally, we describe the general accessibility guidelines and design of the method used during the pilots. We also give some examples of why it is important to use various methods to ensure the accessibility of a system like TeSLA.

Accessibility can be seen from the three following points of view: accessibility by everyone, using any technology and allow access in any environment or location [28, pp. 28–29]. In this chapter, accessibility is discussed by focussing on the user perspective (accessibility by everyone). It should be noted that the use of mobile devices is not included in the TeSLA technology developed during this project.

2 Effects of e-Learning on Students with Special Educational Needs and Disabilities

While discussing the effects of e-learning on students with disabilities, Seale [28] points out many positive outcomes, including flexibility and adaptability, access to inclusive and equitable education, access to learning experience, empowerment, independence and freedom. In contrast, according to this researcher, the main negative element seems to be inaccessible design. Even if students with disabilities access the virtual learning environments (VLEs), there may still be accessibility issues with the content, including activities, resources, collaboration and interaction tools [15]. Accessibility and usability are critical for online student success [3]. Usability can be defined as the 'extent to which a system, product or service can be used by specified users to achieve specified goals with effectiveness, efficiency and satisfaction in a specified context of use' [13].

When sites are correctly designed and developed, all the users have equal access to information and functionality. In terms of accessibility, it is a question of technology and pedagogy, including at least the course design, study materials and assessment. At the beginning of the project, the consortium agreed that the TeSLA system should be developed from its inception to be accessible for all students. This means that the system is regularly tested using automated tests and assistive technology, as well as by a variety of students. In the TeSLA project, technological developers were responsible for developing the accessible instruments and accessible system. Students and staff of the pilot institutions had a role as users, testing and giving the feedback on the technological solutions that were established. Still, the higher education institutions had an important and essential role in building up the context to their VLEs and realising accessible course design.

During the project, the data on disability issues were collected from seven pilot universities, namely the Open University of Catalonia (UOC), Open University of the Netherlands (OUNL), Sofia University (SU), Open University United Kingdom (OUUK), Technical University of Sofia (TUS), Anadolu University (AU) and University of Jyväskylä (JYU). It is worth clarifying that there were some differences between the TeSLA pilot universities: Some were traditional universities, selecting their students in one way or another; others, especially the open universities, were open to all students. Some of the pilot universities were fully online universities, while some combined face-to-face, online, and blended learning modes. The student profiles in the pilot universities also differed. The common element for all the pilot universities was that the pilots were mainly implemented in real educational environments with real assessment activities.

The pilot universities' expectations for the TeSLA system concerning the support of students with disabilities were mainly related to the flexibility of assessment modes through improved student identification and the recognition of authorship of a student's work. In practice, for example, this meant possibilities for taking an exam at home, without traveling to the university campus. The TeSLA system was also expected to improve study opportunities, with the possibility of offering all the activities online and making them available for all, regardless of the student's location. In addition to offering the option for traditional exams, the TeSLA system is considered to provide possibilities for a wider variety of assessment modes (e.g. continuous and formative assessment instead of summative assessment). The psychological stress or discomfort caused to students by recording exams or assessments was mentioned as a possible challenge emerging from TeSLA. Regarding assessment, Ball [2] points out that there are many standards relating to technical aspects of screen assessment and the accessibility and usability of onscreen material but no standards for the accessibility testing of assessments. He underlines the aspects of security and reliability in assessment design. As a part of the e-assessment system, some of the pilot universities highlighted the alternative use of different biometric identification instruments. This seemed to be an important and essential element when developing the TeSLA system for all.

3 Ensuring a Wide Variety of User Experiences During the Pilot

According to the ISO 9241 standard, user experience refers to the perceptions and responses that result from the use and/or anticipated use of a product, service or system. The concept refers to experience in a broad sense, including all the persons' emotions, beliefs, perceptions, preferences, physical and psychological responses, behaviours and accomplishments that occur before, during and after use [13].

The process of ensuring a wide variety of user experiences, meaning especially students with a variety of abilities and using a variety of assistive technologies, can be described as encompassing the following actions:

1. Recognising the SEND students to ensure the variety of use experiences during the pilots (by the local method in every pilot institution);
2. Understanding the context and variety of practices related to accessibility and support for SEND students in pilot universities (by questionnaires for local pilot leaders, who were in charge of the pilot implementation);
3. Taking care of the accessible course design and local VLE (locally in institutions);
4. Implementing accessibility tests by staff (in two pilot institutions);
5. Asking about the students' attitudes and experiences after using TeSLA (questionnaires and local focus groups);
6. Observing and video recording students and screen recording while testing the system (at one of the pilot universities); and
7. Interviewing SEND students during the pilot (at four of the pilot universities).

After all these actions, feedback was given to the technical developers of the TeSLA system.

4 How to Recognise Students' Diversity

It is widely understood that, in the field of inclusive education, there is a challenging dilemma: On the one hand, we know how stigmatising the different categories of SEND are, but on the other, we need them, especially to allocate or prioritise support. Individualistic models of disability are based on the construction that the problems and difficulties that disabled people experience are a direct result of their individual physical, sensory or intellectual impairments. The currently widely accepted social model of disability underlines that it is not the individual with a disability that needs to be changed, but rather, the society. The social model of disability argues that disability is located "in social practice" rather than "an individual body". A person may have a certain impairment, but it is the influence of decisions made by society that causes it to be a disability (see [20, 28]).

Disability is activated differently online compared with face-to-face meetings. On the one hand, impairments that may encounter significant disabling environments in face-to-face meetings may have less of an effect when using the internet. On the other, some impairments may find a different appearance or meaning in online environments, and thus, online environments can be significantly disabling (see [8, 11]).

It is estimated that about 80 million people in the European Union (EU) have a disability that ranges from mild to severe. They are often prevented from fully taking part in society because of the environmental and attitudinal barriers [9]. There are no exact international data on students with disabilities or educational challenges enrolled in higher education. It has been estimated that 10–15% of students in higher education institutions have some disabilities or special educational needs. In online universities or online programmes the number is even higher. As an example, in Finland, according to the results of a national survey among Finnish higher education students, the proportion is 8.2% [16]. Estimated amounts of students with disabilities at the pilot universities varied from 0.8 to 8.5% of all students. Some of the universities did not give any number because no reliable data exist.

In the beginning of the project, there were two main recommendations for the TeSLA system development process, which were as follows:

1. Representatives of SEND students should be included in the TeSLA system development process to help developers see the system and accessibility from the perspective of end users; and
2. Pilots should be designed so that pilot groups of users testing the TeSLA system include students with disabilities. These data on user experience are the only way to obtain the relevant data and feedback for accessibility of the TeSLA system.

The first task was recognising and including these students in the pilots. According to the pilot universities, it was always the student's responsibility to report on his/her special educational needs or other needs for individual study arrangements. In many countries, it is not permitted to ask if students have a disability. Institutions can only share the information for the students in terms of how to notify the organisation and instructors if they require individual arrangements. In some other countries, students are already asked about disabilities during registration. It is clear that there is also a group of students who never disclose their disabilities or special educational needs (see also Crichton and Kinas [5]. They either need no individual arrangements or they have developed successful coping strategies independently; alternatively, they may want to hide their problems for fear of stigma (see [24]). Roberts et al. [26] find that most students with disabilities choose not to reveal their disability. According to these researchers, the students did not even request accommodations to help with

Table 1 SEND students with various disabilities or special educational needs participating in pilot 3

Category	*N*
Blind or partially sighted	25
Deaf or hearing loss	57
Restricted mobility or motor disability	91
Specific learning difficulty (e.g. dyslexia)	77
Chronic illness	101
Psycho-social problems	82
Some other disability, special educational need or exceptional life situation	81
Prefer not to say	34

access to the course material that was presented in an inaccessible format. This phenomenon means that students with disabilities can become invisible online. The problem of discovering those who have special educational needs was also mentioned as a challenge by pilot universities. This challenge is especially evident in situations where the adaptations required are not disclosed or recognised in advance. Some students prefer not to disclose their disabilities even when asked, or they do not want to be recognised as SEND students at all.

There are many ways of categorising impairments and disabilities. During the pilots, the students were asked to describe their special educational needs or disabilities to make sure that a variety of end users participated in the pilots. See Table 1 for the categories used in the pilots and the amount of SEND students participating in the largest pilot, pilot 3.

It is worth recalling that SEND students are a widely divergent group. It was highly important for the project to ensure the participation of diverse students in the pilots, including students with different types of learning difficulties, life situations and ways of communicating, as well as using different types of assistive technology. Because of the wide variety of instruments in the TeSLA system, the experiences and attitudes towards the use of single instruments varied, depending partly on the students' disabilities [19, 22, 23] . There is not one right way to meet the e-learning needs of such a diverse group of people. Moreover, the recent literature has recognised disabled students as individuals who reflect different experiences [5].

5 Importance of Accessibility and Disability Regulations and Practices in Higher Education Institutions

While considering accessibility in e-assessment, Ball [2] states that organisations' managements should ensure a clear accessibility policy and training to ensure compliance; moreover, they should create a process to guarantee that the policy and training will be successfully implemented. Rice and Carter [25] state that many online educators are unaware of their legal responsibilities for students with disabilities. National laws give the boundary conditions for higher education institutions. If laws are binding, it may be unnecessary to create local regulations. All the universities participating in the TeSLA pilots had some local regulations or guidelines concerning accessibility or support to students with disabilities, although they varied significantly. They often contained principles and guidelines for admission examinations, implementation of exams and web accessibility guidelines. In addition, the importance of promoting employee awareness, guidance for faculties and development of the staff's competence was recognised.

The pilot universities described a wide range of persons, groups, teams and services managing the accessibility issue and individualised study arrangements. This partly appears to be a strength, but as Asuncion et al. [1] observed, it may also create disagreement about who is responsible for such tasks—disability service providers, e-learning professionals or professors and lecturers. These researchers ultimately recommended establishing a role for an e-learning accessibility specialist to oversee these elements. At least, educational institutions should have a common understanding of who is responsible and whom students should contact. Creating a role for an e-learning accessibility specialist, adopting e-learning accessibility guidelines and improving staff training are some of the research-based recommendations for universities in terms of improving the accessibility of e-learning environments [1].

All the pilot universities described how accessibility issues are considered in the organisations. In addition to creating organisational accessibility policies and guidelines, universities named persons and teams responsible for accessibility issues at many different organisational levels. They also offered support and services for students with disabilities. Still, there was concern about the accessibility issue involving shared responsibilities and a multidimensional network of agents and teams. As examples of the authority and good practices related to accessibility at the pilot universities, the following were mentioned:

- Managerial support;
- Multilevel responsibilities;
- Contact persons for students with special educational needs and disabilities;
- Centralised student support or advisory services;
- The inclusion of the accessibility perspective in all types of different groups in universities;
- Representatives of students or the student union in teams and groups;
- A wide perspective on the accessibility issue, encompassing the built environment, VLEs and websites;
- A full-time planning coordinator; and
- Utilising the expertise of the whole personnel.

The role of the institution is significant when building up practices related to the accessibility policy. Services and accessible learning environments for SEND students must be guaranteed, not only in the guidelines and recommendations, but also in practice. It has been pointed out that it is sometimes easier to create recommendations and guidelines than make them come true in practice. However, while having some challenges with accessibility in e-learning environments, it is important to make all the necessary guidance, tutoring and support available for students.

The role of the institution seems to be important because it can elicit students' trust, especially when offering new study modes. Levy et al. [17] noticed that there is a need for awareness raising and increased user support when integrating biometrics in e-learning systems in universities. Their study indicated that learners of online courses are more willing to provide their biometric data when provided by their university compared with the same services provided by a private vendor. The same phenomena were recognised during the TeSLA pilots. In terms of students' attitudes and trust in the use of TeSLA instruments, they saw them as quite safe and trustworthy because the biometric data were collected by the local higher education institution.

6 The Wide Variety of Individual Arrangements in the Pilot Universities

To be entitled to individual study arrangements, the pilot universities required a medical certificate from the student. In some cases, an expert report from a psychologist or special education teacher was accepted. In many countries, privacy regulations forbid sharing information about medical diagnoses, but an official document about the disability is needed. Sometimes, a medical certificate was required only if the study arrangements needed complex adjustments. Usually, the certificate only included the medical diagnosis or official document about the disability, not pedagogic suggestions in the study context. A gap between the expert's report and student's wishes was sometimes recognised. To reach a successful learning experience, it is important to listen to the student's perspective whenever possible and when it follows the rules of the university.

The question of how the information about students' disabilities should or could be shared is a sensitive one. Some institutions have developed a process where, to the extent permitted by the student, information about the recommendations of individual arrangements is shared in a centralised manner. Thus, the student does not need to go through the same process in different units and with every teacher.

The universities involved in the TeSLA project offered various adaptations or individualised differentiations for SEND students. The main categories found were as follows:

- Information and study guidance;
- Alternative modes of study (e.g. flexibility in schedules, virtual exams);
- Alternative or adapted study materials (e.g. audio recordings, audiobooks, assistive technology);
- Alternative course completion or exam arrangements (e.g. exams at home, extended time slots);

- Use of an assistant, resource teacher or invigilator;
- Extra support from tutors;
- Alterations made in the physical environment (e.g. ramps, separate exam rooms, special lighting);
- Use of special tools and devices; and
- Discounts in certain cases.

Some universities had specified guidelines or good practices related to certain types of disability (physical disabilities, visual disabilities, hearing disabilities, dyslexia, ADHD, mental disorders).

The pilot universities described their multidimensional practices and large number of staff members connected to the disability issue. They also described the challenges they still had while organising services and responsibilities concerning individual study arrangements. It was stated that more information and knowledge are needed. It seems obvious that there is a need for enhancing staff awareness about the students' diversity and individualised arrangements (including the TeSLA system), but at the same time, clearly allocating the persons responsible for the issue in practice.

OUUK has reported encouraging experiences with the work of the accessibility specialists appointed in every faculty [30]. An individual responsible for accessibility issues has an important role in increasing disability awareness and supporting the staff responsible for curriculum content. Embedding accessibility into curriculum design and production is often the point where help is needed. When an accessibility specialist is named for each unit of the organisation, it is possible to work proactively and focus on the right questions (see [30]).

Students with a disability may perceive their disability to have a negative effect on their ability to be academically successful. They may not disclose the disability because they do not know what accommodation to ask for. Many of the pilot universities considered the TeSLA system as an opportunity for students to have an e-exam at home or any other place. A couple universities mentioned that students may have a personal assistant, resource teacher or invigilator. It was also mentioned that a student using the TeSLA system while taking an e-examination at home may require another person's assistance.

It was clear at the beginning of the pilot that universities using the TeSLA system must provide sufficient instructions and guidelines and offer the required guidance, support and services for users of the TeSLA system. The instructions and guidelines should be accessible for all.

7 Building up the Accessible Online Education and e-Assessment

The European school system has a long history of segregation of students with disabilities. While promoting the importance of individualised arrangements, it is worth remembering that we have a strong commitment to inclusive education [9].

This means that we should avoid segregation, discrimination and useless 'special or individualized arrangements'. This should result in preferring accessibility and design for all principles when planning study and assessment modes for all courses. However, some of the practices are still based on segregating the students with disabilities from others (e.g. studying alone with alternative materials or taking an exam). As an example, the OUUK is committed to inclusivity and they aim to improve accessibility for disabled students and deliver an equivalent study experience to that of non-disabled students. Students' needs are included already at the design rather than when students are already studying [30]. The social construction of knowledge, meaning of interaction on learning and its practical applications, such as peer tutoring, group discussions and co-operational learning, are widely accepted ways of studying, and they are also used as a part of continuous and formative assessment. To support equal study opportunities, students with disabilities should be included and supported to participate in regular student groups. While building up new educational practices, it is good to be aware of one's role as a potential creator of disability.

Roberts et al. [26] suggest that courses should be designed to be accessible from the beginning. Making accommodations for students with disabilities often occurs only after a student has disclosed his/her documented disability. This means adjusting the design of the existing course and is more reactive in nature. This leads to a design–redesign approach [26]. Implementing universal design principles from the beginning avoids costs caused by the redesign and serves to include those students who would otherwise be excluded by an unwillingness to request accommodations.

While evaluating accessibility of the TeSLA system, it is important to understand that all the integrated technology (e.g. web browsers, VLEs) affect the end user experience. Therefore, it is also recommended to regularly evaluate the accessibility of the VLEs. Good educational design and accessibility for study and assessment modes are important for SEND students. At the same time, it is important to keep in mind that taking care of those aspects usually means good education and good practices for other students as well. Macy et al. [18] state that this is a theme that runs throughout the educational literature.

8 Universal Design for Learning

The variety of higher education students' abilities and characteristics emphasises the importance of the universal design for learning (UDL) as a leading principle while developing the digital learning environments and e-assessment procedures. According to Rose and Meyer [27], the UDL provides the framework for creating more robust learning opportunities for all students. It is an inclusive approach to course development and instruction that underlines the access and participation of all students. It builds on the work of Vygotsky and later advances of neurosciences elucidating how the brain processes information (see more [27]). UDL offers three guiding principles for developing curricula that eliminate barriers to learning, build on student strengths and abilities and allow different ways to succeed. For teachers and course design, the UDL method offers three guiding principles, which are as follows:

1. Supporting diverse recognition networks. From the teacher's perspective, this means providing multiple examples, highlighting critical features, providing multiple media and formats and supporting background context. Moreover, it means that various ways of acquiring information and knowledge are recommended and allowed;
2. Supporting diverse strategic networks. This means providing flexible models of skilled performance, providing opportunities to practice with supports and ongoing and relevant feedback and offering flexible opportunities for demonstrating skills. In addition, it represents alternative ways for students to demonstrate what they know; and
3. Supporting diverse affective networks. This means offering choices of learning context, content and tools, offering adjustable levels of challenge and multiple ways to be successful. Moreover, it means engagement to tap into students' interests and appropriate challenges to motivate students to learn [4, 18].

Flexibility and different assessment modes should be described when there are possibilities for adaptations. The aim should be that the needs of all students, including the disabled students, are always considered at the initial stage of course design [30]. Designing a product or system with disability in mind will better serve the needs of all users, including those who are not disabled. It is good to remember that convenience, adaptability and flexibility are some of the reasons why SEND students are looking for online courses as an opportunity to participate in higher institution studies (see [14]). Still, many of the online educators lack the required knowledge related to online accessibility [18].

Usually, students' first contact with a course is the syllabus. The written syllabus has an important role, especially for students studying single courses at open universities. It is important to describe the competences and basic requirements concerning the specific course and program. Several pilot universities underlined that the basic requirements for all students are equal.

Griful-Freixenet et al. [12] find important individual differences regarding learning needs and preferred learning approaches among all students. They report differences among students labelled with the same disability type. Furthermore, they argue that the traditional model of providing retrofitting accommodations depending on the student's disability type is inefficient. Instead of this, they advocate a high number of accommodations being incorporated into the design of the syllabus for all students, regardless of disability, right from the start.

When designing a new syllabus and its material, whenever possible, it is important to choose e-material from the publishers, which offers accessible electronic content. One incoming challenge is open resources, which are increasingly being incorporated in courses. This implies having less control over reviewing and ensuring that these resources accomplish the accessibility standards.

9 Universal Design for Learning Implementation for Online Courses

It is widely recognised that online educators lack sufficient knowledge on how to ensure the accessibility of online courses or online education. Some guidelines recommended as UDL implementation tips are available in the literature (see [6, 18]); these are as follows:

1. Create content first and then design the course;
2. Provide simple and consistent navigation;
3. Include an accommodation statement;
4. Use colour with care;
5. Choose fonts carefully;
6. Model and teach good discussion etiquette;
7. Choose content management system tools carefully;
8. Provide an accessible document format;
9. Convert PowerPoint to HTML;
10. If the content is auditory, make it visual; and
11. If the content is visual, make it auditory.

This list has proven to be useful when giving a wide perspective on course design. The accommodation statement and good discussion etiquette are important tools for supporting students' participation.

Slater et al. [30] highlight that, whenever possible, the aim is to use original course material produced in an accessible way. When this is not an option, an alternative learning material or experience must be provided. The pilot universities reported use of several different VLEs and alternative or adapted study materials (e.g. audio recordings, audiobooks, assistive technology). However, they also experienced a lack of accessible learning materials (e.g. audiobooks involving symbols and different letters, specialised software and hardware for different groups, accessible material including mathematical formulas, guidelines on how to create accessible learning resources). Slater et al. [30] also point out that there are significantly different accessibility issues in different subject areas. In this study, an example mentioned by one pilot university was audiobooks involving mathematical symbols. The developed web content accessibility guideline (WCAG; see Sect. 10) offers detailed guidance on how the four design principles (perceivability, operability, understandability, robustness) should be considered when creating accessible content [33].

In addition to the accessibility of study materials, the TeSLA system and platform for examinations or assignments should be accessible. Thomson et al. [31] present some basic rules for lecturers to follow. Moreover, having reviewed the literature, Macy et al. [18] state that there are strategies that can be easily implemented to promote student success. Elements from the two sets of rules are combined in Table 2.

There are some interesting tools for automated accessibility checking. As already stated, accessibility issues should be considered from the beginning of the product development process. Developers should utilise guidelines and good practices,

Table 2 Elements and recommendations for teachers to follow while designing an accessible online course

Topic	Recommendation
Colour	Do not use colour alone to convey information Ensure that the text colour has sufficient contrast to the background colour (see details in WCAG)
Page content	Structure content semantically in HTML so that assistive technology (screen reader) users can reach the content and navigate effectively Avoid automatic slide transitions and use simple slide transitions when possible. Complex transitions can be distracting
Tables	Add definition of column and row headers into the tables that are used for data. Header attributes can help define table headers
Presentation slides	Check the reading order of the textboxes that are not part of the native slide layout. A screen reader usually reads these last Avoid automatic slide transitions and use simple slide transitions when possible
Images	Add alternative texts to images that alert the student to the image content Add closed captioning. It provides text to visual content
Font	Use easy-to-read fonts Ensure that the font size is sufficient Use one font type throughout Limit the use of bold, italics or CAPS
Audio	If there is embedded audio, ensure that a transcript is included
Multimedia	If there is embedded video, ensure that the video is captioned and the player controls are accessible Captions should include the spoken text and sounds that are important for understanding (laughter, applause, music)
Authentic assessment	Assessments challenge students to demonstrate their ability to apply and synthesise course content Ensure communication with students about what they have learned Use innovative assessment modes, for example, multimedia presentations, oral presentations, etc.
Auto-testing tools	There are many auto-testing tools to integrate with the existing system The system shows the accessibility challenges and offers recommendations for correcting the issue

consult usability experts and use automated tests. It is recommended to employ practical rather than simulated tests, as the simulated environment may not reflect how individuals work in practice [2, 15]. It is also recommendable to include users with disabilities in the development process and testing of products (including software) at the earliest stages. Disability rights organisations are also active on this issue [15].

10 Web Content Accessibility Guidelines

WCAG are widely used as design principles for making web content more accessible. All the pilot universities were familiar with these guidelines. During the TeSLA project, the current version was WCAG 2.0 [32]; version 2.1 was published in 2018. Following the recommendations, the guideline will make web content accessible to a wide range of students with disabilities (including visual, auditory, physical, speech, cognitive, language, learning and neurological disabilities) and more usable for all other users as well [33]. Twelve WCAG design guidelines are based on the following four principles of accessibility:

- Perceivability (users must be able to perceive the information and user interface components);
- Operability (users must be able to operate the interface);
- Understandability (users must be able to understand the information, as well as the operation of the user interface); and
- Robustness (users must be able to access the content as technologies advance).

Testable success criteria are provided for each guideline described above. To meet the variety of needs of different groups and different situations, three levels of conformity are defined, which are as follows: A (lowest), AA and AAA (highest).

The desired level of WCAG for the TeSLA system, including instructions and guidelines for the users, was AA. Some pilot universities have set that level as their standard, so the TeSLA system should align with this and not restrict opportunities for their students. According to the pilot universities, there were differences in achieving the desired WCAG level. For example, one university offered a detailed description of the functionalities implemented to ensure easy access to all the contents. In contrast, another university stated that there were many challenges in achieving accessibility. Some stated that, instead of the desired WCAG level, they had prepared standards for the quality of e-learning resources, including the web content accessibility of all types of e-resources. Every distance learning course had to meet these standards (Table 3).

It is important to remark that even content that conforms at the highest level (AAA) will not be accessible to all users. There are also studies showing the limitations and lack of detailed research, for example, concerning user experiences of problems (see e.g. [21]). Additional information on WCAG levels can be found in the *Guide to understanding and implementing Web Content Accessibility Guidelines 2.1* (W3Cc).

11 Accessibility Test Implemented by Pilot University Staff

While evaluating the accessibility of the TeSLA system, it is important to understand that all the integrated technology (e.g. web browsers, VLEs) affects the end user's experience. While the march of technology is rapid, it is recommendable to evaluate

Table 3 WCAG 2.1 web content accessibility principles and guidelines

Principle/guideline	Content	Description
Perceivability	Text alternative	Provide text alternatives for any non-text content so that it can be changed into other forms needed, such as large print, braille, speech, symbols or simpler language
	Time-based media	Provide alternatives for time-based media
	Adaptable	Create content that can be presented in different ways (e.g. simpler layout) without losing information or structure
	Distinguishable	Make it easier for users to see and hear content, including separating foreground from background
Operability	Keyboard accessible	Make all functionality available from a keyboard
	Enough time	Provide users enough time to read and use content
	Seizures and physical reactions	Do not design content in a way that is known to cause seizures or physical reactions
	Navigable	Provide ways to help users navigate, find content and determine where they are
	Input modalities	Make it easier for users to operate functionality through various inputs beyond the keyboard
Understandability	Readable	Make text content readable and understandable
	Predictable	Make webpages appear and operate in predictable ways
	Input assistance	Help users avoid and correct mistakes
Robustness	Compatible	Maximise compatibility with current and future user agents, including assistive technologies

the accessibility of the VLEs regularly. This was only implemented in a couple pilot universities. More commonly, it was evaluated occasionally or while developing or acquiring new tools. As a good practice, it was also mentioned that every technological project should fulfil the Web Accessibility Initiative (WAI) standards and all learning objects should be evaluated by experts.

During the pilot, there were two options for integrating the TeSLA enrolment tool into the organisation's VLE. Either it could be integrated into the VLE (e.g. Moodle) or it could be used as an external tool via a Learning Tool Interoperability (LTI) integration. LTI is a standard that links content and resources to learning platforms. The accessibility evaluation discussed in this section focussed on LTI enrolment for two reasons. First, TeSLA was running on different versions of Moodle in the pilot institutions, but the LTI enrolment was identical for all institutions and users (see Fig. 2). Second, the accessibility of the LTI enrolment is crucial for SEND students using the TeSLA system. If they cannot access to LTI enrolment, they are not able to access the assignments using e-authentication.

The aim of the accessibility test was to test TeSLA's LTI enrolment version using the following three methods: (1) navigation by tabbing (tab, shift + tab, enter, space bar); (2) using a screen reader (JAWS 2018); and (3) the WAVE Chrome extension, which enables evaluating web content for accessibility issues directly within the browser.

The TeSLA system was tested using Windows 10 and three versions of Chrome. Web browsers are constantly developing software, and new versions are launched frequently. Chrome was selected because, at the point of testing, it was by far the

Fig. 2 LTI user interface of the face recognition instrument's enrolment

most popular web browser across platforms (desktop, tablet, mobile). According to w3counter.com, Chrome's market share was 55.2% in June 2018.

There are several automated tools for evaluating the accessibility of web content (e.g. WAVE, Siteimprove and Axe). WAVE was selected because it is free of charge and easy to use. However, automated tools can only identify a certain amount of errors. Only humans can determine whether specific web content is accessible. Therefore, we also need user testing and accessibility evaluation. TeSLA enrolment was also tested by a usability/accessibility specialist.

The general findings of TeSLA's LTI enrolment's accessibility evaluation concern the enrolment of all instruments. These accessibility tests were conducted at the end of pilot 3 in June 2018. The test report was shared for the whole project, but especially, for the technical developers. The tables below present the test results. The findings and recommendations are copied directly from the WAVE reports (Table 4).

All the instruments were tested separately. The tester recognised some good solutions while testing the enrolment of the face recognition instrument. Elements had individual buttons; this was good because the user could navigate without the mouse. There were many challenges as well. (See Tables 5, 6, 7 and 8 and the comments and recommendations for the enrolment of each instrument.)

These test results are presented with all the details to demonstrate how the accessibility can be tested by a specialist and what types of information these test methods offer. After this kind of feedback, how the recommendations are met will be in the hands of the technical developers.

12 Observing and Recording Students' Test Situations

Using automatic testing systems or consulting a usability/accessibility specialist is not enough to guarantee the best outcome. It was important to have some end users not only using the TeSLA system independently but also testing it under the recorded test and research design conditions. Fifteen students from one pilot university participated in such research. The group of students was highly heterogeneous, including 4 male and 11 female students with a variety of special educational needs, for example, because of limited vision, chronic illness, dyslexia or panic disorder. Two of them used sign language. They were volunteers and tested the system out of the pilot courses, meaning that the test situation was not the real assessment situation with exam stress. Several data collection methods were used for ensuring rich data. The whole test situation was video recorded; the screen was also recorded. Two researchers observed the test and supported the students in case of technological challenges. Observing the test situation made it possible to provide accurate feedback for technical developers. (See Fig. 3 for an example of the test situation.)

Table 4 General findings related to the accessibility of enrolment of the TeSLA instrument and the recommendations for improvements

Feedback provided by WAVE	Recommendation by WAVE
Images miss the image alternative text. Each image must have an alt attribute. Without alternative text, the content of an image will not be available to screen reader users or when the image is unavailable – TeSLA logo on the upper left corner – Buttons (Face Recognition, Voice Recognition, Keystroke Dynamics and Forensic Analysis) – Start and stop buttons – EU flag on the footer	Add an alt attribute to the image. The attribute value should accurately and succinctly present the content and function of the image. If the content of the image is conveyed in the context or surroundings of the image, or if the image does not convey content or have a function, it should be given empty/null alternative text (alt=" ") (High priority)
The instructions are justified. Large blocks of justified text can negatively impact readability due to varying word/letter spacing and 'rivers of white' that flow through the text	Remove the full justification from the text. (Medium priority)
The enrolment page has no headings. Headings (<h1>–<h6>) provide important document structure, outlines, and navigation functionality to assistive technology users	Provide a clear, consistent heading structure, generally one main heading and subheadings as appropriate. Except for very simple pages, most webpages should have a heading structure (Medium priority)
The language of the document is not identified. Identifying the language of the page allows screen readers to read the content in the appropriate language. It also facilitates automatic translation of content	Identify the document language using the <html lang> attribute (e.g. <html lang="en">) (High priority)
Contrast is very low on the icons (Face Recognition, Voice Recognition, Keystroke Dynamics and Forensic Analysis) – Foreground colour: #b5b5ba – Background colour: #eeeeee	Increase the contrast ratio of the icons. (High priority)
Tester's comment	Tester's recommendation
Before finishing the enrolment of all four instruments, the user can exit the enrolment only by clicking the back button of the web browser as many times as needed to get back to Moodle view. For screen reader users or those navigating with tabs, this is too difficult	Add an exit button to the enrolment view (High priority)
When navigating by tabbing, the user interface does not clearly indicate when the Start or Stop button is activated. The thin blue frame does not stand out from the blue button. The frame is so thin that it is impossible to notice the difference	The frame should be of a different colour than the button or much thicker to stand out (High priority)

Table 5 Comments and recommendations for face recognition enrolment

Feedback provided by WAVE	Recommendation by WAVE
Web camera's user interface lacks alternative text	Add an alt attribute to the web camera screen (High priority)
Tester's comment	Tester's recommendation
Web camera is always active when the user is on the face recognition page. This happens even when the user has not activated face recognition	The web camera should be active only when the user has activated it by pressing the Start button (High priority)
The user cannot exit the process of saving the video	User should be able to stop the process of saving the video if it takes too long (High priority)
When the screen reader user navigates by using tabulator, it only reads the names of the instruments at the top of the page and then continues to the selected video device's dropdown menu. The user cannot return to the instructions	The user should be able to view the instructions if needed. Enable accessing them (High priority)
The (meta) information of the web camera's user interface cannot be read properly. This may be caused by several nested div or button elements	Check and remove the nested elements to allow the user to access the meta information (High priority)
The screen reader does not identify the Start button. Users of screen readers cannot complete the face recognition enrolment	Meta-information must be added to the Start button (High priority)
Face recognition works differently with different versions of Chrome – Web camera loops. Version 67.0.3396.87 (Official Build) (64-bit) – Start button returns the user to the start of the page. Version 66.0.3359.181 (Official Build) (64-bit)	The web camera should work properly. It should not loop. The Start button should start the camera (High priority)

As the following examples show, many issues emerged in the testing that were not recognised earlier:

- A student with limited vision was not recognised by the web camera because the student's face was too near to the screen;
- Against the former assumptions, some students using sign language sometimes preferred to use the voice recognition instrument as well;
- It took too many recordings and too much time to complete the enrolment activity, especially among students with slow speech or many breaks in their speech;
- It took a relatively long time to complete the keystroke enrolment if a student was a slow writer or had dyslexia; and
- The TeSLA system seemed to be robust. One student was too 'busy' to read the instructions; pushing many buttons almost at the same time did not break the system.

Table 6 Comments and recommendation for voice recognition enrolment

Tester's comment	Tester's recommendation
When navigating by tabbing the Start button does not indicate that it is active	Start button should indicate clearly that it is active (High priority)
The screen reader does not identify the status voice model information. How does the user of a screen reader know the progress?	Meta-information must be added to the status voice model (High priority)
Voice recognition works differently with different versions of Chrome – User can save the sample but not stop. The system loops trying to get the voice sample but does not inform the user of the loop or any notifications, such as, 'Sample has too long a period of silence preceding it'. Version 67.0.3396.87 (Official Build) (64-bit) – It is not possible to save the sample. Returns to the start of the page. Version 66.0.3359.181 (Official Build) (64-bit)	User should be able to save the sample and stop recording when needed (High priority)

Table 7 Comments and recommendations for keystroke enrolment

Feedback provided by WAVE	Recommendation by WAVE
A form control does not have a corresponding label. If a form control does not have a properly associated text label, the function or purpose of that form control may not be presented to screen reader users. Now the screen reader does not identify the form and user only finds it if he navigates the keystroke dynamics enrolment page by tabbing	Use the element to associate it with its respective form control (High priority)
Tester's comment	Tester's recommendation
Once screen reader user starts typing in the form, the system does not inform him about the progress (increasing percentages) or instructions	The instructions and status keystroke model should be available when typing in the form (High priority)

The main focus of the test situations was collecting user experiences and giving feedback for technical developers of the TeSLA system. In addition to this, the recordings offered rich, interesting data about how students used the keyboards, how they acted in the Moodle environment during the enrolment and follow-up activities and what types of choices they made. All this information will help higher education institution staff generate better solutions while building new online courses with new technology.

Table 8 Comments and recommendations for forensic analysis enrolment

Feedback provided by WAVE	Recommendation by WAVE
A form control does not have a corresponding label. If a form control does not have a properly associated text label, the function or purpose of that form control may not be presented to screen reader users. Now, the screen reader does not identify the form and user only finds it if he navigates the forensic analysis enrolment page by tabbing	Use the element to associate it with its respective form control (High priority)
Tester's comment	Tester's recommendation
The Start button appears under the form when the user has inserted enough text into the form. The system does not inform screen reader users when the Start button appears on the screen. How does the user know there is a sufficient number of words?	The Start button should inform the user when it appears on the user interface (High priority)
The user can navigate the entire page using tabs. During the process, the number of words is read aloud, but this may take several minutes	The number of words should be available all the time on the user interface (High priority)

Fig. 3 Student with limited vision testing the TeSLA system

13 Conclusions

Accessible online education is a salient topic for at least two reasons, namely, the new European legislation and the growth of online education programmes and courses offered by universities. The TeSLA e-assessment system has an important role, creating new possibilities and flexible ways for diverse students to study. By organising accessibility tests and implementing the recommendations in the development of the TeSLA system, it is possible to ensure that the TeSLA system is accessible for diversity of students.

The best result in improving accessibility is reached when using all three methods of end user testing, automated testing tools and tests by accessibility specialists. All these methods were employed during the TeSLA project. Accessibility testing requires time but not necessarily financial investments; for example, using automated testing tools is fast, easy, free of cost and does not require special training. One does not have to be an accessibility specialist to interpret the outcome. The tools also provide clear instructions, with concrete examples of how to fix accessibility issues.

Software, devices and platforms are in constant development. Accessibility issues have become part of technical development and solutions. For example, some programs (e.g. PowerPoint, Word) contain built-in accessibility checks for the end user. Checks are easy to use and advise the user on how to fix accessibility issues. Even mobile devices already have features (e.g. dictation, text to speech) that improve accessibility, and thus, reduce the need for separate accessibility devices.

Considering that even WCAG level AAA does not guarantee accessibility to all, it is important to be vigilant to ensure that the TeSLA system does not become a barrier in itself. As described in this chapter, universities have variety of practises in terms of accessibility, individual arrangements and support for their students. This means that universities must carefully plan their e-assessment modes, choose and use appropriate TeSLA instruments and allow different user profiles for their students. Finally, they should continue to implement accessibility tests and collect user experiences, and if needed, offer alternative and traditional modes of study and assessment.

References

1. Asuncion JV, Fichten CS, Ferraro V, Chwojka C, Barile M, Nguyen MN, Wolforth J (2010) Multiple perspectives on the accessibility of e-learning in Canadian colleges and universities. Assist Technol 22(4):187–199
2. Ball S (2009) Accessibility in e-assessment. Assess & Eval High Educ 34(3):293–303
3. Betts K, Welsh B, Pruitt C, Hermann K, Dietrich G, Trevino JG, Watson TL, Brooks ML, Cohen AH, Coombs N (2013) Understanding disabilities & online student success. J Asynchronous Learn Netw 17(3):15–48
4. Coyne P, Ganley P, Hall T, Meo G, Murray E, Gordon D (2006) Applying universal design for learning in the classroom. In: Rose DH, Meyer A (eds) A practical reader in universal design for learning. Harvard Education Press, Cambridge, pp 1–11

5. Crichton S, Kinash S (2013) Enabling learning for disabled students. In: Moore MG (ed) Handbook of distance education, 3rd ed. Routledge, New York, pp 216–230
6. Dell CA, Dell TF, Blackwell TL (2015) Applying universal design for learning in online courses: Pedagogical and practical considerations. J Educ Online 13(2):166–192
7. Easton C (2013) Website accessibility and the European Union: citizenship, procurement and the proposed Accessibility Act. Int Rev Law Comput Technol 27(1–2):187–199. https://doi.org/10.1080/13600869.2013.764135
8. Ellis K, Kent M (2011) Disability and new media. Routledge, New York
9. European Commission (2010) Communication from the commission to the European Parliament, the Council, the European Economic and Social Committee and the Committee of the Regions. European Disability Strategy 2010–2020: A Renewed Commitment to a Barrier-Free Europe. https://eur-lex.europa.eu/legal-content/EN/TXT/HTML/?uri=CELEX:52010DC0636&from=SL. 5 Apr 2019
10. European Commission (2015) Proposal for a Directive of the European Parliament and of the Council on the approximation of the laws, regulations and administrative provisions of the Member States as regards the accessibility requirements for products and services. https://eur-lex.europa.eu/legal-content/EN/TXT/?uri=celex:52015PC0615. 5 Apr 2019
11. Goggin G, Newell C (2003) Digital disability: the social construction of disability in new media. Rowman and Littlefield, Lanham
12. Griful-Freixenet J, Struyven K, Verstichele M, Andries S (2017) Higher education students with disabilities speaking out: perceived barriers and opportunities of the Universal Design for Learning framework. Disabil & Soc 32(10):1627–1649
13. ISO 9241. Excerpts from international standard ISO 9241 Ergonomics of human system interaction—bibliographic references—part 210: human-centred design for interactive systems. https://www.iso.org/obp/ui/#iso:std:iso:9241:-210:ed-1:v1:en. 19 Mar 2016
14. Jacko VA, Choi JH, Carballo A, Charlson B, Moore JE (2015) A new synthesis of sound and tactile music code instruction in a pilot online braille music curriculum. J Vis Impair & Blind 109(2):153–157
15. Kent M (2015) Dsability and eLearning: opportunities and barriers. Disabil Stud Q 35(1). https://doi.org/10.18061/dsq.v35i1.3815
16. Kunttu K, Pesonen T, Saari J (2016) Korkeakouluopiskelijoiden terveystutkimus 2016. Ylioppilaiden terveydenhoitosäätiön tutkimuksia 48. (Student Health Survey 2016: a national survey among Finnish university students). https://www.yths.fi/filebank/4237-KOTT_2016_korjattu_final_0217.pdf. 5 Apr 2019
17. Levy Y, Ramim MM, Furnell SM, Clarke NL (2011) Comparing intentions to use university-provided vs vendor-provided multibiometric authentication in online exams. Campus-Wide Inf Syst 28(2):102–113
18. Macy M, Macy R, Shaw ME (2018) Bringing the ivory tower into students' homes. Promoting accessibility in online courses. Ubiquitous Learn: Int J 11(1):13–21
19. Noguera I, Guerrero-Roldán A, Peytcheva-Forsyth R, Yovkova B (2018) Perceptions of students with special educational needs and disabilities towards the use of e-assessment in online and blended education: barrier or aid? In: Proceedings of INTED2018 Conference, Valencia, Spain, 5–7 March, 2018, pp 817–828. Available from IATED Digital Library. https://library.iated.org/view/NOGUERA2018PER. 26 Apr 2019
20. Oliver M (1996) Understanding disability: from theory to practice. Macmillan, Basingstoke
21. Petrie H, Kheir O (2007) The relationship between accessibility and usability of websites. Empir Stud Web Interact 3:397–406
22. Peytcheva-Forsyth R, Yovkova B, Aleksieva L (2019) The disabled and non-disabled students' views and attitudes towards e-authentication in e-assessment. In: Proceedings of INTED2019 Conference, Valencia, Spain, March 11–13, 2019, pp 2862–2871. Available from IATED Digital Library. https://library.iated.org/view/PEYTCHEVAFORSYTH2019DIS. 26 Apr 2019
23. Peytcheva-Forsyth R, Yovkova B, Ladonlahti T (2017) The potential of the TeSLA authentication system to support access to e-assessment for students with special educational needs and disabilities (Sofia University experience). In: Proceedings of ICERI2017 Conference Sevilla,

Spain, November 16–18, 2017, pp 4593–4602. Available from IATED Digital Library. https://library.iated.org/view/PEYTCHEVAFORSYTH2017POT. 26 Apr 2019
24. Pirttimaa R, Takala M, Ladonlahti T (2015) Students in higher education with reading and writing difficulties. Educ Inq 6(1):5–23
25. Rice MF, Carter RJ (2015) When we talk about compliance, it's because we lived it. Online educators' roles in supporting students with disabilities. Online Learn 19(5):18–36
26. Roberts JB, Crittenden LA, Crittenden JC (2011) Students with disabilities and online learning: a cross-institutional study of perceived satisfaction with accessibility compliance and services. Internet High Educ 14(4):242–250
27. Rose DH, Meyer A (eds) (2006) A practical reader in universal design for learning. Harvard Education Press, Cambridge
28. Seale J (2006) E-learning and disability in higher education: accessibility research and practice. Routledge, Abingdon, Oxon
29. Seale J, Cooper M (2010) E-learning and accessibility: an exploration of the potential role of generic pedagogical tools. Comput Educ 54:1107–1116
30. Slater R, Pearson VK, Warren JP, Forbes T (2015) Institutional change for improving accessibility in the design and delivery of distance learning—the role of faculty accessibility specialists at the open university. Open Learn 30(1):6–20
31. Thomson R, Fichten C, Budd J, Havel A, Asuncion J (2015) Blending universal design, e-learning, and information and communication technologies. In: Burgstahler SE (ed) Universal design in higher education: from principles to practice, 2nd edn. Harvard Education Press, Boston, pp 275–284
32. WCAG 2.0. Web Accessibility Initiative: web content accessibility guidelines. Updated 2 October 2012 (first published July 2005). 21 Mar 2016
33. WCAG 2.1. Web Accessibility Initiative: web content accessibility guidelines. Recommendation 05 June 2018. https://www.w3.org/TR/WCAG21/. 26 Feb 2019

An Evaluation Methodology Applied to Trust-Based Adapted Systems for e-Assessment: Connecting Responsible Research and Innovation with a Human-Centred Design Approach

Okada Alexandra and Whitelock Denise

Abstract This chapter describes a novel evaluation methodology designed, deployed and refined during the development of the EU-funded TeSLA system which was produced to check student authentication and authorship. This methodology was underpinned by a Responsible Research and Innovation approach combined with human-centred design. Participants were 4058 students, which included 330 with special needs, together with 54 teaching staff and 21 institutional members from seven universities who completed consultation, focus groups, questionnaires and interviews. The findings suggest that the evaluation methodology was able to identify a broadly positive acceptance of and trust in e-authentication for online assessments by both women and men, with neither group finding the e-authentication tools to be either particularly onerous or stressful. The methodology facilitated the development of a framework with five features related to "trust": (1) The system will not fail, (2) be compromised, (3) data will be kept safely and privately, (4) the system will not affect students' performance and (5) the system will ensure fairness.

Keywords Evaluation methodology · Trust-based adapted systems · Responsible research and innovation · Human-centred design approach · Academic integrity

Acronyms

RRI	Responsible Research and Innovation
HCD	Human-Centred Design
NGO	Non-Governmental Organisation
VLE	Virtual Learning Environment

O. Alexandra (✉)
The Open University, Stuart Hall Building 2nd Floor, Milton Keynes MK76AA, UK
e-mail: ale.okada@open.ac.uk

W. Denise
The Open University, Jennie Lee Building, Milton Keynes MK76AA, UK
e-mail: denise.whitelock@open.ac.uk

D. Baneres et al. (eds.), *Engineering Data-Driven Adaptive Trust-based e-Assessment Systems*, Lecture Notes on Data Engineering and Communications Technologies 34,
https://doi.org/10.1007/978-3-030-29326-0_11

LTI	Learning Tools Interoperability
CIC	Computer Integrated Communication
SEND	Special Educational Needs and Disability
API	Application Programming Interface

1 Introduction

Responsible Research and Innovation (RRI) is a recent approach, which was coined by the European Commission at the beginning of this decade. It became a vital approach for funded research projects, particularly by the European Commission, such as the programme Horizon (2014–2020).

RRI has been applied to various fields with the aim to align scientific-technological advances with societal needs and expectations [16, 21]. This constructive alignment occurs through the interaction of all distinctive societal representatives during the all phases of the innovation process: designing, planning, implementation, testing and evaluation. The purpose of RRI is to promote greater involvement of societal members in the process of research and innovation from the beginning to increase knowledge, understanding and better decision-making about both societal needs and scientific innovations [1, 2].

There are various similarities between RRI and the human-centred design approaches for developing and evaluating technological innovations. This chapter presents the evaluation methodology used during the European-funded TeSLA system, which was funded as part of the European Horizon2020 programme: innovation-action for large scale impact. The TeSLA system was designed to check student authentication and authorship through a combination of biometric, textual analysis and security instruments.

- **Biometric instruments** refer to facial recognition for analysing the face and facial expressions, voice recognition for analysing audio structures and keystroke dynamics for analysing how the user uses the keyboard).
- **Textual analysis instruments** refer to plagiarism detection for using text matching to detect similarities between documents and forensic analysis for verifying the authorship of written documents.
- **Security instruments** refer to digital signature for authenticating and timestamp for identifying when an event is recorded by the computer.

This evaluation methodology was developed through an interactive process with all members of TeSLA project. It was implemented in three phases with all stakeholders including 7 Universities located in 6 countries.

This chapter is organised into five sections after the Introduction. In Sect. 2, we present the principles of RRI and the correlations with the human-centred technology approaches for software development which underpinned this work and our research questions. Section 3 illustrates the implementation of this evaluation during 3 pilot studies. Section 4 described the findings which were integrated from all pilots with recommendations for stakeholders. Section 5 discusses the findings from the implementation of the TeSLA system in 7 Institutions and its limitations. Finally, Sect. 6 includes the final remarks and suggestions for future work.

2 Background

The evaluation model implemented for the European TeSLA system for e-authentication and authorship verification was developed in 3 stages and involved five groups of stakeholders: students, teaching staff, pilot coordinators, technical teams, and institutional leaders. The model was conceived, developed and implemented through the continual interaction of technological innovators and end-users. This section presents the Responsible Research and Innovation principles, components and stakeholders who underpinned our work.

The Responsible Research and Innovation (RRI) approach is grounded on previous work developed by the European Commission about Ethics in Science Technology [11]. RRI approach was disseminated at the end of FP7 programme as a vital approach to highlight the importance of promoting scientific technological innovations with and for society and foster scientific advances to ensure security, prosperity and sustainability [8].

Various scholars who have been working with technological and scientific innovation, policies and science with for society have been presenting RRI through various definitions, principles and examples. Von Schonberg [21]'s definition is one of the most influential description of RRI:

> Responsible Research and Innovation is a transparent, interactive process by which societal actors and innovators become mutually responsive to each other with a view to the (ethical) acceptability, sustainability and societal desirability of the innovation process and its marketable products (in order to allow a proper embedding of scientific and technological advances in our society).

Our evaluation approach was conceived to provide transparent procedures and interactive methods to engage Institutions and the TeSLA consortium members to responsibly reflect on the ethical acceptability, sustainability and societal desirability of e-authentication with authorship verification. RRI approach was useful to properly embed the innovative system to ensure scientific and technological advances in terms of technology-enhanced assessment [17] for academic integrity [14].

There are a set of six components [2, 11, 15, 16] that must be taken into account to develop RRI practices:

- **Governance** refers to a set of principles, procedures, instruments and recommendations to foster responsibility and accountability among all actors to ensure acceptable and desirable outputs from scientific innovations. Our evaluation process engaged TeSLA partners and Institutions to reflect and establish the governance collaboratively.

- **Scientific education** aims to equip citizens with knowledge, skills and attitudes for all societal members to participate in R&I debates with evidence-based thinking. Our evaluation instruments included a set of educative artifacts such as TeSLA videoclips, Informed Consent, FAQ and guidelines prepared by the Course and Technical Teams to support students with various issues such as: data privacy and security (including ethics), technical problems, special educational needs requirements (accessibility), e-authentication and authorship verification steps, system interface (usability) and information about cheating and plagiarism.
- **Ethics** supports research integrity through the awareness and prevention of research practices that are unacceptable. It considers principles and procedures to minimise the risks of scientific and technological developments. In TeSLA, our evaluation approach engaged all pilot partners and project leaders to discuss ethical issues, including the General Data Protection Regulation (GDPR), safe procedures for consent forms and guidelines for evaluation data management using common and protected templates with anonymised data. Our questionnaire and focus group guides were designed to include ethical reflection.
- **Open access** contributes to good research practices and knowledge sharing, as well as allowing others to adopt or adapt their approaches and encourage innovation. Our evaluation model led to a set of publications with open access [9, 10] and also academic findings were translated to open educational resources, such as articles published in the OpenLearn platform [7] to support formal and informal education.
- **Gender** equality is key to ensuring diversity of participants, providing gender balance with equal opportunities for all involved in research projects. Our evaluation studies considered gender issues, whose findings were also published and used to inform all stakeholders.
- **Public participation** promotes inclusion, research activities and innovation, in which they need to inform and generate reflection for a better understanding of social, cultural and environmental contexts, thus engaging organizations and society. Our evaluation outputs were presented in public events for large audiences, including events, conferences and social media. We also reached more than 1000 open learners who contributed to our evaluation from different countries.

There are eight principles which guide RRI practices [1, 18], described as follows:

- **Diversity and inclusion**: these principles aim to engage a wide range of participants (innovators with society) at early stage in RRI practice with interactive methods such as deliberation, consultation and collaborative decision-making. This promotes wide access to knowledge and sources of expertise. In TeSLA, our evaluation methods engaged all distinctive groups of stakeholders in all stages of TeSLA project.
- **Anticipation and reflection**: aim to better understand how RRI shapes the future, which means we need to identify impacts, consequences, risks and benefits. In TeSLA, our evaluation procedures were designed to gather valuable insights for increasing our pre-knowledge for better evidence-based decisions.
- **Openness and transparency**: aim to communicate methods, findings, and implications in a meaningful and effective way for enabling societal dialogue. The

visibility and understanding of Research and Innovation through an open and transparent way helped TeSLA to reach very large communities in a considerable number of countries.

- **Responsiveness and adaptation**: aim to be able to respond to changes and modify modes of thought and behaviour in response to new circumstances, knowledge, and perspectives. This aligns the actions with the stakeholders' and public' needs. In TeSLA, our evaluation outcomes provided a set of recommendations to respond and adapt to issues that emerged for all stakeholders, during the three phases of the project.

The RRI projects completed to date highlight five groups of society who should interact in all phases of any development and evaluation of new work [1]. These are:

1. **Research community**: refer to academic researchers, innovative scientists, research managers, public affairs and communication officers. In the TeSLA project, this included all professionals and 'pilot coordinators' involved in the research studies, data collection and data analysis.
2. **Education community**: refer to teachers, teacher trainers, pedagogical coordinators, 'technical teams', course developers and students. In the TeSLA project, this involved all educational members, such as: teaching staff, course teams, assessors, learning designers, lecturers and instructors.
3. **Business industry and companies**: are large, medium and micro-enterprise, professional entrepreneurs' groups, 'technology developers', including transnational organizations and institutions. In the TeSLA project, this group included technology providers and exploitation companies interested in TeSLA system as well as technology developers of the TeSLA instruments: facial recognition, voice recognition, keystroke dynamics, forensic analysis and plagiarism detection system.
4. **Policy makers and policy influencers**: they range from influential policy makers to policymakers, directors of research centres and representatives of scientific societies, whether at European, national or local level. In the TeSLA project, this group involved all those who defined how research and innovation should be carried out in their own area of influence and also 'institutional leaders'.
5. **Civil and Society organization**: are individuals to organizations, including NGOs (Non-Governmental Organisations), communities, media professionals, representatives of civil society. In TeSLA, this group engaged open learners, non-formal learning providers (e.g. OpenLearn community) who contributed to discussions and reflected about how the TeSLA system and instruments meet the needs of society for formal and informal education.

2.1 Correlations Between RRI and Human-Centred Design Approaches

Human-centred design methods [13] are recognised as a significant approach for technology development and system evaluation through continual interaction with end-users to ensure that the innovation will address their needs and expectations. The Human-centred design approach considers human perspectives in all steps of the technology innovation [3]. The interactive process with end-users enables software engineers to examine the requirements more effectively with the end-users. The human interaction is initiated at early stage by discussing the problem within context, brainstorming, conceptualising, designing, developing and evaluating the first technology innovation model. These iterative and cyclic procedures enable innovators to improve the technology system and identify the relevant factors about costumers' acceptance or adoption and product's scalability.

However, some barriers highlighted by the literature might potentially impact on the process of evaluation such as the lack of users' interaction, usage and feedback [19]. These barriers include the users' difficulties and concerns. There are various factors that might affect the user experience and consequently the evaluation process, for example, technical problems, usability and accessibility issues, data privacy and security, lack of digital skills, training and support. One of the difficulties to implement human-centred approaches is there is a high level of novelty, uncertainty and potential risks which might impact on the user experience and may cause dropout.

To explore this challenge, the evaluation model developed and implemented during TeSLA project was refined based Human-centred design approach combined to RRI as described by Table 1. All these approaches, which are iterative in nature enabled the gathering of data together with recommendations and lessons learned related to the challenges (novelty, uncertainty and potential risk) to build expertise throughout the 3 phases of the empirical studies. The Human-centred Design model described through 3 stages (*prototype, deployment, exploitation*) focuses on user requirements to better align with the technology development [12], whereas RRI (presented also with 3 phases (*planning, development and sustainability*) focuses on the societal actors (all stakeholders)' needs and expectations to align more closely to the innovation process. Our model used in TeSLA combines requirements, needs, expectations and trust experience to dove tail with the innovative trust-based system through 3 pilot studies (*small, medium and large*).

3 Implementation of Pilots

Three pilots were conducted to obtain data about the usage of TeSLA system with information feedback and recommendations to refine the instruments and protocols.

1. **Small Educational Pilots**: In this first stage, seven institutions engaged together 500 learners during the first year of the project. In this phase, the TeSLA system

Table 1 Evaluation model for the TeSLA project refined with respect to RRI and Human-centred design

Human-centred design	Responsible Research and Innovation	Evaluation model for Trust-based e-assessment system
Prototype 1. Initial Planning 2. Initial Requirement 3. Initial Design 4. Minimum implementation 5. Small pilot with users 6. Component Tested	*Planning* 1. RRI Introductory Plan 2. Initial participants interaction 3. Needs identified 4. Next steps planned 5. Innovation discussed 6. Initial Feedback collected	*Small Pilot study* 1. Small pilot evaluation plan 2. Consent Form 3. Stakeholders Consultation 4. Questionnaires design 5. Initial Trust data collection 6. Requirements and recommendations
Deployment 1. Detailed Planning 2. Requirement analysis 3. Integrated Design 4. Major Development 5. Integrated System Test 6. Medium pilot with users	*Development* 1. RRI exploitation Plan 2. Multi-Stakeholders engaged 3. Expectations discussed 4. Implementation 5. Innovation improved 6. Feedback collected	*Medium Pilot study* 1. Medium pilot evaluation plan 2. Consent Form 3. Focus groups 4. Pre- Questionnaire 5. Instruments-Trust feedback 6. Post-questionnaires
Exploitation 1. Plan for scalability 2. Requirement for new users 3. Expanded Design 4. Development/Deployment 5. Large-scale testing 6. Large scale-evaluation	*Sustainability* 1. RRI sustainable Plan 2. Public Engagement 3. Next priorities identified 4. Final steps planned 5. Innovation consolidated 6. Final Feedback collected	*Large Pilot study* 1. Large pilot evaluation Plan 2. Consent Form 3. Focus groups and Dropout analyse 4. Pre- and Post-questionnaire 5. Trust System feedback and reports 6. Stakeholder Interviews

was under development, therefore no technology to be tested, but the defined protocols and data flows between the project members and stakeholders (learners, teachers, auditory, …) were implemented and evaluated. The learning and assessment activities to be used with TeSLA system were tested and evaluated. A first critical risks guideline was also defined by the project team.

2. **Medium Test-bed Pilots**: This second phase was conducted during the second year of the project with the TeSLA system with the five instruments. Approximately 3500 learners used TeSLA instruments.
3. **Large Scale Pilots**: This final phase was conducted during the third year of the project. Two rounds were performed during this phase, involving a total of more than 17,000 learners. The goals of this phase were: (1) To test the TeSLA system's integration and scalability. (2) To test the refinement of the TeSLA e-assessment Model in a large-scale scenario. (3) To test the reliability of authentication and authorship mechanisms.

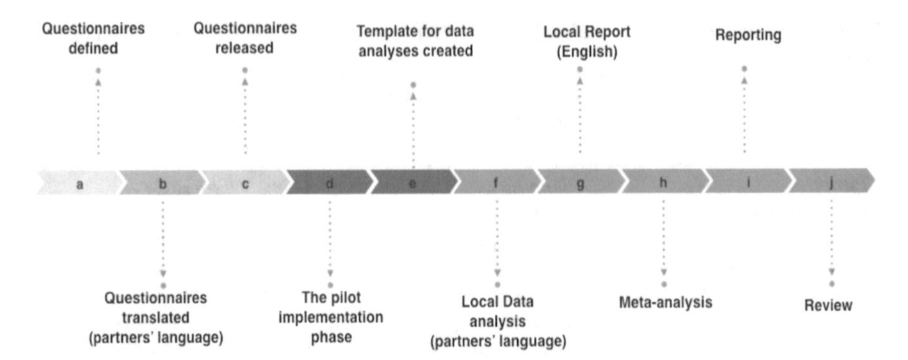

Fig. 1 Steps used to implement our methodological approach

In each pilot a set of 10 steps (see Fig. 1) were implemented to support the evaluation process.

The pre-pilot and post-pilot questionnaires for students and teaching staff were implemented by means of an online template set-up using the Bristol Online Survey system (https://www.onlinesurveys.ac.uk/). This survey system has now been transferred to JISC. All partners used the same survey system with an identical set of questions. Each of the seven partners was responsible for translating questionnaires into their local language. In addition, the three questionnaires for the pilot coordinator, the institutional and technical leaders were constructed and ran in English.

Each partner implemented their data collection and statistical analysis with respect to their local context. The partners' data analyses were then integrated and interpreted before the final steps, which included writing the evaluation report with recommendations and receiving peer-review feedback.

3.1 Evaluation Questions

Table 2 presents the nine thematic categories grouped the nine overarching thematic questions for the TeSLA pilots, through stakeholders' deliberation and consultation events.

3.2 Participants

There were seven Higher Education organisations in total who participated in the three pilots: two online learning institutions (Universities 4 and 7) and five universities with blended learning courses (Universities 1, 2, 3, 5 and 6). The number of students who completed the evaluation questionnaires were 336 during pilot1-2016, 1085 during pilot2 in 2017 and 4428 during pilot 3 in 2018. This chapter focuses

Table 2 Overarching principle questions used in Pilot-3

Thematic categories	Key thematic questions
1. Student perspectives	What are student's perceptions about e-authentication systems and TeSLA tools?
2. Staff perspectives	What are educators' views on TeSLA interface and students' experience?
3. Technology development	What are the technical team' views about the data & system integration?
4. Effectiveness of the authentication and authorship	How effective is TeSLA system in authentication and authorship verification?
5. Assessment design and pedagogy	How engaged are teachers and managers in assessment design, teaching, training and support with TeSLA system?
6. Award bodies and policy makers	Does TeSLA contribute to and support national education policy and social aspirations?
7. Staff, resource and financial costs	What was the technological readiness of staff?
8. Methodology	What are the issues related to data collection and analysis, students' consent, SEND participation?
9. Trust	Do users feel informed, comfortable and confident with the TeSLA system?

on data about pilot3 final phase, which refers to the final evaluation of the TeSLA system. The participants included a large group from the universities: 67 teaching staff, 7 pilot coordinators (research role), 7 technical teams (IT department), and 7 institutional leaders (director role) who contributed with their views.[1]

In terms of students' participation in the pilots, there were a total of 11,102 who used TeSLA system in pilot 3 (final stage). The total of students who replied the pre-questionnaire was 3528 and the post-questionnaire was 2222. The total of teaching staff who also completed pre- and post- questionnaires was 67. There were 7 technical teams' coordinators, 7 pilot' coordinators who were interviewed and 7 institutional leaders (Table 3). The seven institutional leaders from the 7 institutions who replied a questionnaire were: (1) Dean, (2) Director of Distance Education Centre, (3) Vice manager of the Learning Technologies and R&D Department, (4) Manager educational logistics, (5) Associate Director Quality Enhancement Prof Technology Enhanced Assessment and Learning, (6) Director, (7) IPR manager.

[1]The role of the different actors within the project can be reviewed in Chap. 8, Sect. 4.3.

Table 3 Summary of students data—Pilot 3 from questionnaires

Universities	1	2	3	4	5	6	7	Total
Target—(expected number of students)	1500	1500	1500	1500	1500	1500	1500	10,500
Total of students (unique participants)	2325	1844	417	1617	1457	1574	1868	11,102
Students who used facial recognition	2684	100	9	25	644	1163	1116	5741
Students who used voice recognition	247	54	0	0	117	370	235	1023
Students who used KeyStroke dynamics	247	250	29	0	46	407	915	1894
Students who used forensic analysis	53	1661	229	126	189	150	686	3094
Students who used plagiarism detection	48	1586	365	1541	814	321	1674	6349
Students who replied pre-questionnaire	240	167	84	853	232	783	1169	3528
Students who replied both the pre- and post- pilot questionnaires	171	115	57	574	226	452	627	2222
Students dropout rate (%)	29	31	32	33	3	42	46	29
Teachers who replied both the pre- and post- pilot questionnaires	8	8	3	4	4	6	34	67
Pilot coordinators interviews	1	1	1	1	1	1	1	7
Technical teams Interviews	1	1	1	1	1	1	1	7
Institutional leaders interviews	1	1	1	1	1	1	1	7

3.3 *Limitations*

During the three pilot studies there were a few limitations/caveats. Some institutions did not have access to system instruments results of e-authentication and authorship verification. Two institutions used the TeSLA system in a separate VLE (Virtual Learning Environment) universities 3 and 4. In addition, all students who used TeSLA and signed the informed consent form were volunteers and two universities engaged a low number of participants (universities 2 and 3).

4 Findings

Our results were grouped based on the overarching principle questions presented in Table 2.

4.1 Student Perspectives: What Are Students' Perceptions About E-Authentication Systems and TeSLA Tools?

In terms of the benefits of e-assessment with e-authentication most students (more than 70% of participants from each institution) selected various advantages (Fig. 2). The most popular reasons chosen for using a system such as TeSLA were to prove that their work is authentic, to improve the rigour of assessment, to ensure trust and prevent cheating.

Only a few of them mentioned that there was no advantage. Students from Universities 1, 4 and 7 provided other examples about the advantages of e-authentication with e-assessment, such as:

- University 1: Reliable and fair evaluation, location and time independence.
- University 4: Less stressful and more adapted to my mental problems, to not have to travel to an examination centre and avoid all the associated logistics, (transport, time off, childcare, school pick-up, etc.). Realistically, most real-life applications

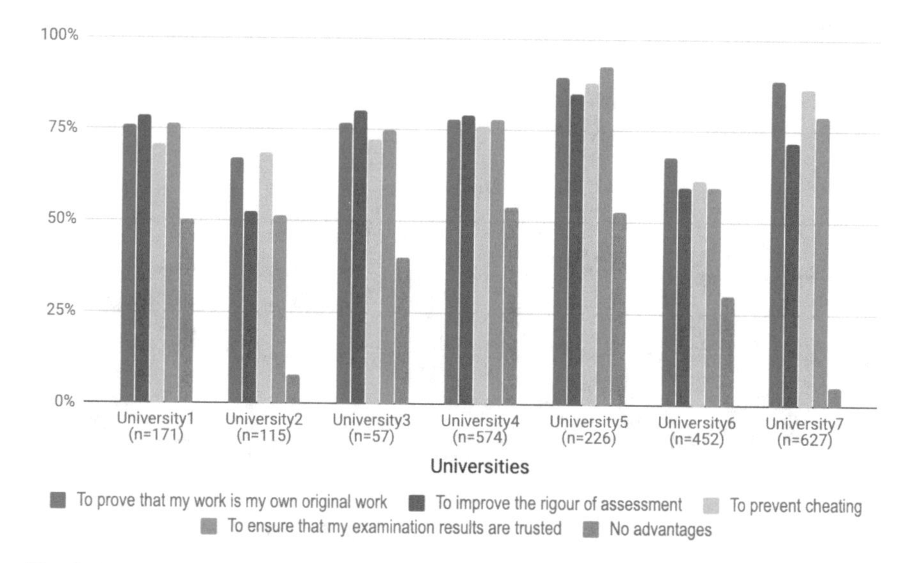

Fig. 2 Students' opinions about the main advantages of e-authentication in e-assessment—data from post-questionnaire

of what we have learned would allow for the source material to be available for consultation.

- University 7: To Avoid face-to-face exam, to help SEND students (reduced mobility), and less stressful.

There were also students who pointed out disadvantages from all institutions (Fig. 3). More than 20% from each institution mentioned that e-authentication in e-assessment can be intrusive. More than 55% of students from universities 1, 3 and 5 mentioned that have to share personal data and e-authentication can involve more work than traditional assessments. The other reasons were: reliability about outcomes, technical issues, more time and challenging. Qualitative data from students from three institutions also indicate other difficulties:

- University 2: Internet connection problems, blackouts during the process, or lack of technical skills. In online assessment, personality features and communication do not affect interpretation and evaluation. It is impossible to or difficult to ask clarifying questions.
- University 3: Reducing interaction with other people too much reliance on technology; too many factors that can go wrong; the system might not work properly or might not adapt to changing model, not suitable for type of assignments/software to be used, e-authenticated exams not clear or not applicable.
- University 4: There may be always faults that make system difficult to be used, for example, a new computer with a different kind of a keyboard, new glasses, flu etc. The system might not recognise the person with a different appearance. The system may be unstable and work insecurely, e.g. logging out suddenly and losing all the work done.

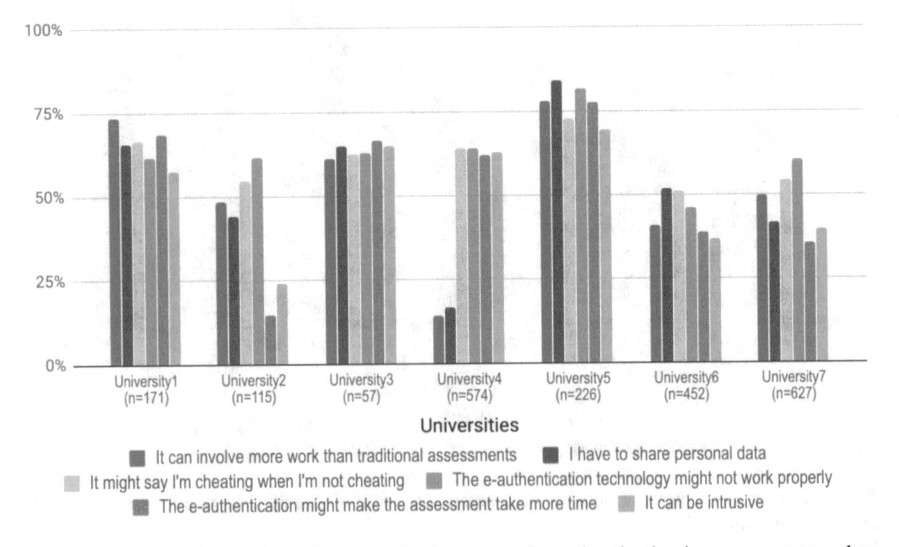

Fig. 3 Students' opinions about the main disadvantages of e-authentication in e-assessment—data from post-questionnaire

4.2 *Staff Perspectives—What Are the Educators' Views on TeSLA System and Students' Experience?*

Most teaching staff found TeSLA an user-friendly and relevant system. They indicated three main advantages for using an e-authentication based e-assessment system: to avoid having to take an examination under formal examination conditions, to have assessments better adapted to students needs and to allow anytime anywhere assessments. However, they also highlighted three key issues: all students must share their personal data, the system must work properly, and it might take more time for students to complete e-assessment. In addition, more than 30% of teachers from four institutions (universities 1, 4, 5, 6 and 7) listed some concerns about TeSLA system: it can involve more work than traditional assessments and it might say a student is cheating when they are not.

Pilot leaders were asked what their institution expect the TeSLA system to do to assist them with e-assessment. They listed four key factors:

- A system which triggers warnings about cheating and dishonesty behaviours. Then, the teachers only have to check these potential cases.
- Clear and accurate feedback for student authorship and authentication in the assessment activities.
- Additional information to assist teacher in determining authorship.
- Secure platform and reliable results.

Pilot leaders were also asked whether they would recommend TeSLA to another colleague. Five pilot leaders said yes because:

- The opportunities it gives for e-assessment.
- It will enable secure e-assessment and provide opportunities for using different assessment activities.
- With some restrictions. So far, we have little insights that shows that TeSLA works.
- It may contribute to increase the trust by reducing cheating and dishonesty behaviours.
- It has a great potential to make e-assessment more secure for teachers, and for students to become more responsible in the assessment process.

4.3 *Technology Development—What Are the Technical Team' Views About the Data and System Integration?*

The directors of the technical teams of each institution who were responsible for the TeSLA system integration were also interviewed. Four institutions (Universities 2, 3, 5 and 6) integrated TeSLA system in their institution VLE. Two institutions integrated it in an external system linked to their Institution VLE (Universities 1 and 4). Another institution (University 7) used both approaches.

They were asked whether they faced issues with the TeSLA system integration with the VLE and student records system. Only one technical team reported no problems (University 7). The other six technical teams described various issues, such as:

- In transversal assessment tools, like a classroom forum, there is no way in TeSLA to define time periods to restrict the audit data to the content generated for each learning activity.
- There were some issues revealed during the initial tests which were solved quickly with help of the TeSLA technical team. After the start of the pilot Keystroke Dynamics enrolment has stopped working and it was cancelled in all the planned activities. The problem was resolved in some weeks ago.
- We found that integration would be impossible, given project and university constraints, we focused on a standalone system. We developed custom student-data import functionality for Moodle.
- VLE and student records system are 2 different systems. We only integrated TeSLA with our VLE. There were many issues to solve, first of all because proper documentation was missing.
- Some installation issues. Some documentation problems.
- We did not get answers for some problems that we reported.

The technical teams also mentioned problems faced by students with service disruption, delays or disconnection:

- **Browser**: *Sometimes we got issues with some browsers.*
- **Various implementation issues**: *The system is working but with lot of changes and small issues solved during the implementation.*
- **Enrolment**: *Yes, [we faced problems] especially when enrolments were performed in face-to-face mode. The system was not stable and could not send all the information simultaneously. The strength of the internet connection was also very important (Wi-fi or mobile).*
- **Technical skills**: *Students were invited (not mandatory). Some of those who responded struggled.*
- **Upgrades**: *Some issues related to infrastructure upgrades, performed during the project.*
- **Instability and delays**: *The TeSLA system was not stable enough in the beginning of pilots, so there were delays on the courses. Some course activities had to be skipped.*

Five institutions reported significant extra workload for the institutions' technical teams due to problems faced by students with service disruption, disconnection and instability of the system. However, five institutions have confirmed their intention and capacity to continue using the TeSLA system after the project's completion.

4.4 *Effectiveness of the Authentication and Authorship—How Effective Is TeSLA System in Authentication and Authorship Verification Instruments?*

The overall students' experience with the TeSLA instruments was positive for more than 50% of the students from all partner universities. More than 70% participants from all universities found that the instruments were easy to use. More than 60% were comfortable with the system and would be willing to use it in future online assessments. Three main factors contributed for participants to feel confident with the instruments: very clear and detailed instructions; familiarity with the system and tutorials with guidelines. Users provided positive feedback from all instruments:

- **Facial recognition** (Fig. 4) was used by students from most universities apart from University 3. Half of the participants (50%) considered it not intrusive, they were comfortable and would be willing to use it again, apart from university 6 (only 25% found it did not take too much time and were willing to use it again).
- **Voice recognition** (Fig. 5) was used by students from five universities (universities 1, 2, 5, 6 and 7). Approximately half of the participants (50–60%) were comfortable to use this instrument, and willing to use it again apart from University 1 (20%) and University 5 (27%).
- **Keystroke dynamics** (Fig. 6) was used by students from six institutions (except university 4). Most students were comfortable to use this instrument, and willing to use it again particularly from universities 3 and 7 (75%).

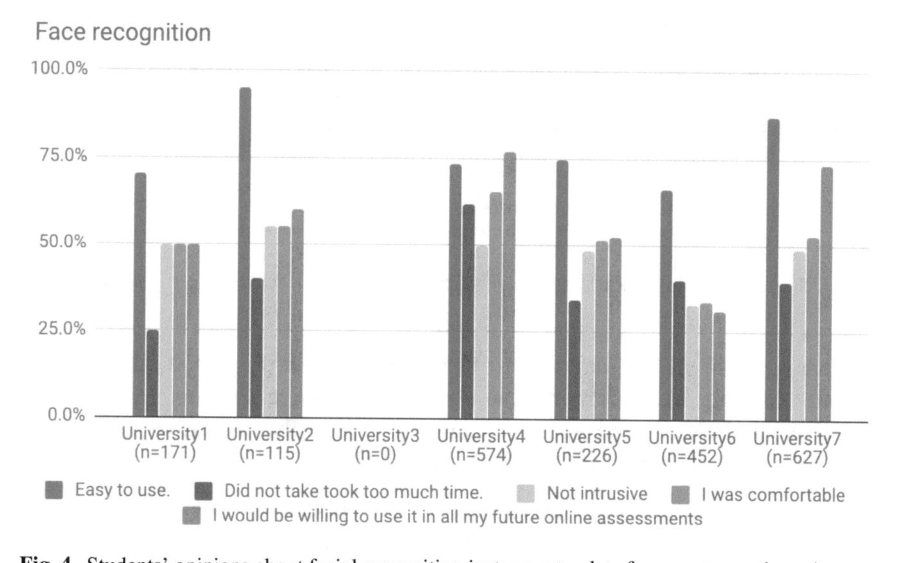

Fig. 4 Students' opinions about facial recognition instrument—data from post-questionnaire

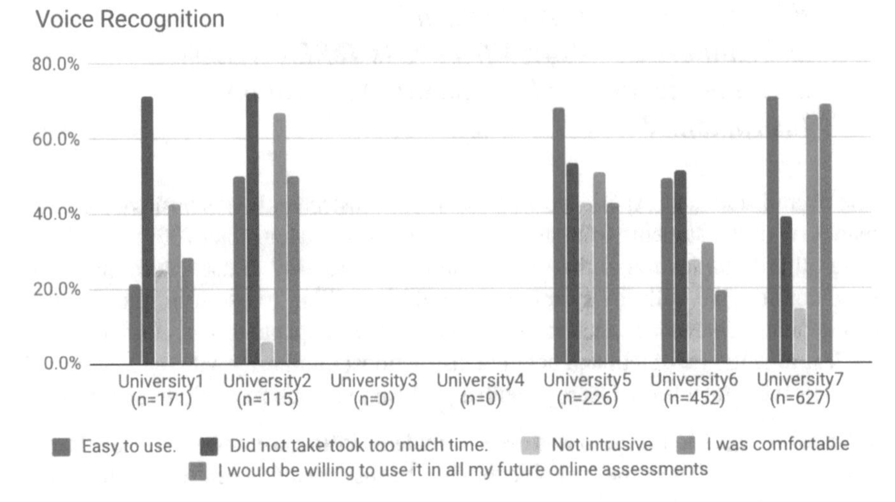

Fig. 5 Students' opinions about voice recognition instrument—data from post-questionnaire

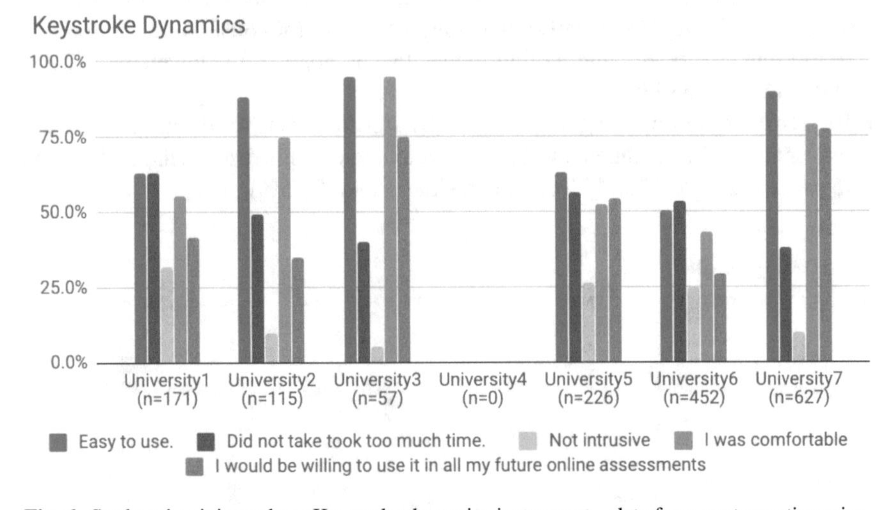

Fig. 6 Students' opinions about Keystroke dynamics instrument—data from post-questionnaire

- Students from all institutions used **Forensic Analysis** (Fig. 7) and **Plagiarism Detection** (Fig. 8). Their opinion across institutions were very similar (apart from University 6). Many students were comfortable (more than 70%) and half of them willing to use it again (particularly Universities 3 and 7).

Pilot Coordinators also presented their views about the effectiveness of the *TeSLA* system for authentication (successful and failed attempts, inauthentic and inappropriate uses; disruption and invasion). Pilot Coordinators presented some benefits of using e-authentication such as new types of assessments and the opportunities for

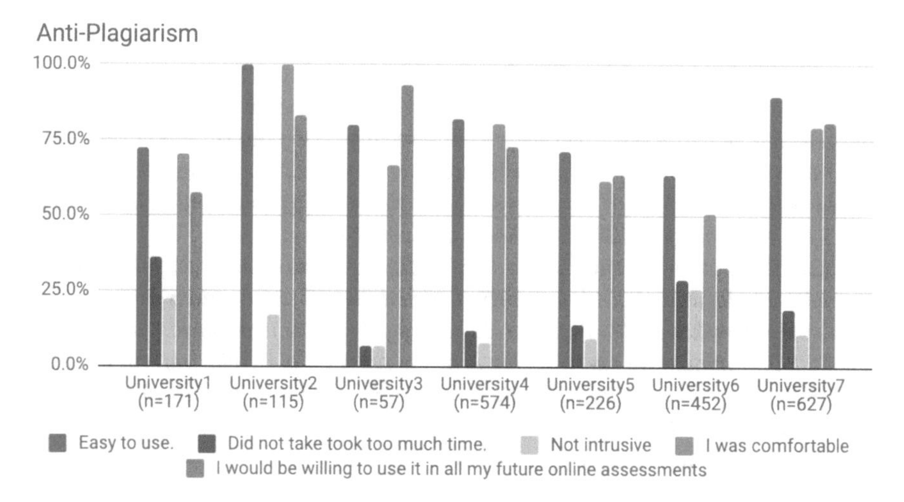

Fig. 7 Students' opinions about plagiarism detection instrument—data from post-questionnaire

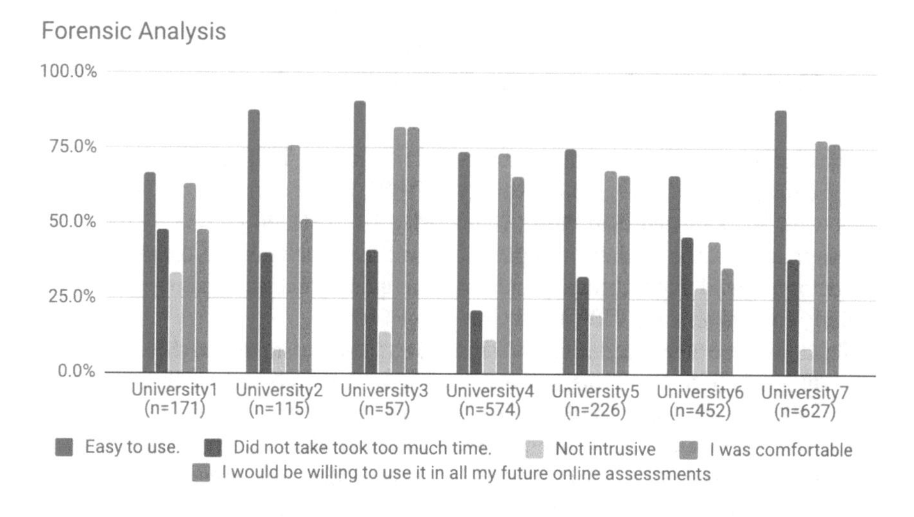

Fig. 8 Students' opinions about forensic analysis instrument—data from post-questionnaire

increasing trust by reducing cheating and academic malpractice. Five institutions would recommend TeSLA to another colleague. Two institutions who reported that they were not sure highlighted that the technological implementation was difficult.

As the results of e-authentication were not available during pilot 3, all Pilot Coordinators reported that they were not sure whether the TeSLA instruments assisted the Institution in checking e-authentication.

The seven Institutional Pilot Coordinators were asked if the suite of TeSLA instruments assisted their Institution with checking e-authentication and whether teaching staff had been able to review the outcomes promptly. There were two institutions who were not sure and five who mentioned that they did not. Pilot coordinators from universities 1, 3 and 4 replied that they could not answer this question, because teaching staff have not seen the outcomes. Universities 5 and 6 believe that TeSLA assisted in e-authentication, but they were also unable to confirm it. Universities 2 and 7 also mentioned that they are not sure "*Results for some instruments just begin to be available*", but they "*have not analysed the feedback yet*" *and* "*did not know how well or properly TeSLA instruments are working*".

University 7 reported that "*the results of Facial Recognition were useful. Keystroke dynamics results were difficult to interpret and additional explanations from the technical colleagues have been required. Regarding the other instruments, and considering the information we currently have, our impression is Plagiarism will be easily understandable, and Voice Recognition may suffer similar problems of interpretation to those described for Keystroke Dynamics. The results for Forensic Analysis are unknown (the technical team is working on that).*"

4.5 Assessment Design and Pedagogy—How Engaged Are Teachers and Managers in Assessment Design, Teaching, Training and Support with TeSLA System?

Various teaching staff agreed they were satisfied with the TeSLA experience particularly University 6 (70%) and University 5 (100%). Participants from two institutions were less satisfied than others (University 7—50% and University 4—25%). Although most of the participants agreed they received technical guidance, a smaller percentage agreed that technical problems were quickly and satisfactory solved particularly for University 3 (50%) and University 4 (0%).

Teachers were able to redesign and recreate new e-assessment activities supported by e-authentication. For some universities (3, 4, 7) the process of integrating the TeSLA instruments started at the course design stage. However, some teaching staff and course teams would have liked to obtain more guidance from the system about how best to combine the instruments and analyse their outcomes in order to use the instruments more effectively. In addition, some Institutions did not have access to the results and would have liked additional support to interpret the instruments' outcomes. The negative factors that impact on their staff's experience with TeSLA was that the system failed or stop working properly (Universities: 1, 2, 4, 5 and 7) where the workload was greater than expected particularly for University 1 (75%) and University 4 (71%) and University 5 (40%).

4.6 Award Bodies and Policy Makers—Does TeSLA Contribute to and Support National Education Policy and Social Aspirations?

All pilot coordinators also mentioned that the university has in place procedures to deal with cheating and plagiarism, however, teaching staff must receive some guidance to interpret the feedback and know how to solve technical issues with the results.

In many institutions, policy makers confirmed that the implementation of TeSLA will be an opportunity to raise the profile of academic integrity within the institution. So that this might draw attention both to academic integrity policies and to quality assurance for e- learning.

Local regulations including policy development, education of staff and students, assessment practice and technology support will be important for all institutions interested in e-authentication. The technical team provided a few comments about the TeSLA technology providers' support:

- **Reducing delays for solutions**: *TeSLA technical team were very supportive during the whole pilot. They made a lot of efforts and spend a lot of time resolving our issues.*
- **Providing more documentation**: *Key developers were responsive* via *issue-tracker, e-mail and Skype. However, only two of the components had proper CHANGELOGs/release notes (I requested), and there was not consistent and low-level information within the code-base—READMEs, API.md* etc.
- **Improving communication with schedule**: *There was an over-reliance on verbal communication in meetings, with no minutes, and instructions buried in communication forums. The development schedule and methodology were not well communicated to Pilot institutions.*

4.7 Staff, Resource and Financial Costs—What Was the Technological Readiness of Staff?

Technical teams were asked whether they have capacity to carry on using the system in their Institution after the TeSLA completion. Four institutions replied yes and provided comments:

- My institution is working in a continuity plan.
- We need to test the final release and documentation to be sure about all the answers. They are now based on the current experience.
- We do have the expertise.
- We are able to use/test the TeSLA system after the project if the final version is technically stable enough and the technical support and version delivery are arranged properly.

Five institutions (apart from 1 and 7) mentioned that there was more effort from the technical team than they expected during the pilot implementation because of the technical issues, delays and long period without getting an answer. The reasons were:

- **Lack of stable version before the pilot implementation**: *Technical teams are deploying TeSLA versions every week. This is not what was expected. Efficient way to proceed is having a stable version before the pilot implementation.*
- **Workload to report and solve issues with the system**: *When issues were raised with the instruments (especially with the enrolments) it needs a lot of efforts to report and resolve them. At the beginning of both phases of the 3rd pilot we put much efforts every day in order the pilot to start successfully.*
- **Lack of clear tasks for the technical team**: *My role was not well defined at the outset. As our Pilot studies were ALL remote and at a distance, I needed to: 1. customise core Moodle & TeSLA plugin language texts; 2. customise LTI-enrollment language texts; 3. deploy Moodle-TeSLA plugin, and TeSLA backend components; 4. liaise with developers, report bugs …; 5. test the TeSLA plugin and functionality. One of the biggest challenges was reporting and mitigating bugs found when software was delivered late, with little or no time before pilots were due to start (late).*
- **Lack of plugin for a different VLE**: *The project promised to deliver a plugin that could be used in any VLE but could not do so. Consequently, we had to develop our own implementation on top of the planned resources. Even then, we had to develop quite some additional features.*
- **Lack of information and translated guidelines**: *We had problems with version management, translations and getting information of new features.*

The technical teams of the seven institutions provided support to pilot coordinators during pilot 3 which included: technical guidelines, solutions for endusers' technical problems, FAQ to support teaching staff and technical support for teaching staff with SEND.

The comments about their support were:

- We had spent more resources that it was planned originally.
- We worked very close with the pilot leader and the technical guidelines and user manuals were result from our collaborative work.
- We tested together the system and we were supporting teachers and students collaboratively as well.
- As technical lead, I supported a lot of the Moodle course and activity configuration, supported writing Pilot-specific course content, documented and showed colleagues how to access TeSLA data, configured the TeSLA components, supported testing by the team.
- As far as I know, we had no way to specifically invite SEND students to participate in the Pilots.

In terms of Institutional Leaders' views about the *TeSLA* system in authentication (staff, resource and financial costs, methodology and trust), six leaders would be willing to adopt an e-authentication system (e.g. TeSLA) for their institution. Three leaders would buy an e-authentication and plagiarism detection system for providing more flexibility and possibilities of e-assessments that are trustful. Their expectations and suggestions are a user-friendly system, a usable product, well-documented references, information about how the tools work and guidelines for interpreting results and detecting cheating.

They mentioned some potential benefits: increase the opportunities for online teaching with e-assessment as part of their curricula, offer more possibilities for SEND students to complete assessment at home, improve quality and security of e-assessment.

They also indicate some potential challenges for their institution with the adoption of TeSLA such as: changing university regulations, potential resistance of some students or teachers, technical support and readiness, increasing capacity related to resource allocation and administrative staff support.

4.8 Methodology—What Are the Issues Related to Data Collection and Analysis, Students' Consent, SEND Participation?

Many students (more than 50%) indicated that they found the consent form (which gives permission by the student for the system to use their data which, in the case of TeSLA, can also include biometric data) easy to understand in all institutions. They also indicated that the form provided enough information and they were confident with the way their personal data was being used by TeSLA. However, some students (20% to 40%) from most of the universities indicated a negative experience, i.e., 15% to 20% of students were unsatisfied with the amount of feedback that they received.

An average of 35% did not understand how the TeSLA system was used to e-authenticate their identities and checking their authorship. An average of 15% of students faced technical problems, apart from University 3 who used only a few instruments in pilot 3 (Fig. 6).

The number of students who were unwilling to share any type of personal data was very small (less than 5%, apart from University 6 which was 15%). However, the number of students who were willing to share personal data are less than 25% in four institutions (1, 3, 4 and 5). This means that most students in these institutions are not sure if they want to share their personal data (See Fig. 7). In contrast, there are two institutions whose 75% students are more willing to share their personal data (Universities 2 and 7).

4.9 Trust—Do Users Feel Informed, Comfortable and Confident with the TeSLA System?

Participants found in general that TeSLA system will increase trust in e-assessment. Students from all institutions provided their views about e-authentication and trust showing that they were informed, comfortable and feeling confident.

Most of them consider that e-authentication and authorship verification will increase trust on e-assessment particularly from Universities 2, 4, 6 and 7.

University 2: Preventing cheating is a good thing. Even though you're honest yourself, there's no guarantee that others are. *Of course e-authentication creates a sense of surveillance, but I do not think it is a negative thing*

University 4: *I think that participants will trust online assessment more than before.*

University 6: *The security measures are very important in spite of the connected to the internet devices. The results will be more secure.*

University 7: *It is important for the university to increase the trust between university, industry and students. Industry and society continue to mistrust fully online assessment mostly by preconception about online assessment*

There were also students that were more resistant and consider that e-authentication and authorship verification will not be enough to ensure trust particularly from Universities 1, 3 and 5.

University 1: *The measures taken will not be sufficient. There may be some difficulties in terms of reliability, validity and usability. Face-to-face evaluation is more appropriate for now.*

University 3: *Online assessment can be an option for certain situations. Too many variables, too much doubt if system will work reliable, not clear what is meant by online, not able to judge.*

University 5: *It is very difficult. Not any additional comments.*

In terms of Teachers' views about positive factors to promote trust on e-authentication and authorship verification, data from pre- and post-questionnaire data were very similar. Most participants from all institutions (more than 60%) selected positive factors particularly that it will increase the trust among universities and employers and it will help participants trust the outcomes of e-assessment.

The seven technical teams confirmed that the TeSLA technology providers supported them with information and guidance, including security and data protection information; apart from the extra documentation with support for a different platform (very little assistance).

In terms of impact, pilot coordinators were asked whether the TeSLA had impact on students who were previously unable to participate in assessments (e.g. for reasons of location, disability, lifestyle). Although many of them mentioned that there was not an impact yet; they presented a few benefits:

- It will have a positive effect when the system is completed and ready to use.
- TeSLA system will provide new kind of possibilities and alternatives for students including SEND.

- TeSLA has been only used during continuous assessment.
- Most of the SEND students, as well as many students who live far from the university building were very happy that they could conduct their activities from home by using the TeSLA system. Their attitude towards the usage of such system was very positive before and after their experience with TeSLA.

Five Institutional leaders mentioned that very frequently there is a need to authenticate students during assessment in their Institution. In addition, six leaders would be willing to adopt an e-authentication system (e.g. TeSLA) for their institution.

They were also asked whether there was any other system or approach that they would use instead of TeSLA. Some of them suggested: e-proctoring and National IT-solution.

Three leaders mentioned that would buy an e-authentication system for their institution. Their views about the potential benefits of using e-authentication in their institution were:

- To increase the cases of e-assessment and its quality.
- The opportunity to offer more flexibility for students.
- More secure; more possibilities for e-assessments; assessments at home.
- SEND students using the system at home without supervision.
- Trustful authentication and plagiarism detection system.
- Opportunity to introduce online teaching as part of the curricula offered on more regular basis.

They were also asked about the potential challenges of using e-authentication in their institution. All of them mentioned various issues, such as:

- Implement it to the entire university.
- Changing the university regulations.
- Resistance of some students or teachers for adopting a new system.
- Training needed for students and teachers to learn how to use a new system.
- Extra workload for establishing such a system, creating new procedures and guidelines.
- Technical issues (e.g. link to LMS and grading systems), governance (e.g. accreditation), trust (e.g. by teaching staff).
- Technical support form IT-services.
- Technical readiness of students to use TeSLA.
- Fanatical resources allocation and administrative staff support.

Some Institutional leaders also presented their final comments about their expectations and suggestions:

- A usable product/system including well-documented references.
- More information about how the tools work to devise appropriate assessments.
- Information about tools that are new for institutions such as Forensic analysis.
- The confidence values for detecting cheating.
- A very user-friendly system.

5 Discussion and Conclusions

This work presented a novel approach: an Evaluation methodology for trust-based adapted systems by connecting Responsible Research and Innovation with Human-centred design approach. This methodology was applied and refined throughout three pilot studies which enabled the project team to measure and evaluate trust during the development of the innovative technology TeSLA.

The EU-funded *Adaptive Trust-based e-Assessment System for Learning* (*TeSLA*) (http://tesla-project.eu) was developed to check student authentication and authorship through a combination of various instruments, such as: facial recognition, voice recognition, keystroke analysis, plagiarism detection and forensic analysis.

The findings (Table 4) suggest a broadly positive acceptance of and trust in e-authentication for online assessments by both women and men, with neither group finding the e-authentication tools experienced to be either particularly onerous or stressful [10].

Table 4 summarises the findings based on our key thematic questions. All online distance universities (4 and 7) trusted the system more and had less difficulty accepting and working with the tools comparing to the other institutions. Many students (more than 70%) considered that examination results will be trusted and that the essay's authorship can be verified. Only a few students (5% to 19%) faced technical problems in all institutions. Various teaching staff were satisfied with the system and highlight the importance of having technical issues faster and satisfactory solved. Technical teams recommended sufficient capacity including cloud solution and training. Course coordinators found that e-authentication enabled new types of assessments and opportunities to reduce academic malpractice. Institutional leaders who would be willing to adopt an e-authentication system expect user-friendly and usable system with guidelines for interpreting results.

Five features related to a "trust-based e-assessment system", which emerged during the medium-test-bed [9], were confirmed during the large-study:

1. **The system will not fail or be compromised**: participants who faced technical problems received support and were able to complete the assessment tasks independently.
2. **Data will be kept safely and privately:** participants were informed about data security, privacy and safety which helped them share personal data and become more confident with the system.
3. **No adverse impact on assessment experience**: procedures were discussed including technical and pedagogical support in case the system did not recognise students' identity and authenticity.
4. **The system will not affect performance:** teaching staff provided alternatives about instruments particularly for students with special educational needs.
5. **The system will ensure fairness**: the e-authentication and authorship verification system provided opportunities for flexible, supportive and trustful e-assessment

Table 4 Summary of Findings related to the key thematic questions

Thematic categories	Key thematic questions
1. Student perspectives	Most of the student's perceptions about e-authentication systems and TeSLA tools were positive. They selected various advantages in particular to prove that their work is authentic, to improve the rigour of assessment, to ensure trust and prevent cheating. Students from universities 4 and 7 listed various benefits including SEND participants
2. Staff perspectives	The educators' views were positive, most of them would recommend it to other colleagues. However, there were some concerns in terms of students' experience: personal data provided by all students; system working properly; enough time for students to complete e-assessment. In terms of TeSLA interface, they expected clear and accurate feedback, guidance and reliable results
3. Technology development	Most of the technical team' views about the data and system integration presented various issues apart from University 7: time, audit data, technical problems, institutional constraints, documentation, settings, infrastructure, and technical support from the technology providers
4. Effectiveness of the authentication and authorship	The results of e-authentication were not available to the Institutions, all Pilot Coordinators reported that they were not sure about the reliability of the system and whether the TeSLA instruments assisted the Institution in checking e-authentication
5. Assessment design and pedagogy	Teachers and managers were satisfied in supporting the assessment design, teaching and training with TeSLA system, in particular University 5 and 6. Most of the teachers agreed they received guidance. However, only a smaller percentage agreed that technical problems were quickly and satisfactory solved
6. Award bodies and policy makers	Institutions mentioned that they have in place procedures to deal with cheating and plagiarism. However, teaching staff need some guidance about e-authentication to interpret the feedback and know how to solve technical issues with the results. The key requirements for policy makers at institutional level are: reducing delays for solutions, providing more documentation, and improving communication with schedule
7. Staff, resource and financial costs	Four institutions replied that they have capacity to carry on using the system, but there was more effort from the technical staff than they expected because of the technology problems and the lack of guidelines. Six leaders would be willing to adopt an e-authentication system (e.g. TeSLA) for their institution, three of them would buy an e-authentication system. The potential challenges reported were: changing university regulations, potential resistance of some students or teachers, technical support and readiness, increasing capacity related to resource allocation and administrative staff support
8. Methodology	The methodological approach was accepted and understood by various participants (more than 50%) in terms of data collection and analysis including consent forms. The key issues were the lack of feedback about e-authentication results, technical problems faced and uncertainties about sharing personal data for e-authentication

(continued)

Table 4 (continued)

Thematic categories	Key thematic questions
9. Trust	Various participants considered that were informed, comfortable and confident with the TeSLA system. There were only a few comments presenting resistance to trusting an e-authentication with authorship verification system, particularly from the universities that are not distant education institutions

Our findings show that TeSLA system might address the concerns highlighted by the literature on academic integrity in the digital age. Universities will not be compromising the public trust by allowing incidents of plagiarism to go unchecked [6] when using a trust-based e-assessment with e-authentication. "*One of the casualties of academic misconduct is the general sense of broken trust; students, faculty members, university administrators, potential employers, and the general public agree on very little when it comes to plagiarism, but all seem to share the sense that their trust in some aspect of university has been violated*" Evans-Tokaryk [5]:1).

To conclude, the RRI with a human-centred design approach was designed to support the scalability, sustainability and societal desirability of a technological innovation [21]. This methodological approach enabled the evaluation of the European TeSLA system during its development through a set of studies. Findings revealed that this approach was vital to examine the perceptions and needs of distinctive users about the e-assessment with e-authentication and authorship verification system. Future work will be important to address the key issues reported by participants for increasing trust of e-assessment in Higher Education [4, 20]. In particular, more studies will be necessary to examine the reliability and accuracy of the TeSLA system including technology integration and technical support with new institutions.

References

1. EC (2017) Responsible research and innovation. European Commission. Retrieved from https://ec.europa.eu/programmes/horizon2020/en/h2020-section/responsible-research-innovation
2. EC (2012) Assessment of key competences in initial education and training: policy guidance. Commission staff working document. European Commission. Retrieved from http://eose.org/wp-content/uploads/2014/03/Assessment-of-Key-Competences-ininitial-education-and-training.pdf
3. Leveson NG (2000) Intent specifications: an approach to building human-centered specifications. IEEE Trans Software Eng 26(1):15–35
4. ESG (2015) Standards and guidelines for quality assurance in the European higher education area Brussels, Belgium. Retrieved from http://www.enqa.eu/wp-content/uploads/2015/11/ESG_2015.pdf
5. Evans-Tokaryk T (2014) Academic integrity, remix culture, globalization: a Canadian case study of student and faculty perceptions of plagiarism. Across Discipl 11(2):1–40

6. Gulli C, Kohler N, Patriquin M (2007) The great university cheating scandal. Maclean's. Retrieved from http://www.macleans.ca/homepage/magazine/article.jsp?content=20070209_174847_6984
7. Okada A (2018) What are your views about technologies to support and assess writing skills? OpenLearn Retrieved from https://www.open.edu/openlearn/education-development/what-are-your-views-about-technologies-support-and-assess-writing-skills
8. Okada A, Rodrigues E (2018) Open education with open science and open schooling for responsible research and innovation. In: Texeira CS, Souza MV (eds) Educação Fora da Caixa: tendências internacionais e perspectivas sobre a inovação na educação (Org), vol 4. Blucher, São Paulo, pp 41–54
9. Okada A, Whitelock D, Holmes W, Edwards C (2019b) e-Authentication for online assessment: a mixed-method study. Br J Educ Technol 50(2):861–875. Retried from https://onlinelibrary.wiley.com/doi/full/10.1111/bjet.12608, https://doi.org/10.1111/bjet.12608
10. Okada A, Noguera I, Aleksieva L, Rozeva A, Kocdar S, Brouns F, Whitelock D, Guerrero-Roldán A (2019b) Pedagogical approaches for e-assessment with authentication and authorship verification in Higher Education (http://oro.open.ac.uk/id/eprint/58380). Br J Educ Technol (early access)
11. Owen R, Macnaghten P, Stilgoe J (2012) Responsible research and innovation: from science in society to science for society, with society. Sci Public Policy 39(6):751–760
12. Salah D, Paige RF, Cairns P (2014) A systematic literature review for agile development processes and user centred design integration. In: 18th International conference on evaluation and assessment in software engineering (EASE)
13. Seffah A, Gulliksen J, Desmarais MC (eds) (2005) Human-centered software engineering-integrating usability in the software development lifecycle, vol 8. Springer Science & Business Media
14. Simon B, Sheard J, Carbone A, Johnson C (2013) Academic integrity: differences between computing assessments and essays. In: Proceedings of the 13th Koli calling international conference on computing education research. ACM, pp 23–32
15. Stahl BC (2013) Responsible research and innovation: the role of privacy in an emerging framework. Sci Public Policy 40(6):708–716
16. Stilgoe J, Owen R, Macnaghten P (2013) Developing a framework for responsible innovation. Res Policy 42(9):1568–1580
17. Stödberg U (2012) A research review of e-assessment. Assess Eval High Educ 37(5):591–604. https://doi.org/10.1080/02602938.2011.557496
18. RRI-TOOL (2016) RRI-TOOL project. https://www.rri-tools.eu/
19. Rubin J, Chisnell D (2008) Handbook of usability testing: how to plan, design, and conduct effective tests. Wiley, Indianápolis, Indiana
20. Van den Besselaar PAA (2017) Applying relevance-assessing methodologies to Horizon 2020
21. Von Schomberg R (2011) Prospects for technology assessment in a framework of responsible research and innovation. Springer, pp 39–61. Retrieved from https://link.springer.com/chapter/10.1007/978-3-531-93468-6_2

Ethical, Legal and Privacy Considerations for Adaptive Systems

Manon Knockaert and Nathan De Vos

Abstract The General Data Protection Regulation (GDPR) is the new EU legal framework for the processing of personal data. The use of any information related to an identified or identifiable person by a software will imply the compliance with this European legislation. The objective of this chapter is to focus on the processing of a specific category of personal data: sensitive data (mainly face recognition and voice recognition) to verify the user's identity. Indeed, the GDPR reinforces requirements for security measures to ensure the integrity and confidentiality of these personal data. We analyze three privacy aspects: the possibility to obtain a valid consent from the user, how to ensure the transparency principle and the implication of openness and the framework to implement in order to use the feedback given by the system to the user. From an ethical point of view, the request for consent is legitimized by the existence of a real assessment alternative left to the student. Then the different components of the right to transparency are illustrated by examples from the field. Finally, the question of feedback is expressed in the form of a dilemma highlighting the possible risks of poorly justified decisions due to the way feedback is exposed.

Keywords Biometrics · GDPR · Privacy · Personal data · Ethics · Consent · Transparency · Right to information

Acronyms

GDPR General Data Protection Regulation
HEI Higher Education Institution
LTI Learning Tools Interoperability
TEP TeSLA e-Assessment Portal

M. Knockaert (✉) · N. De Vos
CRIDS/NADI Faculty of Law, University of Namur, Rue de Bruxelles 61, 5000 Namur, Belgium
e-mail: manon.knockaert@unamur.be

N. De Vos
e-mail: nathan.devos@unamur.be

D. Baneres et al. (eds.), *Engineering Data-Driven Adaptive Trust-based e-Assessment Systems*, Lecture Notes on Data Engineering and Communications Technologies 34,
https://doi.org/10.1007/978-3-030-29326-0_12

TIP TeSLA Identity Provider
VLE Virtual Learning Environment

1 Introduction

As members of the Research Centre in Information, Law and Society (CRIDS), we have worked on both the legal and ethical aspects of the TeSLA (Trust-based authentication and authorship e-assessment analysis) project.[1] As the software is using personal data of the students from the involved institutions (face recognition, voice recognition and keystroke dynamics), it was necessary to adopt a reflection on both the legal and ethical aspects of data processes involved—and this, from the beginning of the project, to ensure that the outcomes are legally valid and well received by society.

Our Research Centre is used to dealing with these two aspects both in research and development projects. Based upon our experience in the field and long-standing reflection on how to make legal and ethical interventions in this context, we have adopted a particular stance towards the other members of the project. Indeed, being neither engineers nor pedagogues, our expertise could tend to place us on sort of an external position. However, our credo is to try to collaborate as much as possible with all the partners, despite the diversity of the profiles we have to deal with. This modus operandi avoids the trap of an overhanging attitude that would consist in prescribing from the outside a series of directly applicable injunctions. We worked together with the project members to stimulate collective reflection on a number of issues that we found necessary to address.

The legal approach intends to put in place the safeguards required by the law to legalize the processing of personal data. The major objective was to implement the technical measures to ensure the proportionality, the confidentiality and the security of the personal data processes. The first guarantee implemented is that the system can only process the data of the students that give their consent. The second major guarantee concerns the rules of access. The system must be designed to allow each institution to have access only to its own data and only to the relevant data. Furthermore, the system must be designed to facilitate deletion or anonymization of the personal data when it is no longer necessary to store the data in a form permitting the identification of the student. Next to the technical considerations, the GDPR also foresees key principles surrounding the processing of personal data. In this chapter, we will mainly focus on this second aspect of the legislation, with a particular attention to the transparency principle.

Conversely, the ethicist's contribution is less related to operational aspects. It aims more to open the project, to bring a broader point of view to it, to see how the object created would interact with the rest of society. The ethical approach highlights

[1] To learn more about the project: https://tesla-project.eu/.

hypotheses on the various potential impacts of the technology, sheds light on the societal choices embedded in the object and invites other partners to think about them. The product of ethical work therefore takes the form of a reflection rather than a recommendation.

For a project such as TeSLA, the legal aspects already provide a solid basis to be taken into account to ensure the possibility of implementing the system. However, the Law will not necessarily enlighten us about all the critical technical choices (especially given the novelty of the object, some legal gaps may exist). The ethical approach can also be used to complete the legal approach when it does not have the resources to effectively advise partners. In this sense, but also because of the two aspects mentioned above, the legal approach and the ethical approach are complementary.

The first part of this chapter is dedicated to the consent of the students to process their personal data. As TeSLA is using biometrics data, it is important to obtain an explicit consent. We will analyze the legal conditions to obtain a valid consent and how the ethical considerations could reinforce the requirements to have a free, specific, informed and unambiguous consent. The second part is dedicated to the transparency principle. This principle involves a right for the students to receive specific information and a corollary obligation for the data controller to provide that information. In this second part of the chapter, we will analyze what information has to be given and how they can be adapted to the user. Finally, as a third part, we will focus on the result given by the TeSLA system. The result of matching or non-matching is personal data but ethical considerations could help the teachers to work with these indicators. For each part, we will firstly expose the GDPR requirements, and secondly explain the ethical considerations.

2 Ethical Preliminary Remarks

Three methodological approaches were undertaken in order to question the project from an ethical point of view.

First of all, a state-of-the-art report was produced regarding the potential social effects TeSLA could have on the learning and teaching experience. It was based on an interdisciplinary approach, ranging from academic texts on e-learning to the sociology of technology. This state of the art aimed to introduce initial reflections with a more global aim into the project (both through reports and direct discussions) and to get familiar with the subject of study. To this end, our analytical framework was informed by the work of key authors in the field of sociology of techniques such as Feenberg[2] or Akrich.[3]

Following this theoretical approach to the above concerns, an ethical fieldwork was developed, based on individual semi-structured interviews of teachers and learners

[2]Feenberg [6].

[3]Akrich [1].

Table 1 Interviews conducted in each pilot university

Number of participants	UOC	OUNL	SU	TUS
Learners	6	5	1	5 (focus group)
Teachers	3	3	2	3

involved into the pilots (which consist of the experimental and gradual integration of the TeSLA system in 7 consortium member universities). These interviews aimed to help us to concretely and pragmatically understand the social effects the TeSLA project could have on teaching and learning experiences of the participants. Four pilot institutions were implicated in the ethical fieldwork: two full distance universities (the Open University of Catalunya UOC and The Open University of the Netherlands OUNL) and two blended ones (Sofia University SU and the Technical University of Sofia TUS). Dutch OUNL stakeholders were interviewed at the end of Pilot 2, the other pilot stakeholders at the end of pilot 3A (not earlier because we needed to interview users who tested a sufficiently advanced version of the system). The choice of these universities was made to collect contrasted social contexts with respect two criteria (distance versus blended Higher Education Institutions HEI—technical backgrounds versus humanities backgrounds of teachers and learners) (Table 1).

Each interview lasted an average of 1.30 h. Two specific features characterized those interviews: first of all, they were based on the TeSLA experience of the interviewees, i.e. their understanding of the system and the activities undertaken with the system (what we call 'situated interviews'). Secondly, they were explorative and not representative: the ambition was just to help us to pragmatically understand concerns 'theoretically' pointed out by the literature or through the discussions experienced during the project. The number of interviewees is too low to expect some statistical representativeness and this is not the point of this study. The aim was to discuss, with users of the system (professors and students), the uses they have made of a new system. The reactions should allow the technical partners to see how their system is received by the public.

In this respect, the interview methodology is inspired by sociologists such as Kaufmann, who claims a very open approach to interviews.[4] This openness allows interviewees to address the aspects that they consider most relevant and not to confine the debate to the interviewer's pre-conceptions. This is how our fieldwork differs from other fieldworks accomplished in the context of this project: this methodology allows us to report the sense of the practices, the deep reasons why users act a specific way. It enables us to consider diversity among users and to ask specific questions according to the issues they are more sensitive about. This wider freedom in the way to answer questions presented by our methodology is complementary to the pilot partners' questionnaires which are more formalized and gather more standardized quantitative data.

To do this, interview grids were designed: they include a set of sub-themes to explore in order to deal with the system in a comprehensive way and examples

[4]Kaufmann [7].

Table 2 Teachers' interview grid

Topic	Why approaching it?
Background of the teacher (both global and more specifically about e-learning)	Allows a very contextualized analysis
Understanding of TeSLA	Observing whether they really understand what is TeSLA and how it is supposed to be used is a prerequisite to analyse other issues
Pedagogical integration of TeSLA	Their feeling about the principles of use of the system, the different instruments, why did they use this/these instrument(s) in place of others …
Feedback interpretation	Their use of the numbers sent back by TeSLA. Are they a sufficient prove? Are they reliable?
Monitoring aspect	The surveillance induced by TeSLA, their eventual fear of "slides"
Defence against a decision coming from the system	What would they do if the system accuses them of cheating? Would they like to be added in the system to allow this defence?
Regulation of the judgement	The relation between the use of TeSLA and the internal regulation of institutions, the role allocated at the teacher in the management of TeSLA

of questions that can help the conversation to move forward (interviewees are not required to answer each of them if they find them uninteresting). Here are simplified versions of these grids both for teachers interviews and students' interviews (Tables 2 and 3).

Table 3 Learners' interview grid

Topic	Why approaching it?
Background of the learner (both global and more specifically about e-learning)	Allows a very contextualized analysis
Understanding of TeSLA	Observing whether they really understand what is TeSLA and how it is supposed to be used is a prerequisite to analyse other issues
Explanation of the use	Their feeling about the principles of use of the system, the different instruments, and their integration with the activities they are asked to do …
Monitoring aspect	The surveillance induced by TeSLA, their eventual fear of "slides" …
Defence against a decision coming from the system	What would they do if the system accuses them of cheating? Would they like to be added in the system to allow this defence?
Regulation of the judgement	How are the teachers supposed to deal with the feedbacks sent by TeSLA?

The information collected during the ethical fieldwork have been compared (where possible) with the observations from other partners, particularly the data collected in the surveys and focus groups organized to assess the pilots experiences, to widen our point of view.

3 Legal Preliminary Remarks

Privacy law concern TeSLA. Indeed, the General Data Protection Regulation[5] (hereafter the "GDPR") defines personal data as "*any information relating to an identified or identifiable natural person ('data subject'); an identifiable natural person is one who can be identified, directly or indirectly, in particular by reference to an identifier such as a name, an identification number, location data, an online identifier or to one or more factors specific to the physical, physiological, genetic, mental, economic, cultural or social identity of that natural person*".[6] TeSLA implies the processing of personal data when using face recognition, voice recognition and keystroke dynamics to verify students' identity during distance exams and pedagogical activities. One of the objectives is to develop strong privacy features that will prevent unnecessary identification of persons during the e-assessment. The project began under the rules of the Directive 95/46/EC[7] and continued under the GDPR which is applicable from 25 May 2018. For the purpose of this chapter, we only focus on the processing of face and voice recognition. Indeed, even if keystroke dynamics are personal data, they do not fall into the specific category of sensitive personal data (see Sect. 5.1.2 of this chapter).

In order to establish the responsibilities of each partner and the architecture of the system, a first step was to identify the data controller and the data processor.

Let us indicate that the *data controller* is the natural or legal person that determines the means and the purposes of the personal data processing.[8] When using TeSLA, the institutions are the solely responsible for the processing of personal data of their students.

On the other hand, the *data processor* is the natural or legal person, which processes personal data on behalf of the controller.[9] TeSLA is the third party, which develops the tools needed to perform the e-assessment. It will be the interface executing the *privacy filters* and sending alerts of fraud detection to the institution. The

---

[5]Regulation (EU) 2016/679 of the European Parliament and of the Council of 27 April 2016 on the protection of natural persons with regard to the processing of personal data and on the free movement of such data, and repealing Directive 95/46/EC (General Data Protection Regulation), OJ 2016, L 119/1, p. 1.

[6]Article 4.4 of the GDPR.

[7]Directive 95/46/EC of the European Parliament and of the Council of 24 October 1995 on the protection of individuals with regard to the processing of personal data and on the free movement of such data, OJ 1995, L 281, p. 0031.

[8]Article 4.7 of the GDPR.

[9]Article 4.8 of the GDPR.

privacy filters are the guarantees developed in the system to ensure the compliance with GDPR requirements.

Each institution willing to use TeSLA is in charge of the major part of the obligations provided by law and is the main contact for data subjects. The obligations are mainly:

- Getting the consent from students. The plug-in, in the university domain, verify that the student signed the Consent Form;
- The internal determination of who can have access to what data, and send these rules of access to the TeSLA system. The combination of the VLE (Virtual Learning Environment) and LTI (Learning Tools Interoperability) provider establish this privacy filter. The infrastructure has to be conceived in a way that limits access to the relevant teacher and only to the relevant data;
- The period of retention of personal data by the system;
- To keep the table of conversion between the TeSLA ID and the real identity of students. The TeSLA ID is a unique and secure identifier assigned to each student in TeSLA.

The processing by the data processor must be governed by a binding legal act that sets out the subject matter and duration of the processing, the nature and purpose of the processing, the type of personal data and the obligations of each party. The data processor can only act if the data controller provides documented instructions.[10]

The project uses a hybrid model of cloud. The TeSLA system has student's biometric data but only through the coded identity of the students. The institutions, on their servers, have the uncoded samples and the table of conversion between the TeSLA ID and the real identity of the students. The TIP (TeSLA Identity Provider) converts the identity of the student into a pseudonymized TeSLA ID. The main function is to reduce as much as technically possible the identification of unneeded information. TeSLA only works with pseudonymised data and it sends the coded data to the relevant institution only when there is an abnormality detected during the exam. Pseudonymisation means the processing of personal data in such a manner that the personal data can no longer be attributed to a specific data subject without the use of additional information. This additional information needs to be kept separately and be subject to technical and organisational measures to ensure that the personal data are not attributed to an identified or identifiable natural person. On the contrary, anonymization means that there is no possibility, with reasonable means, to identify a person.

Each instrument has its own database. It permits to separate the data (face recognition, voice recognition, and keystroke dynamics) in order to send only the relevant data to verify the cheating. The combination of TEP (TeSLA E-assessment Portal) and the RT (Reporting Tool) constitute service brokers that gather-forward requests and learner's data. Furthermore, the system is designed to facilitate deletion or anonymization of the data when it is no longer necessary to store them in a form permitting identification of the corresponding person. Another filter is applied over

[10]Articles 28.1 and 28.3 of the GDPR.

the results. If the instrument does not detect a dishonest behavior, the teacher will not have access to the audited data, but only to the score itself. Furthermore, both data controller and data processor are in charge of data accuracy. In case of incorrectness, they have to take reasonable steps to rectify or erase the concerned personal data.[11]

The data processor is liable for damages caused by processing which do not comply with the GDPR requirements. TeSLA will also be liable if it has acted outside or contrary to lawful instructions of the controller. Nevertheless, a data processor will be exempt from any liability if it proves that it is not responsible for the event giving rise to the damage.[12]

The GDPR considers also the situation where there is more than one controller or processor involved in the same processing. Each controller or processor will be held liable for the entire damage in order to ensure a full and effective compensation for the data subject. The person who has paid full compensation has the possibility to claim back from the other controllers or processors involved in the same processing the part of the compensation corresponding to their part of responsibility in the damage.[13] The division of responsibility is tackled after the full data subject's compensation.

4 Personal Data Processing Key Principles

In this section, the aim is not to explain all the legislation but to re-examine the general principles of protection in order to understand the next steps in the development process. Therefore, we focus on the obligation of transparency, purpose limitation, minimization principle, storage limitation and security. All these principles related to the processing of personal data ensure the lawfulness of each software development.

4.1 Lawfulness, Fairness and Transparency

Article 5.1, (a) of the GDPR provides that: "*Personal data shall be processed lawfully, fairly and in a transparent manner in relation to the data subject*".

The GDPR provides for six legitimate assumptions for data processing. Note that the TeSLA project uses biometric data, such as face and voice recognition tools. According to the GDPR, biometric data means "*personal data resulting from specific technical processing relating to the physical, physiological or behavioural characteristics of a natural person, which allow or confirm the unique identification of that natural person, such as facial images or dactyloscopic data*".[14]

---

[11]Article 5.1, d) of the GDPR.

[12]Articles 82.2 and 82.3 of the GDPR.

[13]Articles 82.4 and 82.5 of the GDPR.

[14]Article 4.14 of the GDPR.

These kinds of personal data fall in the special categories of personal data. Consequently, we identified the consent of the student as a way of legitimizing the processing, according to article 9 of the GDPR.

The obligation of transparency requires that the information about the recipient or category of recipients of the data must be disclosed to the data subject. It means that the student has to be informed about TeSLA as the data processor and has the right to access the data processing contract.[15]

4.2 Purpose Limitation

Article 5.2, (b) of the GDPR provides that: "*Personal data shall be collected for specified, explicit and legitimate purposes and not further processed in a manner that is incompatible with those purposes*". For example, biometric data can be useful to identify the user of the TeSLA system or to check if the user is cheating or not.[16] In addition, the GDPR prohibits further processing in a manner that is incompatible with the original purpose. To determine the lawfulness of a further processing, the GDPR establishes a list of factors that should be taken into account. Here are some examples of relevant factors (this list is not exhaustive[17]):

- The existence of a link between the original purpose and the new one,
- The context in which the data was collected,
- The nature of the data and the possible consequences of the processing,
- And the existence of safeguards such as encryption or pseudonymisation.

Concerning TeSLA, the purpose could be defined as follow: the personal data from students who gave their consent (face recognition, voice recognition and keystroke dynamics) will be collected and processed by the institution in order to certify the real identity of the student.

4.3 Minimization Principle

Article 5.1, (c) of the GDPR states that: "*Personal data shall be adequate, relevant and limited to what is necessary in relation to the purposes for which they are processed*".

To analyze the necessity to use biometrics data, we consider whether there are less privacy-invasive means to achieve the same result with the same efficacy and if the

[15]See article 12 of the GDPR.

[16]Notice that the traceability of the learner and the correlation between data collected are a process by themselves and need to respect all privacy legislation. Consequently, the student must be informed and consent.

[17]Recital 50 of the GDPR.

resulting loss of privacy for the students is proportional to any anticipated benefit.[18] To minimize the risk, we put in place one database for each instrument to avoid the use of a centralized database that could lead to a single point of failure, and we use the technique of pseudonymisation.[19]

Pseudonymisation means "*the processing of personal data in such a manner that the personal data can no longer be attributed to a specific data subject without the use of additional information, provided that such additional information is kept separately and is subject to technical and organization measures to ensure that the personal data are not attributed to an identified or identifiable natural person*".[20]

The GDPR encourages the use of this method because it is a technique to implement privacy by design and by default.[21] Indeed, the GDPR implements these two notions. It implies that the principles of data protection have to be taken into account during the elaboration and conception of the system. Furthermore, it may contribute to meeting the security obligations.[22]

We established that the TeSLA project would adopt a hybrid system, which would consist of having pseudonymized (or coded) data in the cloud and uncoded data in institution servers (data controller).

4.4 Storage Limitation

Article 5.1, (e) of the GDPR states that: "*Personal data shall be kept in a form which permits identification of data subjects for no longer than is necessary for the purposes for which the personal data are processed*". The retention period could be the time required to verify the fraud alert if any or the extinction of recourse by the student. For each type of collected data and in consideration of the relevant purposes, it is necessary to determine if the personal data needs to be stored or whether it can be deleted.

4.5 Security

Finally, article 5.1, (f) of the GDPR states that: "*Personal data shall be processed in a manner that ensures appropriate security of the personal data, including protection against unauthorised or unlawful processing and against accidental loss, destruction or damage, using appropriate technical or organisational measures*". In other words, personal data should be processed in a manner that ensures appropriate data security.

[18]Article 29 Data Protection Working Party-WP193 2012, p. 8.

[19]Belgian Commission for the protection of privacy 2008, pp. 14–16.

[20]Article 4.5 of the GDPR.

[21]Article 25 and Recital 58 of the GDPR.

[22]See article 6.4 (e) of the GDPR.

The data processor is also responsible for the security of the system, in particular for preventing unauthorized access to personal data used for processing such data. The security system should also prevent any illegal/unauthorized use of personal data. The GDPR imposes no specific measure.

According to article 32 of the GDPR, the data controller and data processor shall implement appropriate technical and organizational measures to ensure security of personal data, taking into account the state of the art, the costs of implementation, the type of personal data, and the potential risks. According to the GDPR, the data processor should evaluate the risks inherent in the processing such as accidental or unlawful destruction, loss, alteration, unauthorized disclosure of, or access to, personal data transmitted, stored or otherwise processed, which may lead, in particular, to physical, material or non-material damages. The GDPR gives some examples of security measures:

- Pseudonymisation and encryption of personal data;
- The ability to ensure the ongoing confidentiality, integrity, availability, and resilience of processing systems and services;
- The ability to restore the availability and access to personal data in a timely manner in the event of a physical or technical incident;
- A process for regularly testing, assessing and evaluating the effectiveness of technical and organizational measures for ensure the security of the processing.

5 Consent

In the TeSLA system, each student has to consent to the processing of his/her biometrics data. In order to be legally valid, the consent needs to fulfil legal obligations. These are mainly the obtention of a free, specific, informed and unambiguous consent. Ethical considerations strengthen and reinforce legal obligations arising from the GDPR. After describing each of the conditions, we will expose their implementations in the TeSLA system.

5.1 Legal Considerations

5.1.1 Definition and Objectives

The GDPR provides that: "*Processing shall be lawful only if and to the extent that at least one of the following applies: (a) the data subject has given consent to the processing of his or her personal data for one or more specific purposes*". The consent of the data subject is: "*any freely given, specific, informed and unambiguous indication of the data subject's wishes by which he or she, by a statement or by a clear affirmative action, signifies agreement to the processing of personal data relating to him* or her".

The GDPR intends to stop the abusive use of uninformed and uncertain consent. In order to ensure a true reflection of individual autonomy, the GDPR has strengthened its requirements.[23] Consent must now be free, specific, informed and unambiguous.[24]

Consent is also enshrined in Convention 108+. This is currently the only international legislation on personal data protection.[25] Having in mind the increase and globalization of the use of personal data as well as the vast deployment of technologies, the Convention 108 wants to give more power to citizens. As stated in the Explanatory Report: "*A major objective of the Convention is to put individuals in a position to know about, to understand and to control the processing of their personal data by others*".[26]

5.1.2 Conditions

The data controller must therefore prove compliance with four conditions.

Free Consent

The objective is that the data subject has a real choice to consent. To ensure a real expression of willingness, three elements must be taken into account: the imbalance of power, the granularity and the detriment.[27] Recital 43 states that: "*Consent should cover all processing activities carried out for the same purpose or purposes. When the processing has multiple purposes, consent should be given for all of them*". It is up to the controller to determine and specify the different purposes pursued and to allow an opt-in of the data subject for each one. Furthermore, if the data subject feels compelled, afraid to face negative consequences in case of refusal or to be affected by any detriment, the consent will not be considered as freely given. As stated in Recital 42, "*consent should not be regarded as freely given if the data subject has no genuine or free choice or is unable to refuse or withdraw consent without detriment.*"[28]

Specific Consent

Secondly, consent must be specific. In case of multiple purposes for several data processing, the data controller must receive a separate consent for each purpose.

Informed Consent

Thirdly, the consent must be informed. Articles 13 and 14 of the GDPR list the information that the controller must provide to the data subject. At least, it concerns

[23] de Terwangne et al. [4], p. 306.

[24] Article 4.11 of the GDPR. See also Article 29 Data Protection Working Party-WP259 2018.

[25] Convention for the Protection of Individuals with regard to Automatic Processing of Personal Data, signed in Strasbourg the 28 January 1981, ETS No.108.

[26] Explanatory Report, p. 2, pt. 10. Available at: https://rm.coe.int/16808ac91a.

[27] Article 29 Data Protection Working Party-WP259 2018, p. 5 and seq.

[28] Article 29 Data Protection Working Party-WP259 2018, pp. 5–10.

information about the data controller's identity, the purpose of each of the processing operations for which consent is needed, the collected and processed data, the existence of the rights for the data subject, notably the right to withdraw the consent and the transfers of personal data outside the European Union if any.

The objective is that the person who consents must understand why, how and by whom his or her personal data will be collected and used.[29] The information must be concise, transparent, intelligible and easily accessible. The information has also to be given in a clear and plain language, avoiding specialist terminology. Therefore, the data controller should adopt a user-centric approach and to identify the "audience" and ensure that the information is understandable by an average member of this audience.[30] As best practice, the Article 29 Working Party encourages the data controller to provide an easy access to the information related to the processing of personal data. The data controller has also to take into consideration the possibility to provide express reminders to data subject about the information notice.[31]

Unambiguous Consent

Fourthly, the consent must be unambiguous and must be the result of a declaration by the data subject or a clear positive act. Consent is therefore not presumed. In addition to the quality requirements, consent may be withdrawn at any time and the GDPR specifies that it must be as easy to withdraw consent as to give it.[32]

Implementation in TeSLA

In TeSLA, we process biometrics data. As it is information related to an identified or identifiable person, biometrics data are personal data in the meaning of the GDPR. In addition, biometrics data enter into a particular category of personal data: sensitive personal data. Sensitive data is defined as data relating to specific information on the data subject. These are mainly listed in Article 9 of the GDPR and include the personal data revealing racial or ethnic origin, political opinions, religious or philosophical beliefs, or trade union membership, and the processing of genetic data, biometric data for the purpose of uniquely identifying a natural person, data concerning health or data concerning a natural person's sex life or sexual orientation.

For the processing of this special category of personal data—which is in principle prohibited—the GDPR requires an explicit consent. It implies that the data subject must give an express statement of consent.[33] As examples, the Article 29 Working Party recognizes the use of an electronic form, the sending of email or using electronic signature. The explicit consent implies for the data controller to be careful when providing information to the data subject. The institution willing to use TeSLA, as

[29] Article 29 Data Protection Working Party-WP259 2018, p. 13.

[30] Article 29 Data Protection Working Party-WP259 2018, p. 14. The Article 29 Working Party gives some methods to target the audience, such as panels, readability testing and dialogue with concerned groups; Article 29 Data Protection Working Party-WP260 2018, p. 7.

[31] Article 29 Data Protection Working Party-WP260 2018, p. 18.

[32] Article 7.3 of the GDPR.

[33] Article 9.2, (a) of the GDPR.

data controller, has to ensure that all the necessary information are given to the data subject to understand the processing and the personal data used, without submerging the data subject with information.[34]

TeSLA's particularity is the collection of both personal data such as students' surnames and first names and special categories of personal data such as face and voice recognition. In view of the factual circumstances, consent seems to be the only basis for legitimization.

The consent form elaborated during the project includes the following information: the identity of the controller and of the processor, goals of the project, personal data collected, purposes, data retention, privacy policy signature and rights of the data subjects. We have been vigilant in balancing the obligation of transparency with the obligation to obtain valid consent and the requirement not to inundate the person concerned with information.[35] We suggest that the period of conservation is the duration of the project TeSLA. In case of duration of TeSLA over the end of the project, the retention period for both types of personal data could be the time required to verify the alert of fraud.

This consent form must be completed and approved by each student. This form is offered to them by a visual display on the user's computer screen at the beginning of the service's use to respect the timing for provision of information.[36] This method has also the advantage to include the information in one single document in one single place.[37] This is intended to facilitate the accessibility and the communication of information by ensuring that the data subject has at least once taken note of all information relating to the processing of his or her personal data.[38]

The consent is also specific because the student receives a clear information about the privacy policy, which is separate from other text or activities and the student has to click in order to give his/her consent. In the text of the consent form, we avoided to use unclear terms like "may', "might", "possible" as it is pointed out by the Article 29 Working Party. These terms are subject to interpretation and do not allow the data subject to have a concrete understanding of what is done with his or her data.[39]

5.2 *Ethical Considerations*

In order to enrich the definition of consent given in the law, we will take up one by one the different characteristics it cites: consent must be free, specific, informed and unambiguous.

[34] Article 29 Data Protection Working Party-WP259 2018, p. 18.

[35] Article 29 Data Protection Working Party-WP260 2018, p. 18.

[36] Article 13 GDPR and Article 29 Data Protection Working Party-WP260 2018, pp. 14–16.

[37] Article 29 Data Protection Working Party-WP260 2018, p. 11.

[38] Article 29 Data Protection Working Party-WP260 2018, p. 11.

[39] Article 29 Data Protection Working Party-WP260 2018, p. 9.

The necessity of freedom implies that a request for consent to provide private data must include a real possibility of alternative in case of refusal. This may stand to reason, but what would be the point of asking for a student's consent to provide biometric data if he or she were unable to take the test if he or she refused?

Ethical analysis makes it possible to enrich the contribution of Law at this level. As we have said, ethics can help to refine our response to technical partners when the law remains somewhat vague. Indeed, the latter informs us of the obligation to provide a viable alternative but remains vague as to the definition of what constitutes a real alternative able to guarantee the free aspect of a consent.

What is actually the situation for TeSLA? The system asks the student if he/she wishes to be authenticated by the system using the modalities chosen upstream by the professor. It is therefore a question of agreeing on a combination of biometric/textual analysis instruments, in other words on a set of samples of data to be made available. TeSLA, as it was originally conceived, provides two possibilities:

- The teacher provides the possibility of an examination completely outside the context of TeSLA (for example, a classic face-to-face examination).
- The teacher allows the student to take the exam with another combination of TeSLA instruments.

We noted that this approach derived by the technical partners had certain shortcomings.

First, it consists in delegating a large part of the very functioning of the system to the teacher. They are faced with a system that would require a fairly substantial logistical effort in cases of massive refusals, whereas the purpose of the system is precisely to avoid the organizational burdens caused by face-to-face examinations. This aspect needs to be clarified for reasons of transparency towards future TeSLA user institutions. Indeed, it is important that they know what they will incur if consent is not given by the student. One thing that can be done at this level is to further clarify these aspects (the necessity to develop an alternative way of exam and the limitations of the internal alternative options) in the description of TeSLA to future customers, as well as clear warnings in the operating instructions. For the institutions, it is a question of being able to perceive how the system could be implemented in their own regulation.

Second, it is clear from our ethical investigation that a significant proportion of the students are unconvinced by the alternative system embedded in TeSLA. They believe that if they have little confidence in the way the system processes their data or in the reliability of the instruments, switching from one combination of instruments to another will not necessarily reassure them. Concerning this problem, we had suggested to the partners to standardize and clarify the uses that would be made by teachers by giving indications on the following aspects: Who can access to which data? For which purpose? How do TeSLA's instruments process data? These indications go beyond the simple question of consent but they would make consent more informed if they were put forward at the time of the request.

Finally, this second way of considering an alternative is ethically questionable because it does not guarantee equivalence criteria between students. Indeed, on which legitimacy is based an examination that has been passed by students with different monitoring criteria? Various abuses are conceivable: students could consciously choose the method that makes it easier for them to cheat, or on the other hand they could feel aggrieved by having to take an exam with a heavier supervisory method than their classmates. This would impact the very legitimacy of the past examination, and by extension the reputation of the institution that proposes it.

The specificity of consent implies that the nature of the data and the way in which the data covered by it are used must be explicitly indicated. We stressed the importance of clearly defining how feedback is treated (more explanation about feedbacks in the next section).

The question of feedback and its processing will be dealt with more precision in the seventh section, but it is important to specify that the definition of this feedback processing affects in particular the quality of the request for consent. In this respect, we have insisted on the need to extend this concern for clarity in the request for consent to the way feedback are exposed and used (and not only basic biometric samples).

Regarding the need to have an informed consent, we fed the reflection with findings of our survey about students' concerns about their understanding of the system. These must be taken into account when writing the consent request in order to make it both user-centred and easier to access. Highlights from our findings include:

- The activation period of biometric instruments.
- The type of data that is collected (picture, voice recording…).
- The way the data is collected (the continuous aspect).
- And so, certain behaviors can be avoided to allow the instruments to correctly collect the necessary and sufficient data.

Consent is therefore an important issue to be addressed because it conditions the very use of the system. Indeed, the very purpose of TeSLA is to generate trust between professors and students. To do this, it uses biometric recognition instruments to identify students remotely. However, these instruments do not provide confidence in a purely mechanical or automatic way through their use. It is worth considering the potential drift of such a system if it claims to bring confidence to users through a strong and widespread monitoring system. Working on the transparency of consent makes it possible to avoid justifying ever greater and deregulated surveillance based on the need for trust.

However, a good request for consent is not limited to a transparent list of the private data collected and the work done with it. The system itself must operate clearly and unambiguously in order to submit a meaningful request for consent. This principle probably applies to any system requiring private data, but is all the more important for TeSLA, whose biometric instruments provide feedback that teachers must interpret.

6 Transparency

The transparency obligation is a core requirement of the GDPR. After explaining this principle and its objectives, we focus on how this obligation imposes to the software developers the openness of their system. Finally, ethical considerations give a good overview of possible difficulties in users' understanding and therefore of the transparency efforts needed to guarantee the right to information.

6.1 Legal Considerations

6.1.1 Principle and Objectives

The objective of the transparency obligation is to strengthen the citizens' control over the processing of their personal data. The transparency of the data controller permits data subjects to understand the use of their information. In addition, this key principle allows individuals to effectively exercise the rights granted by the GDPR.[40]

To respect the obligations for Member States to process the personal data fairly, we assisted to the emergence of long and unintelligible general conditions of use. Needless to say, every citizen accepted them without understanding, or even knowing, what he/she was engaging with, without knowing that he/she had just exchanged aspects of private life for a service.

The GDPR intends to reinforce the transparency requirement. It is now clearly written in the text that transparency is an obligation for the data controller. Article 5 states that: "*Personal data shall be processed lawfully, fairly and in a transparent manner in relation to the data subject*".

Recital 39 adds: "*Any processing of personal data should be lawful and fair. It should be transparent to natural persons that personal data concerning them are collected, used, consulted or otherwise processed and to what extent the personal data are or will be processed. The principle of transparency requires that any information and communication relating to the processing of those personal data be easily accessible and easy to understand, and that clear and plain language be used. That principle concerns, in particular, information to the data subjects on the identity of the controller and the purposes of the processing and further information to ensure fair and transparent processing in respect of the natural persons concerned and their right to obtain confirmation and communication of personal data concerning them which are being processed. Natural persons should be made aware of risks, rules, safeguards and rights in relation to the processing of personal data and how to exercise their rights in relation to such processing. In particular, the specific purposes for which personal data are processed should be explicit and legitimate and determined at the time of the collection of the personal data. The personal data should be adequate, relevant and limited to what is necessary for the purposes for which*

[40] de Terwangne [5], pp. 90–94.

they are processed. This requires, in particular, ensuring that the period for which the personal data are stored is limited to a strict minimum. Personal data should be processed only if the purpose of the processing could not reasonably be fulfilled by other means. In order to ensure that the personal data are not kept longer than necessary, time limits should be established by the controller for erasure or for a periodic review. Every reasonable step should be taken to ensure that personal data which are inaccurate are rectified or deleted. Personal data should be processed in a manner that ensures appropriate security and confidentiality of the personal data, including for preventing unauthorised access to or use of personal data and the equipment used for the processing."

Transparency promotes and develops the empowerment of citizens and is interpreted as a mean of enhancing the privacy of users and facilitating the exercise of their rights.[41] Furthermore, the obligation of transparency is intrinsically linked to trust. As stated by Article 29 Working Party, "*It is about engendering trust in the processes which affect the citizen by enabling them to understand and, in necessary, challenge the processes*".[42]

6.1.2 Transparency and Openness

The obligation of transparency is often presented as the cornerstone of the data subject's right to information and a duty of communication of the controller. It includes:

- Providing information to the data subject in a clear and plain language, at the beginning and all along the processing of personal data. In addition, the data controller must adopt a proactive approach for providing information and not waiting for an intervention of the data subject[43] (see as detailed above and Fig. 1.).
- The data controller has to give a permanent access to privacy information, even if there is no modification in the terms and conditions. In addition, the Article 29 Working Party encourages the data controller to provide express reminders, from time to time, to the data subject about the privacy notice, and where this is accessible[44] (see as detailed above).

However, the obligation of transparency indirectly entails other obligations for the controller and a certain openness of the system in order to permit the verification of compliance with the GDPR.

The duty of transparency has two faces. On the one hand, the data subject have some rights, such as the right to be informed, the right to have access to some information and the right to erasure. On the other hand, the data controller has to prepare himself/herself by technical and organisational measures, to answer to the

[41] Article 29 Data Protection Working Party-WP260 2018, p. 4.

[42] Article 29 Data Protection Working Party-WP260 2018, p. 4.

[43] Article 29 Data Protection Working Party-WP260 2018, p. 6 and p. 18.

[44] Article 29 Data Protection Working Party-WP260 2018, p. 18.

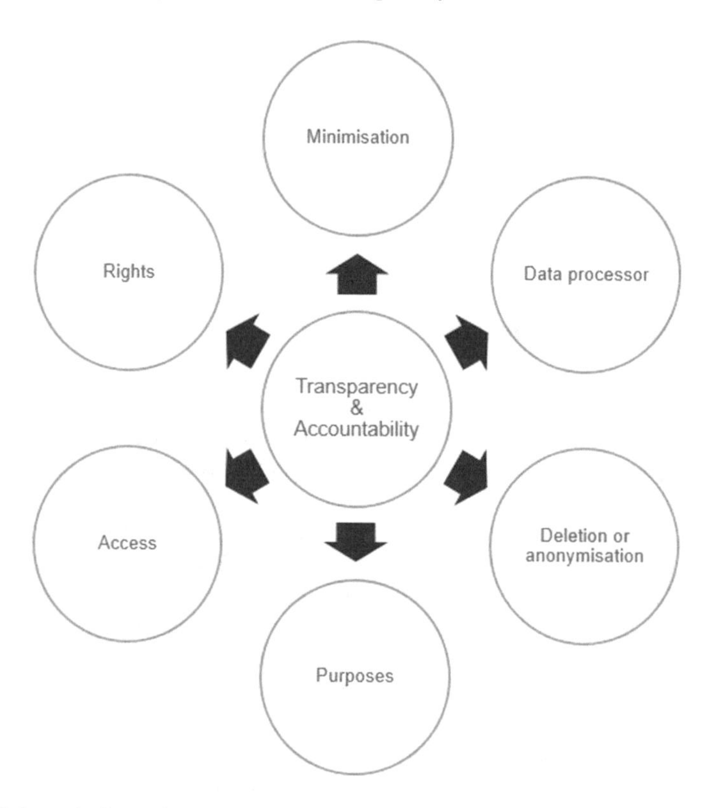

Fig. 1 Information's requirements

data subject. It implies de facto to open the system to the data subject. These two faces are illustrated by Table 4.

This duty of transparency could be a constraint on the shoulders of the data controller, who must identify the type of persons concerned and the most intelligible way of fulfilling the obligation to inform. However, it could be also a decision-making tool for the data controller in the use and management of personal data. In order to

Table 4 Right for the data subject and corresponding obligations for the data controller

Rights for data subject	Duty for data controller
Right of access	To give a secure and online access to one's personal data
Right to be informed	Transparency about the logic and how the system works
Right to have a human intervention and contest the decision	• Procedural rules • To receive a clear and comprehensive feedback from the tools
Right to erasure	To facilitate the deletion in all servers

fulfil its obligation of information, the data controller has to understand how the system works. Thereby, the data controller must know the limits of the tools used and can therefore take an enlightened decision on the basis of the outputs of the TeSLA system. Transparency then becomes, not only a legal constraint, but a real contribution in the academic procedure of decision-making in the case of suspicion of cheating by a student.

6.2 Ethical Considerations

The GDPR gives the right to information as a fundamental prerogative, but this dimension needs to be further explored ethically. This will then make it possible to provide a richer contextual dimension, in a way to relate the standard legal definition of this right into a practical reality.

The GDPR's definition of the right to information is not totally satisfactory because it cannot (and this is normal for a legal definition) take into account the diversity of human behaviors. Indeed, the right to information implies a need of understanding on the side of stakeholders. Our ethical fieldwork gives us a good overview of possible difficulties in users' understanding and therefore of the transparency efforts needed to guarantee the right to information.

The comprehension problems demonstrated by our survey participants are diverse. On the student side, these concerns lie both in the understanding of biometric instruments and the architecture of the system and how it affects the teacher's judgment behind the scenes.

Although many factors may have influenced this lack of understanding regarding how the TeSLA instruments work (the lack of explanations received upstream, the sometimes confusing context that characterizes a test phase…), it is worrying to note how difficult it is for some to access the basic principles of these instruments. The capabilities of these instruments were sometimes overestimated, considered at a level that goes beyond simple authentication. For example, many students believed that the face recognition instrument was able to interpret their actions which is way more than only matching enrolment biometric samples with data samples collected during assessment activities. Some even thought that someone was watching them live via webcam. These interpretations led these students to ask themselves what behaviors were expected of them (should we stay in front of the screen during the entire activity or not? Should we remove the presence of sheets of paper on the desk?). These uncertainties potentially cause a form of anxiety about the system. Taking these understanding shortcomings into account is necessary to ensure the right to be informed in the context of TeSLA.

These aspects concern the understanding of the algorithmic underlying the system (the software attached to the instruments are black boxes from the TeSLA system perspectives), but the opacity that potentially blurs the vision of student users is not limited to them. The feedback that these instruments send back to the teachers is also a veil of opacity towards the students. Legal and ethical issues regarding the feedback will be addressed in the following section. We would like to stress the potential lack of access by students to these data (cfr: the "right of access"). Indeed, these feedback data are subject to the right of information. Students are seeking an opportunity to see how professors manage them and whether they are able to justify their possible charges with supporting evidence.

Another legal aspect that also concerned our research was the right to a human intervention and to challenge the decision taken on the basis of TeSLA's indications. The students' concerns, or even defeatism, about the difficulty to challenge a decision that would have been taken on the basis of the system led us to highlight the importance of implementing TeSLA according to the rules specific to each university. Of course, we could not study all the possibilities of interaction between TeSLA and internal regulations on a case-by-case basis, but we are contributing to the implementation of an instruction manual. This will contain general principles of use allowing universities to easily set up a clear modus operandi. By clarifying TeSLA's role in the teacher's judgment of his or her students, the notice should enable each student to challenge the charges against him or her and hold the teacher accountable if necessary.

The issue of transparency is crucial because it highlights a common dilemma in the development of such systems. On the one hand, it must be easily accessible in order to allow as many people as possible to use it (and this aim of "democratizing" university courses is an argument claimed by TeSLA project) and on the other hand the use of technical tools (here biometric instruments) necessarily leads to a certain complexity of the system. This complexity can hinder users with the least technological literacy (both from the teacher and student point of view).

In this case, the ethical approach raises questions that go beyond the scope of the Law. Beyond the processing of private data, there is the question of the usability of the system and its potentially discriminatory nature. Ethically speaking, it is obvious that it is necessary to ensure the greatest possible accessibility to it, whatever the level of technological literacy of the person. This is of course also linked to the question of trust that we have previously mentioned: how the system can claim to generate trust if it excludes some of the users to whom it potentially addresses? We have tried to favor an approach that would not simply consider that "people have to adapt to technologies, period" by the comprehension of their fears.

The use of TeSLA by teachers must be understood in terms of "clues". As TeSLA's feedbacks do not constitute self-evident evidence, a possible accusation of cheating must be based on more solid materials than the mere presence of a litigious number returned by one of the biometric instruments.

As a result, TeSLA interacts with the cheating regulations previously in place in each university. On the institutional side, it is important to understand what TeSLA produces before integrating it. The latter already includes, in a way, a specific "definition" of what is considered as cheating. Authentication using face recognition implies, for example, that a third party cannot take the place of the student "in front of the screen" for the duration of the evaluation activity. However, TeSLA does not regulate the interactions that these two people would have outside the duration of the assessment, or even during the assessment (if speech recognition is not enabled, there is nothing to prevent them from talking during the activity).

Before any integration of TeSLA, the following two questions must therefore be asked by the institutions:

- What are the typical fraud situations facing our institution? After detailing them field by field (the way of cheating is not the same depending on the field of study, of course), institutions must look at which of these situations TeSLA potentially thwarts or not.
- Are the detections made by TeSLA instruments relevant to our institution? It is in fact the opposite question. The system user guide should contain a detailed description of each instrument and what it is capable of reporting. On the basis of this, the institutions will be able to see which instrument is potentially useful for them or not.

7 Feedback

The aims of this section are to explain the legal framework surrounding automated decision-making and the related rights for the students, in particular the right to obtain, on request, knowledge of the reasoning underlying data processing. A particular point of attention is the possibility of using the results given by the TeSLA system to profile students that are repeatedly cheating. It is also important for teachers using the software to understand the delivered indicators as well as how to interpret and integrate them into internal rules of each institution.

7.1 Definition

As showed in Fig. 2, Feedback consists of confidence indexes comparing enrolment sample(s) (of biometric or text-based data) with samples collected during an activity for each learner. The feedback is classified by instrument (a tab by instrument) and showed chronologically for each activity in the case of biometric instruments (one index is sent for each sample collected every 30 s/1 min more or less).

Each metric included in tables appear in green, yellow or red according to their value in comparison with a threshold. This aims to facilitate teachers' interpretation of these abstract numbers, giving a simple indication. However, even if the teachers are supposed to interpret freely these numbers, the thresholds and the coloration of feedback will influence his/her later decision and perception of learners' work.

Evaluation result	Start date	End date	Audit
0.0%	febrer 15è 2019, 3:18:12 pm	febrer 15è 2019, 3:18:18 pm	More info
59.0%	febrer 15è 2019, 3:18:15 pm	febrer 15è 2019, 3:18:19 pm	
67.0%	febrer 15è 2019, 3:18:18 pm	febrer 15è 2019, 3:18:21 pm	
72.3%	febrer 15è 2019, 3:18:21 pm	febrer 15è 2019, 3:18:23 pm	
69.8%	febrer 15è 2019, 3:18:24 pm	febrer 15è 2019, 3:18:54 pm	
73.1%	febrer 15è 2019, 3:18:27 pm	febrer 15è 2019, 3:18:56 pm	
72.2%	febrer 15è 2019, 3:18:30 pm	febrer 15è 2019, 3:18:57 pm	
69.6%	febrer 15è 2019, 3:18:33 pm	febrer 15è 2019, 3:18:59 pm	
70.6%	febrer 15è 2019, 3:18:36 pm	febrer 15è 2019, 3:19:01 pm	
71.3%	febrer 15è 2019, 3:18:39 pm	febrer 15è 2019, 3:19:02 pm	
70.7%	febrer 15è 2019, 3:18:42 pm	febrer 15è 2019, 3:19:03 pm	
71.4%	febrer 15è 2019, 3:18:45 pm	febrer 15è 2019, 3:19:04 pm	
70.5%	febrer 15è 2019, 3:18:48 pm	febrer 15è 2019, 3:19:05 pm	
70.3%	febrer 15è 2019, 3:18:51 pm	febrer 15è 2019, 3:19:06 pm	
70.3%	febrer 15è 2019, 3:18:54 pm	febrer 15è 2019, 3:19:06 pm	
70.7%	febrer 15è 2019, 3:18:57 pm	febrer 15è 2019, 3:19:07 pm	
70.0%	febrer 15è 2019, 3:19:00 pm	febrer 15è 2019, 3:19:08 pm	
69.4%	febrer 15è 2019, 3:19:03 pm	febrer 15è 2019, 3:19:09 pm	
68.7%	febrer 15è 2019, 3:19:06 pm	febrer 15è 2019, 3:19:10 pm	
71.8%	febrer 15è 2019, 3:19:09 pm	febrer 15è 2019, 3:19:10 pm	

Fig. 2 Screenshot of Feedback interface

7.2 Legal Considerations

The result given by TeSLA is a personal data in the sense that it is information relating to an identifiable person.[45] Consequently, all regulations relating to the protection of personal data must be respected.

[45] Article 4.1 of the GDPR.

7.2.1 Legal Guarantees

The Article 29 Working Party defines the concept of automated decision-making as "*the ability to make decisions by technological means without human involvement. Automated decisions can be made with or without profiling*".[46]

The Article 29 Working Party is sensitive to a decision that has "*the potential to significantly influence the circumstances, behaviour or choices if the individuals concerned*".[47]

If by an algorithmic method the TeSLA system is able to determine a result of cheating, or at least a percentage of probability, of non-matching between the exam and the enrolment process, it seems that this can potentially affect significantly the student's situation in that it may lead to a refusal to grant the diploma or certificate.

In the TeSLA project, the collection and use of personal data from the learners are based on their freely given, specific and informed consent. The consent, that needs to be confirmed by an express statement and a positive action, is an exception that permits the use of automated decision-making. There are several safeguards. Firstly, the data controller has to inform properly the data subject about the existence of an automated decision-making.[48] They have the right to be informed about the logic involved, the explanation of the mechanism and how the system works. It is not mandatory to enter into details on the functioning of algorithms.[49] These information must be provided by the data controller when the data are collected. As a consequence, it does not concern information about how the decision was reached by the system in a concrete situation.[50]

Secondly, students have the right to obtain a human intervention, to express their opinion and to contest the decision.[51] This is of a crucial importance considering the fact that special categories of personal data from students are used in the project. Recital 71 of the GDPR specifies that they need to have the possibility to obtain an explanation of the decision reached. It goes beyond a simple right of information as it requires to give an *ex post* information on the concrete decision. It implies for the data controller to understand and to be able to explain in a comprehensive way the algorithmic functioning of the system.[52] In the context of TeSLA, the teacher has to receive a clear and comprehensive feedback from the tools.

The possibility to obtain an explanation of the decision reached is not mandatory as it is in a recital and not in the article itself, but permits a real and meaningful possibility to express an opinion and to decide to contest the decision or not. The

---

[46] Article 29 Working Party-WP251 2017, p. 8.

[47] Article 29 Working Party-WP251 2017.

[48] See articles 13 and 14 GDPR

[49] Article 29 Working Party-WP251 2017, p. 14.

[50] Wachter et al. [8], pp. 82–83.

[51] Article 22.3 of the GDPR.

[52] Wachter et al. [8], pp. 92–93.

Article 29 Working Party insists on the role of the controller in the transparency of the processing.[53]

Thirdly, the data controller must ensure that someone who has the authority and ability to remove the decision will review the automated decision.[54]

Fourthly, the Article 29 Working Party advises the data controller to carry out frequent assessments on the personal data collected and used. The curacy of the personal data and the quality of the tools used for the processing are core elements of the lawfulness of the system. This is particularly relevant when special categories of personal data, such as biometrics, are processed. The data controller, with the help of the data processor, should not only check the quality and prevent errors or inaccuracies at the time of the collection and the technical implementation of the system but in a continuous way as long as the data is not deleted or anonymized.[55]

Beside the right not to be subject to and automated decision, the right of access for the student includes the right to know "*the existence of automated decision-making (…) and, at least in those cases, meaningful information about the logic involved, as well as the significance and the envisaged consequences of such processing for the data subject*".[56] It implies for the data controller to understand and to be able to explain in a comprehensive way the algorithmic functioning of the system but is not a plenary algorithmic transparency.[57]

In the TeSLA context, the teacher has to receive a clear and comprehensive feedback from the tools. So there is a right for the students to know the data being processed, the criteria and the importance of each criteria but it is not a real full algorithmic transparency because we need also to take into account potential intellectual property rights on the software and trade secret on algorithms.[58]

The modernization of the Convention 108+ gives the right to obtain, on request, knowledge of the reasoning underlying data processing where the results of such processing are applied to him or her.[59]

In conclusion, if the teacher or the institutional committee have suspicions, it is important that they are able to explain to the students how the system works and to give, on student's request, knowledge of the reasoning underlying the data processing and the possibility to obtain an explanation of the decision reached.

To respect the right to rectification and to object, it is also important that the student can demonstrate that his or her behaviour was not optimal for the functioning of the system but it is not for that much a case of fraud.

---

[53] Article 29 Working Party-WP251 2017.

[54] Article 29 Working Party-WP251 2017, p. 10.

[55] Article 29 Working Party-WP251 2017, pp. 16–17.

[56] Article 13.2 of the GDPR.

[57] See Wachter et al. [8].

[58] de Terwangne [3], p. 107.

[59] Article 9.1 (c) Convention 108+.

7.2.2 Feedback and Profiling

The aim of this subsection is to consider the possibility to use the feedback given by TeSLA system for profiling activities.

According to article 4.4 of the GDPR, profiling means "*any form of automated processing of personal data consisting of the use of personal data to evaluate certain personal aspects relating to a natural person, in particular to analyse or predict aspects concerning that natural person's performance at work, economic situation, health, personal preferences, interests, reliability, behaviour, location or movements*".

If the data controller wants to collect the repeated cases of fraud from the same student with TeSLA (the biometrics data will be used to certify the identity of the student), there is no explicit and clear mention of profiling or collection of personal data in order to combine them with others to establish a certain profile of a student in the consent form. Furthermore, it is important to keep in mind that the consent can never legitimate a disproportionate processing.[60]

We can take a second hypothesis: If the envisaged activity is the establishment of blacklists[61] with identities of students that are considered as cheaters, there is no prohibition as such for that. However, these listing need to be done in accordance with the GDPR; transparency, data minimization, exercise of the right of access, information to the data subject on the fact that he/she is on the list, limited time of conservation, accuracy and mechanisms to avoid errors in the identification of students included and errors in the information mentioned and a secured access.[62]

Nevertheless, if the database needs to contain also biometrics data to prove the identity of the learner, there is no explicit consent from the students. Consequently, the data controller has no ground to legitimate such activity. Moreover, this is subject to the condition that the processing of personal data is not contrary to the exercise of the fundamental right to education and therefore could not be regarded as unlawful.[63]

7.3 Ethical Considerations

Addressing the feedback system is important because we do not know in advance how teachers will use this feature. This raises questions in terms of fair treatment of students. How can an accusation of cheating be justified in front of them if the

[60] Belgian Commission for the protection of privacy 2008, p. 10.

[61] The Article 29 Working Party defines a blacklist as: "*the collection and dissemination of specific information relating to a specific group of persons, which is compiled to specific criteria according to the kind of blacklist in question, which generally implies ad-verse and prejudicial effects for the individuals included thereon and which may be discriminate against a group of people by barring them access to a specific service or harming their reputation*"; Article 29 Working Party-WP65 2002, pp. 2–3.

[62] See article 5 and Recitals 39 and 50 of the GDPR. See also, Article 29 Working Party-WP65 2002, pp. 8–12.

[63] Burton and Poullet [2], p. 109.

teacher is not able to explain how he used the system? How can we avoid a totally hazardous use of the system that could produce uncontrollable effects of judgement? This is a major issue regarding the second aspect cited by the legal approach: the right to have a human intervention.

One of the topics we discussed with the professors during our ethical survey was their understanding and use of these index tables. Five out eight teachers found the feedback hard to understand and had difficulties to answer the following questions: What does each number mean? How are you supposed to interpret them concretely? What does the green/yellow/red coloration of the feedback really mean? Their answers were often unclear, and based largely on the idea that this has to be judged on a case-by-case basis. Some others, however, complained head-on about the way feedback was presented in the interface. The comments of some professors even suggested that they would use this feedback in a way that goes beyond the framework initially imagined in the project, for example by developing a long-term student profiling based on the confidence indexes returned by the TeSLA instruments. Then, it was necessary to intervene ethically on this aspect of the system.

A dilemma has arisen in the face of this observation: what should we do to make the use of feedback less ambiguous and what is the status that should be attributed to them? Should they have decision-making and justification powers or are they just indicators that need to be informed by other evidence?

This dilemma is materialized in the way trust indicators are presented in the teacher's interface. These are exposed in a very raw way, as we have explained. However, some professors imagined that they would actually be presented in a more aggregated and summarized form, allowing for more immediate use and requiring less interpretation work. The advantage of this second way to perceive feedback would be to reduce the random aspect of teachers' understanding of the indexes. On the other hand, this version of the system would tend to "pre-decide" who is a cheating student and who is not in the teacher's place, and so contribute to an even more automated decision process. It is to avoid this pitfall and leave each institution free to decide how to use the system that the project has kept a very rough exposure of the data.

However, it is our ethical responsibility to insist on the limits of these indexes (that they must be assisted with other means to justify a suspicion of fraud) and to give advice on the use of these index tables. These should help to answer the following questions: What to consider when there is a whole spot with numbers written in red on the interface (which would correspond to a long while without a correct matching)? What could be the reasons for red numbers outside the attempt to mislead the system (e.g. a non-optimal biometric data collection environment)? On the other hand, which cheating scenarios seem to be difficult to deal with via TeSLA?

TeSLA's feedback raises the question of how to interpret the results returned by biometric instruments in the context of continuous authentication. Indeed, the specificity of TeSLA is that biometrics is not used from time to time (for example, in an airport to authenticate passengers when crossing the border) but throughout an activity. Biometrics, in the context of TeSLA, goes beyond its usual role of authentication, in fact: student behaviour can be attributed to certain feedbacks. What does a negative result mean? That the student was replaced? That someone else passed in front of the webcam field? That the student went away for a few moments?

This diversion from the more "traditional" uses of biometrics raises many questions. Such as the fact that use of biometrics in decision-making could lead to random and arbitrary forms of decisions: it is difficult to know how professors will deal with such feedbacks if they are not provided with a minimum of clarification.

In this regard, it should however be noted that an argument in favor of the project is its flexibility and the fact that it does not impose a way of using it. The system provides "clues" that the student may have cheated (the teacher must then still prove it with means that exceed TeSLA). The feedbacks do not therefore say head-on "this student cheated", teachers are free to interpret feedbacks differently depending on the context. Nonetheless, this argument in favor of TeSLA may mask the potential pitfalls of a completely deregulated use. If no minimal standardization of the use of TeSLA feedback (e.g. through a user manual for teachers) is developed, how can students trust the decisions that will be made using the system?

8 Conclusions

TeSLA involves the processing of personal data when using face recognition, voice recognition and keystroke dynamics to verify student's identity during distance examinations and pedagogical activities. In this chapter, we focused on three privacy aspects: the need to have a valid consent, the transparency principle and the feedback given by the system to the user.

Firstly, as TeSLA is working with a special category of personal data, the GDPR requires an explicit consent. The consent form established during the project includes information that helps the students understand what will be done with their personal data. The consent form must be completed and approved by each student. The consent is also specific because the students receive a clear information about the privacy policy, which is separate from other texts, and the students have to click to give their consent. From ethical surveys, we learned that students require clear information about who can have access to their personal data, the activation period of biometrics instruments, the continuous or non-continuous collection of personal data and the behaviors to be avoided to ensure the proper functioning of the system. This information has to be given to the students to have a real and informed consent and to adopt a user-centric approach.

Secondly, the transparency principle involves a right, for the data subject, to receive information, and a corollary obligation for the data controller to provide information in a clear and plain language. The information should be given at the beginning and all along the processing of personal data. While the GDPR establishes a list of information to be provided, ethical analysis could highlight some particular requirements from the data subject. In TeSLA, the students seem to have difficulties to really understand how the system is working and what the TeSLA possibilities are.

Ethical considerations may therefore justify the provision of additional information to that required by law, in order to ensure a non-discriminatory use of the system.

Thirdly, we focused on the feedback given by TeSLA and the teachers that used it. A concern is that the students cannot effectively exercise their right to object if they do not receive information about how the system works and how the decision was made by the professor. Accordingly, a decision had to be made between giving raw information and leaving the interpretation of these feedback to teachers, on the one hand, and giving immediately interpretable results, on the other hand. As this second option would lead to pre-decide for the teachers, it has been decided to keep a very rough exposure of the data.

From the above, it stems that if the law regulates the processing of personal data, ethical considerations make it possible to concretize the rules imposed by the GDPR in order to facilitate the integration of TeSLA into an institution and its pedagogical activities.

References

1. Akrich M (1987) Comment décrire les objets techniques? Tech Cult 9:49–64
2. Burton C, Poullet Y (2006) Note d'observations à propos de l'avis de la Commission de la protection de la vie privée du 15 juin 2005 sur l'encadrement des listes noires. Revue du droit des technologies de l'information, 102–122
3. de Terwangne C (2015) La réforme de la convention 108 du Conseil de l'Europe pour la protection des personnes à l'égard du traitement automatisé des données à caractère personnel. In: Castets-Renard C (ed) Quelle protection des données personnelles en Europe?, 425–451, Larcier
4. de Terwangne C, Rosier K, Losdyck B (2017) Le règlement européen relatif à la protection des données à caractère personnel: quelles nouveautés? J Droit Eur, 302–316
5. de Terwangne C (2018) Les principes relatifs au traitement des données à caractère personnel et sa licéité. In: de Terwangne C, Rosier K (eds) Le règlement général sur la protection des données (RGPD/GDPR): analyse approfondie, 87–142, Larcier
6. Feenberg A (2012) Questioning technology. Routledge
7. Kaufmann JC (2011) L'entretien compréhensif. Armand Colin, Paris
8. Wachter S, Mittelstadt B, Floridi L (2016) Why a right to explanation of automated decision-making does not exist in the general data protection regulation. Int Data Priv Law, 76–99

Further Reading

9. Article 29 Working Party, Working Document on Blacklists, 3 October 2002, WP65
10. Article 29 Working Party, Guidelines on Automated individual decision-making and Profiling for the purposes of Regulation 2016/679, 03 October 2017, WP251
11. Article 29 Data Protection Working Party, Guidelines on consent under Regulation 2016/679, 10 April 2018, WP259

12. Article 29 Data Protection Working Party, Guidelines on transparency under Regulation 2016/679, 11 April 2018, WP260
13. Article 29 Data Protection Working Party, Opinion 03/2012 on developments in biometrics technologies, 27 April 2012, WP193
14. Belgian Commission for the protection of privacy, Opinion 17/20008 on biometrics data processing, 9 April 2008

Underpinning Quality Assurance in Trust-Based e-Assessment Procedures

Paula Ranne, Esther Huertas Hidalgo, Roger Roca, Anaïs Gourdin and Martin Foerster

Abstract Assessment is a key aspect of curriculum design and is fundamental to the learning process. The methods used for assessment are of prime pedagogical importance as they have a direct bearing on both teaching and learning provision and the learning experience. Assessment methods must be planned so that they align well with intended learning outcomes in e-assessment procedures in particular, in order to ensure reliable and secure learner authentication and authorship in online and blended learning environments. A framework for quality assurance in trust-based e-assessment processes was developed in accordance with these objectives. The framework has been designed in line with the Standards and Guidelines for Quality Assurance in the European Higher Education Area (ESG) to assure the applicability of the e-assessment system to all online and blended learning environments while respecting the shared European framework for quality assurance (QA). This chapter provides contextual information on the quality assurance of higher education in the European Higher Education Area (EHEA) and presents in more detail the rationale

P. Ranne (✉) · A. Gourdin
European Association for Quality Assurance in Higher Education, Avenue de Tervueren 36-38, 1040 Brussels, Belgium
e-mail: paula.ranne@enqa.eu

A. Gourdin
e-mail: anais.gourdin@enqa.eu

E. H. Hidalgo
Agència per a la Qualitat del Sistema Universitari de Catalunya, C. dels Vergós, 36-42, 08017 Barcelona, Spain
e-mail: ehuertashidalgo@aqu.cat

R. Roca
Clavelles 8, 08560 Manlleu, Barcelona, Spain
e-mail: rogersroca@gmail.com

M. Foerster
European Quality Assurance Network for Information Education, Mörsenbroicher Weg 200, 40470 Düsseldorf, Germany
e-mail: foerster@asiin.de

D. Baneres et al. (eds.), *Engineering Data-Driven Adaptive Trust-based e-Assessment Systems*, Lecture Notes on Data Engineering and Communications Technologies 34,
https://doi.org/10.1007/978-3-030-29326-0_13

behind the development of the framework, together with aspects that need to be taken into consideration when designing and implementing the quality assurance of e-assessment.

Keywords e-assessment · e-learning · Framework · European standards and guidelines · Higher education · Quality assurance

Acronyms

EHEA	European Higher Education Area
EI	Education International
ENQA	European Association for Quality Assurance in Higher Education
ESG	Standards and Guidelines for Quality Assurance in the European Higher Education Area
ESU	European Students' Union
EUA	European University Association
EURASHE	European Association of Institutions in Higher Education
EQAR	European Quality Assurance Register for Higher Education
FQAeA	Framework for the Quality Assurance of e-Assessment
GDPR	General Data Protection Regulation
HEI	Higher Education Institution
IMS	Information Management System
IT	Information Technology
KPI	Key Performance Indicator
QA	Quality Assurance
VLE	Virtual Learning Environment

1 Introduction and Motivation

The Standards and Guidelines for Quality Assurance in the European Higher Education Area [8] set the requirements for the quality assurance of higher education and include all approaches to quality assurance such as the accreditation, audit and evaluation on programme or institutional level at higher education institutions (HEIs). The ESG are generic standards that apply to all higher education across the 48 countries of the EHEA, regardless of the mode or place of delivery, and therefore include cross-border provision, distance learning and all different forms of technology-enhanced learning from full online programmes to blended learning.

Although not all national higher education systems have fully developed systematic approaches to address the quality assurance of e-learning, one approach adopted by QA agencies in certain countries has been to establish specific criteria, indicators and quality assurance methods to address this form of delivery. It has been estimated that around 23% of systems have established special QA requirements for e-learning. In the majority of countries, e-learning is integrated into an overarching framework designed to cater equally and appropriately for all forms of delivery [4].

Taking into account the increasing supply of e-learning provision, it is important to address specific ways in which general QA procedures can be used to evaluate different forms of e-learning and e-assessment, and how these can be integrated into general QA practices. Regardless of how it is conducted, the adequate quality assurance of both e-assessment and e-learning is crucial, not least for gaining public confidence in the provision of technology-enhanced education within a larger context [7]. The framework for the quality assurance of e-assessment (FQAeA) [9] was developed as part of the TeSLA project to accommodate the specific characteristics of e-learning and e-assessment, in particular, with the main objective of improving and enhancing educational standards in the various different forms of e-assessment.

In this chapter, Sect. 2 provides some contextual information on the quality assurance of higher education in the EHEA. In Sect. 3, the motivation for the development of the FQAeA is explained. Section 4 describes the steps taken for the development of the framework in the different pilot stages of the TeSLA project.[1] The framework itself is presented in Sect. 5. The analysis of the results and conclusions stemming from the application of the framework are presented in Sects. 6 and 7, respectively. The final Sect. 8 provides some thoughts on the further implementation and dissemination of the FQAeA.

2 Background

The Bologna process is a cooperation initiative between 48 countries in the EHEA that guides the collective efforts of stakeholders in higher education on how to improve, among other things, the internationalisation and modernisation of higher education. As the main objective of the Bologna Process from the time of its inception in 1999 was to create the EHEA, which was meant to ensure more comparable, compatible and coherent systems of higher education across Europe.

Quality assurance in higher education has been a key line of action in the Bologna process right from the very beginning. In 2005, the Standards and Guidelines for Quality Assurance in the European Higher Education Area (ESG) were adopted by the Ministers responsible for higher education following a proposal prepared by the European Association for Quality Assurance in Higher Education (ENQA), the European Students' Union (ESU), the European Association of Institutions in Higher Education (EURASHE) and the European University Association (EUA). The ESG were designed to form the overarching framework for the formation and development of quality assurance systems in all higher education institutions in the EHEA. The ESG also provide the standards and guidelines for quality assurance agencies

[1] More details about the TeSLA pilot execution can be found in Chap. 8.

for a variety of external quality assurance activities carried out by agencies, including evaluation, review, audit, assessment, accreditation or other similar activities at programme or institutional level to verify the effectiveness of internal QA systems and act as a catalyst for improvement [1].

Since the adoption of the ESG in 2005, considerable progress has been made in quality assurance as well as in other Bologna lines of action, such as the recognition of qualifications and the promotion of the use of learning outcomes. In order to follow up on the rapidly changing context of higher education, in 2012 the Ministerial Communiqué invited the E4 Group (ENQA, ESU, EUA and EURASHE) in cooperation with Education International (EI), BUSINESSEUROPE and the European Quality Assurance Register for Higher Education (EQAR), to prepare an initial proposal for a revised and updated version of the standards and guidelines in accordance with the changing landscape of higher education and to improve their applicability and usefulness [1]. The revised ESG, which were adopted in 2015, place more emphasis, among other issues, on the measurement of learning outcomes and the concept of student-centred learning, both of which are essential in e-learning provision. In addition, the revised ESG explicitly state that all types of higher education provision must take account of and adhere to QA, regardless of the mode of delivery or length of activity. Due to the fact that all e-learning provision and related activities are subject to QA in the same way as traditional higher education activities, together with the increased e-learning provision, learner e-assessment has become an increasingly important aspect of the teaching and learning process [6].

As pointed out by Gaebel and Zhang [3], the vast majority of HEIs in the EHEA offer or plan to offer online or blended programmes, which reflects an important change in the context of higher education in Europe. Other studies show that equal consideration has not been given to the QA of online provision compared to traditional face-to-face education [4]. Certain recent initiatives have begun to fill this gap, for example, a report published by ENQA on the considerations for quality assurance of e-learning provision, which introduces a set of considerations for the QA of online provision [5].

The objective of the FQAeA is to increase consistency and transparency in all forms of e-assessment applied in evaluating and measuring learning outcomes. One key element is to ensure that assessment processes are student-centred, relevant, authentic, reliable and trustworthy. The framework seeks to provide minimum standards and guide users towards the thorough implementation and reliable and secure practice of e-assessment.

The main principle underlying the development of the FQAeA was compliance with the ESG. The intention was not to formulate a binding subset of standards, but rather a supporting framework to address the specific characteristics of e-assessment and underpin and enhance external QA procedures involving e-assessment in particular. One of the main aims was to provide guidance and support to higher education institutions in e-assessment design and implementation and to provide particular recommendations to the institutions participating in the TeSLA project to further enhance their e-assessment procedures. One further aspect was the identification of good practices.

3 Problem Statement

Due to new innovations and the rapid advancement of technology, the concept of e-learning is constantly changing. The challenge therefore lies in developing appropriate pedagogical models, together with technologies to support learning and performance.

Institutions need to ensure that programmes are delivered in a way that encourages students to take an active role in creating the learning process, and that the assessment of students reflects this approach, as clearly laid out in ESG standard 1.3. Ways to enable student-centred learning in the e-learning context include flexible learning pathways, addressing the diversity of students and their needs, supporting lifelong learning, equipping students with skills for future labour market, among others. E-learning is also free from the geographical constraints of classroom learning, as well as often being free from time constraints, which adds to the flexibility and adaptability to suit the individual needs of learners.

The report Considerations for quality assurance of e-learning provision [5] suggests that institutions reflect on the pedagogical model best suited to ensuring that the teaching and learning process supports the achievement of the intended learning outcomes. In e-assessment, it is recommended that assessment methods be particularly clear to students and suitable for the measurement of achieved learning outcomes. Methods should be technologically sound and include appropriate tools to ensure authenticity and work authorship. In addition, learners should be adequately informed about the rules of academic citation and referencing, together with rules pertaining to plagiarism, in order to maintain and strengthen academic integrity and trust between the institution and the student community in the changing settings of technology-enhanced learning and assessment.

4 Methodology Used to Define the e-Assessment QA Framework

The FQAeA is designed to be applicable in all online and blended learning environments that include e-assessment, with the ESG functioning as the core document for the framework. This applies both to the development of the internal quality assurance aspects and the design of the external review methodology.

The TeSLA pilots provided an interesting context in which to assess the feasibility of the FQAeA given that the participating partner universities represent a variety of contexts in terms of their geographical scope, size, QA practices and virtual learning environments. In order to best capture the specificities of each pilot case, a set of external review procedures were organised. As in any standard QA procedure, these review procedures were designed fit for purpose, specifically bearing in mind the pilot nature of the exercise (ESG 2.2); they were carried out by groups of external experts including student members (ESG 2.4); and all reporting was required to be clear and accessible (ESG 2.6). These standards have been specifically adapted to the context of the TeSLA system as follows:

- **Designing methodologies fit for purpose (ESG 2.2)**
  "External quality assurance should be defined and designed specifically to ensure its fitness to achieve the aims and objectives set for it while taking into account relevant regulations. Stakeholders should be involved in its design and continuous improvement".

Applicability in the context of TeSLA: this standard lies at the very heart of the process in that external QA procedures need to be specifically adapted and designed for the TeSLA context. The FQAeA-based external QA procedure was designed specifically for this purpose.

- **Peer-review experts (ESG 2.4)**
  "External quality assurance should be carried out by groups of external experts that include (a) student member(s)".

Applicability in the context of TeSLA: The experts selected for the peer-review were individuals of recognised standing in either an academic and/or professional field, as well as students, all of whom were selected on the basis of their independence and no conflict of interests. Geographical and gender balance were also taken into account.

In order to address the distinctive features of the TeSLA pilots, two different types of review panels of experts were established, one main review panel, which was involved in the meta-analysis process and carried out a systematic review of the FQAeA, and a group of regular review panels that carried out the reviews (including interviews with the various stakeholders) in each partner university.

- **Reporting (ESG 2.6)**
  "Full reports by the experts should be published, clear and accessible to the academic community, external partners and other interested individuals. If the agency takes any formal decision based on the reports, the decision should be published together with the report".

Applicability in the context of TeSLA: The reports produced by the regular panels identified good practices and areas of improvement. These reports were used by the institutions as the basis for their respective follow-up action plans.

As regards the formulation of specific standards for the FQAeA, existing practices in the QA of e-learning were identified and a thorough analysis made of the applicability of the ESG in relation to e-assessment. On the basis of this analysis, a first version of the standards and indicators was drafted. This document was used throughout the different stages of the external review procedure in the pilots, during which the experts applying the standards and the participating institutions provided feedback on the applicability of the standards and indicators and identified minimum evidence requirements for the draft framework.

After the review panels completed the evaluations, a workshop was organised to gather further feedback from the experts where the focus was mainly on the subject of terminology and improving the clarity and applicability of the document.

5 Framework for the Quality Assurance of e-Assessment

The FQAeA was designed in line with the ESG and consists of eight different standards that are described below:

5.1 Policies, Structures and Processes for the Quality Assurance of e-Assessment

Standard

The institution has appropriate policies, structures and processes to ensure that e-assessment conforms to ethical and legal standards and is embedded in the organisational culture and values. In addition, the e-assessment proposal is aligned with the institution's pedagogical model and academic and legal regulations and ensures its objectives are achieved on a constant basis.

Indicators

1. The institution has appropriate policies, structures and processes in place to provide guidance on:
   a. E-assessment organisation and protection against academic fraud.
   b. Accessibility to learners with disability, illness and for other mitigating circumstances.
   c. Proper and timely technical support for e-assessment platform use for learners and teaching staff.
2. The institution uses a clearly articulated policy framework and governance structure when deciding on the adoption of new technologies to ensure the expected quality of e-assessment.
3. The e-assessment system is aligned with the educational objectives, regulations and pedagogical models of the institution and it has mechanisms and processes in place for the review and continuous improvement of e-assessment methods.
4. Quality assurance procedures and security measures are in place for external partners that provide e-assessment systems and/or services.
5. The institution has a policy and code of practice for electronic security measures and the use of learner data that cover privacy, security, consent and the purposes for which learning analytics are carried out. This policy and code of practice must ensure information integrity, validity and data protection.
6. The institution has an e-assessment strategy that includes a clear description of responsibilities, roles and procedures. The e-assessment strategy is part of the institution's development plans.

The minimum evidence requirements are as follows:

- Transparent definition of the quality expectation for e-assessment methodology and outputs from processes, in line with institutional assessment regulations and

quality assurance procedures (i.e. mechanisms, instruments and responsibilities to check the quality of system functionality).
- Policy for the sustainable provision of the technological system with regulations for data security, data and privacy protection that are in line with European and national regulations.
- Assessment regulations, including regulations for learners with disabilities and/or mitigation circumstances. Rules for alternative digital assessment methods and pedagogical models. Outputs from transparently defined processes, instruments and allocated responsibilities for e-assessment methodology reviews and updates, based on a cyclical approach.
- Policy and guidelines for external sourcing of the technological system and contractual relationships with external providers and partners.
- Evidence of institutional oversight of assessment procedures and outcomes including reports to institutional committees responsible for academic standards and quality.
- Evidence of institutional oversight of an e-assessment strategy with description of responsibilities, roles and procedures.

5.2 *Learning Assessment*

Standard

E-assessment methods are varied, they facilitate pedagogical innovation and they rigorously determine the level of achievement of learning outcomes. They are designed to assure the timely and fair assessment of learning. As such, they are authentic, transparent and consistent with learning activities and resources. Digital assessment should also promote the participation of learners and adapt to the diversity of both learners and educational models.

Indicators

1. Stakeholders, in particular teaching staff and learners, are informed of the e-assessment methods and grading criteria.
2. E-assessment methods are consistently applied and allow learners to demonstrate the extent to which the intended learning outcomes have been achieved; they reflect innovative pedagogical practices and an understanding of the diversity of learners and groups of learners; these methods are in place and encourage the use of a variety of evaluation and assessment methods (formative, continuous, summative).
3. Mechanisms are in place to ensure that feedback on constructive and developmental learning is given to learners and that it is timely.
4. Feedback on the e-assessment methodology and technical arrangements is collected from learners, teaching staff and managers.
5. Learner appeals processes are in place to ensure fairness.

6. Processes and mechanisms are in place for reviewing new and existing e-assessment and traditional assessment methods that are based on feedback from stakeholders (especially teaching staff and learners) and state-of-the-art developments in pedagogy and technology.

The minimum evidence requirements are as follows:

- Guidelines for teaching staff on available assessment methods and for the design of e-assessment criteria.
- Guidelines or policies for the alignment of assessment methods, teaching methodologies and intended learning outcomes. Guidance on assessment methods and criteria for learners are provided (i.e. learning guides). Details of learning unit accreditation and approval processes.
- The outcomes of feedback surveys (learners, teaching staff, academic managers, etc.) or any other method used to evaluate satisfaction with e-assessment procedures.
- Definition of key areas and quality indicators for the collection of data from stakeholders as input for the subsequent steps of analysis, review and renewal of e-assessment methodology.
- Register of appeals.
- Reports from review panels/groups analysing feedback from stakeholders and providing suggestions, papers and reports on new pedagogical models and new technological developments.

5.3 Authenticity, Transparency and Authorship

Standard

The development and implementation of e-assessment include protective measures that guarantee learner authentication and work authorship. The e-assessment system is secure and fit for purpose.

Indicators

1. The institution has an all-inclusive fail-safe technology development plan, including learner authentication and anti-plagiarism technologies that guarantee learner identity and work authorship, as well as procedures for data protection and privacy requirements.
2. The institution has an all-inclusive fail-safe technology development plan, including a system that provides support to the building and maintenance of the infrastructure for e-assessment and processes for the ongoing review of learner authentication and anti-plagiarism technologies.
3. The code of conduct for learners includes specific elements related to cheating and plagiarism and the sanctions (consequences) that may be imposed, as well as guidance on good practice.

4. The system has the capacity to operate with the maximum number of users in the learning units.
5. Higher education institutions assure the integrity of data collected and used, together with the system's security.
6. Assessment procedures must follow national and international regulations on personal data protection.

The minimum evidence requirements are as follows:

- Guidance for learners on technologies that monitor their behaviour.
- Outputs from clearly defined processes to upgrade the technological system, as and when necessary.
- Code of conduct on academic integrity and regulations including sanctions.
- A record of incidences and mitigation actions regarding system capacity (for instance the number of failed connections, the number of requests for technical support, etc.).
- A record of security incidents that have been detected and mitigated.

5.4 Infrastructure and Resources

Standard

The institution utilises appropriate technologies that match intended learning outcomes and enhance and expand opportunities for learning.

Indicators

1. Procedures are in place for feedback (from learners, teaching staff and managers) with the learning environment and the educational digital technologies (including the institutional virtual learning environment) used and include the following aspects:
   a. Ease of use for all learner profiles (including SEND learners and learners with a different technical background or diverse hardware profiles).
   b. Ethical and legal (privacy in relation to personal data, legal requirements and ethical aspects involved).
   c. Constant updates to reflect technological changes.
   d. Support of a variety of methods and tools.
2. The technical infrastructure and operating systems ensure sufficient coverage and alignment with the different e-assessment methods. This has been sufficiently tested prior to use.
3. The technical infrastructure ensures accessibility to the e-assessment system for SEND learners. This has been sufficiently tested prior to use.

4. Sufficient resources are allocated to ensure the uninterrupted running of the system (technicians, updates, maintenance of running systems, etc.).
5. Compatibility with institutional virtual learning environments.

The minimum evidence requirements are as follows:

- Guidance for learners is provided on use of the learning environment and educational digital technologies.
- Definition of infrastructure requirements, taking into account the system coverage (net coverage) and testing of the technical functionality of assessment methods and different access technologies.
- Feedback surveys, including aspects related to ease of use, privacy, etc. Resource plans indicate the allocation of resources for the maintenance of technology in the respective learning environment.
- Evidence on compatibility with institutional virtual learning environments.

5.5 *Learner Support*

Standard

Learners are aware of, have access to and use effective and well-resourced support services for counselling, orientation, tutoring and facilitation in order to increase student retention and success. Learner support covers pedagogical, technological and administrative related needs and is part of established institutional policies and strategies.

Indicators

1. Procedures are in place to identify the support requirements of learners, including SEND learners.
2. Institutions implement learner support policies and strategies by:
   - Providing access to support services including tutoring and facilitation, technical help-desk, administrative support and choice advice.
   - Ensuring that support services are timely and adequate to learners' profiles and needs.
   - Learner IT skills are taken into account.
3. Feedback procedures are in place.
4. Learners are provided with adequate guidance on digital literacy and academic integrity.

The minimum evidence requirements are as follows:

- Public information on institutional learner support policies and strategies, support resources and contact points for different user groups.
- Qualified pedagogical, technical and administrative support staff.

- Figures on learner retention and achievement.
- KPI for technical support.
- Feedback from learners on learner support provision and the accessibility of advice and guidance.
- Report on the number of learners who have registered for the specific course of study (for the monitoring of platform use).

5.6 *Teaching Staff*

Standard

Teaching staff are skilled and well supported in relation to the development of the technological and pedagogical requirements and e-assessment methods.

Indicators

1. Procedures are in place to identify the support requirements of teaching staff.
2. Teaching staff are trained and proficient in the use of digital learning technologies and e-assessment methods.
3. Technical and pedagogical support services for teaching staff are adequate, accessible and timely.
4. Feedback procedures are in place for teaching staff in relation to institutional support that they have received.
5. Support is available for the development of innovative practices in e-learning delivery and e-assessment.
6. Teaching staff are provided with adequate references on digital literacy and academic integrity.

The minimum evidence requirements are as follows:

- Responsibilities and outputs from procedures dealing with pedagogical and technical aspects are clearly allocated among support staff.
- Training and continuing professional development in the use of e-assessment are provided and widely subscribed to by teaching staff.
- Qualified pedagogical and technical support staff, with clearly allocated responsibilities.
- Feedback survey outcomes.
- Innovative learning and assessment practices are in place and have been evaluated.

5.7 *Learning Analytics*

Standard

The institution has an information management system that enables agile, complete and representative collection of data and indicators derived from all aspects related to e-assessment methodology and authenticity and authorship technologies.

Indicators

1. The institution ensures the collection and dissemination of relevant information from stakeholders (learners, academic staff, support staff, etc.) for effective management and enhancement of the e-assessment methodology (including authenticity and authorship technologies).
2. The institution analyses relevant information for effective management of the e-assessment methodology.
3. The institution uses relevant information for effective management of the e-assessment methodology to promote improvements in the learning experience of learners.

The minimum evidence requirements are as follows:

- The institution has effective processes in place for systematic data collection and management.
- Available data are analysed for output. Indicators to monitor learner performance are defined and accessible to teachers and students.
- Learning analytics are consistent with the institution's pedagogical approach and are used to monitor progress and promote continuous improvement.

5.8 *Public Information*

Standard

The institution appropriately informs all stakeholders of e-assessment methods and resource requirements. Learners are informed of hardware requirements, learning resources technology and the provision of technical support.

Indicators

1. The institution publishes and makes available reliable, complete and up-to-date information to learners prior to and after enrolment on:
   - The e-assessment methods (criteria, regulations and procedures).
   - The pedagogical model on which these are based.
   - The minimum hardware requirements to make full use of the system.
   - Learning and technical support from the institution.

The minimum evidence requirements are as follows:

- Open access pages on the institution's website.
- Feedback from users on the institution's website.

6 Analysis of Results

The following section introduces the main results and findings obtained from the implementation of the FQAeA at the TeSLA partner universities, according to each of the eight standards.

6.1 Policies, Structures, Processes and Resources for e-Assessment Quality Assurance

In general, all of the TeSLA partner HEIs have QA procedures in place. Concerning the issue of specific e-assessment policies, however, different scenarios are evident. In some cases, the national legal framework is more restrictive and online exams are not permitted, whereas in other more flexible cases there are no restrictions at all on e-assessment. In overall terms, several issues stand out involving internal and external policies and regulations that need to be reassessed in order for institutions to properly adapt their activities concerning online learning and e-assessment applications. Such cases usually involve the more traditional universities that have only recently included blended and online provision into their learning processes.

In cases where e-assessment is not permitted or where e-learning is a new way of delivering courses, there still remains considerable room for improvement at policy level because HEIs are required to design new procedures and policies covering the different aspects of e-learning, e-assessment, definitions for cheating and plagiarism, ethical and legal issues, and security.

Fully online universities are seen to be fully compliant with this standard because the reason they were established in the first place was to offer online learning. This is reflected in appropriate policies that clearly focus on addressing the distinctive features of e-learning. In addition, good practices are often followed in these institutions as a consequence of the alignment of e-assessment procedures with well implemented pedagogical models.

The introduction of the TeSLA instruments will in principle provide all HEIs the opportunity to analyse in greater depth and reflect on their policies, structures and resources for quality assurance in e-assessment and to better adapt to national and European regulations (e.g. General Data Protection Regulation [EU] 2016/679). In order for this to succeed, however, such developments will need support at institutional level, and synergies will need to be sought in general QA procedures.

6.2 Learning Assessment

As demonstrated in the pilot testing, an external review can help identify good practices in learning assessment at online institutions, especially as applied practices often take into account a student-centred pedagogical approach. This is often accompanied with increased flexibility in learning design and delivery. Access to higher education is also broadened and increased as HEIs offer more diversified options for the assessment of learners with special educational needs and disabilities (SEND students).

Analysis of the assessment methods applied in the different TeSLA partner HEIs shows that collaborative assignments are still a challenge as most assessment procedures are designed to be performed individually. In all cases, however, chosen assessment methods were seen to be in line with expected learning outcomes.

There were only a few reported cases of innovative practices involving either pedagogical models or new e-assessment activities. In order to support the full integration of e-assessment procedures, HEIs are encouraged to focus on the dissemination of e-assessment information. Pilot testing has shown that institutions have so far been unable to implement a system that can be fully integrated into the overall learning process. Teachers and learners may have a role to play as agents of change in this, which will entail their involvement in planning and development processes right from the very beginning.

6.3 Authenticity, Transparency and Authorship

Although the HEIs participating in the study were to a large extent aware of the technical and security implications surrounding the implementation of a new e-assessment system, certain issues concerning e-assessment security were identified, together with several areas for improvement, such as the need to implement a register of cyber-attacks and technical problems. Corrective actions in this regard would lead to a more structured approach regarding system security.

Despite the fact that TeSLA partner HEIs already address issues concerning academic integrity and ethical practices, especially against plagiarism, one relevant aspect that has been identified relates to the detection of cheating (authentication and authorship issues) by means of a defined threshold level for what is considered normal behaviour compared to suspicious behaviour. As teaching staff are responsible for detecting fraud, the definition of such a threshold would assist them in the identification of unethical academic behaviour.

One major challenge was noted regarding the sharing of personal data, with students being reluctant to share the personal data necessary for the TeSLA system to operate correctly. Although the system fully complies with the General Data Protection Regulation (GDPR) and national legislations on data privacy, students need to be provided with sufficient information and guidance on how the system deals with the issues of privacy and security.

6.4 Infrastructure and Resources

TeSLA partner HEIs use different virtual learning environments (VLEs) according to their specific needs. As a result, integration of the TeSLA system into each VLE was to some extent a challenge. Aside from the technical requirements necessary for integrating any kind of new technology, which in this case was the TeSLA system into different VLEs, it was noted that integration also calls for other forms of support. For example, it is important for institutions to have in place a centralised technical support service, a ticketing system and procedure guidelines and specifications available to technical staff. In addition, all HEIs should gather feedback from their learners and other main users in the VLE as this information can play an important role in the continuous improvement of both the infrastructure and the system itself, as well as contribute to increased ownership of the system by all users.

6.5 Learner Support

Learner support is one of the most important standards in this QA framework. In general, TeSLA partner HEIs have well-established and readily available support mechanisms in place to meet the students' needs, including administrative, technical and pedagogical support, and in particular support with expertise in e-assessment. In some cases, students were unable to find all the necessary information and so they sought assistance from their teachers. HEIs should therefore allocate specific responsibilities to teaching and support staff.

SEND students are another key stakeholder group that support services must take into account. In general, the TeSLA partner HEIs provide appropriate support to SEND students, which includes a broader range of support services and adapted learning resources. However, this support is only made available if students declare themselves as SEND, which is not always the case.

It is important that institutions pay attention to their feedback processes to ensure that all relevant aspects relating to support services are covered and feedback is gathered and used accordingly for the further development of support services.

6.6 Teaching Staff

As the composition of HEI teaching staff is diverse, with teachers having different roles and responsibilities, institutions need to cater to the professional and skills development of teachers. For example, teaching staff need to be trained in the innovative aspects of pedagogical practice, which includes e-assessment, and the technical aspects of the e-assessment system. In order to achieve these objectives, HEIs should design and implement a professional development plan that includes these aspects, and provide training opportunities to improve teaching skills. In addition, a support system for teaching staff should always be available.

Furthermore, teaching staff should be provided with updated information and guidelines to interpret the results of e-assessment, as well as guidelines and well-defined procedures to deal with issues relating to academic integrity.

Some institutions also need to include procedures to receive regular feedback from the teaching staff on e-assessment procedures.

6.7 Learning Analytics

In general, the added value of having an information management system (IMS) in place was clear to the HEIs taking part in the project. IMS are collectively seen as being an important aspect in the effective management of e-assessment methodology. Furthermore, data that are properly analysed can be used to improve the grading system and in the development of current and future learning material. To this end, it is essential that HEIs clearly define the procedures involving the use and analysis of personal data in order to not inhibit the building of trust in the system. All HEIs in the project collect data on student performance, but only a few apply learning analytics to monitor and improve teaching skills and the learning experience of students.

HEIs may lack or need to enhance an IMS for the systematic collection of data related to the QA of e-assessment. Institutional strategies to expand the use of learning analytics (through the setting up or modification of data collection and analysis systems) should therefore be accompanied by the corresponding regulations and policies to ensure that IMS application and use are well justified and clear to all parties.

6.8 Public Information

In general, all the TeSLA partner HEIs have well-established systems (mainly institutional websites) that provide all stakeholders with accurate information on their e-assessment practices. In some cases, due to the large amount of available information, the websites are not very user-friendly. They should therefore be designed to be more user-friendly, with information better structured to the needs and requirements of different users.

Most of the websites included the main regulations on e-learning. Nonetheless, in TeSLA partner HEIs where e-assessment is not a regular practice, there was no available evidence of public information on e-assessment. It is therefore recommended that institutions update their public information to reflect changes and new methods used in assessment and ensure greater transparency in the use of e-assessment.

Detailed information on the software and hardware requirements should also be provided for appropriate use of the VLE and e-assessment resources.

7 Conclusions

The Framework for the Quality Assurance of e-Assessment (FQAeA) introduces a set of minimum quality standards for reliable and secure digital assessment practice with the aim of preventing academic dishonesty and misconduct. In other words, the framework enhances trust in digital assessment.

The FQAeA has been proven to improve QA procedures in institutions where e-assessment is implemented. Nevertheless, continued efforts by HEIs will be necessary to ensure the better integration of e-assessment practices in the overall learning process, including necessary information and awareness raising throughout the HE community to further encourage the comprehensive use of e-assessment.

Depending on the context and background of each HEI (brick-and-mortar vs. online, on-campus vs. off-campus, etc.), and taking into consideration European and national regulations, an institution's ability and resources to handle the QA requirements of e-assessment may vary. For instance, if the national legal framework does not permit the use of e-assessment, institutions may need to revise aspects such as the pedagogical and assessment models, policies and regulations, etc., and develop approaches to convince regulators to make necessary changes to the respective regulations.

In overall terms, fully online universities comply with most of the QA aspects of e-assessment, whereas traditional universities that offer new blended learning programmes need to pay more attention to aspects that, within their context, are new (e.g. the pedagogical model, VLE, learner and teacher support, etc.).

E-assessment should be perceived from a holistic perspective. As it is not a standalone process, peer-review experts should evaluate how it is integrated into the regular processes and organisational context of each institution.

Reporting on e-assessment practices should be a regular part of an institution's internal QA system and therefore be included in regular self-evaluation processes. HEIs should ensure that external peer-review experts are provided with all the relevant necessary information.

Lastly, the FQAeA is also a valuable instrument for QA agencies. In this regard, QA agencies should take into consideration the following:

- Given the specific and evolving nature of e-assessment, it is crucial to include experts with e-learning expertise in the evaluation panels. In addition, QA agencies should ensure that all peer-review experts are trained accordingly to ensure that they understand the specificities of e-learning and e-assessment.
- It should be stressed that the evaluation of e-assessment practices calls for the involvement of a wider range of stakeholders during the different stages of the

QA procedure. In particular, learning and technical support staff, IT managers and representatives of outsourcing partners need be heard and/or interviewed.
- Examples of good practices should be recorded and disseminated openly in the evaluation procedure, in order to facilitate the implementation of appropriate and effective e-assessment practices.

8 Future Work

The aim of quality assurance in higher education is to ensure that all aspects relating to education provision are implemented with care, taking into consideration the needs of all actors and stakeholders. In this respect, there is clearly a need to provide guidance on new technologies that are appearing in the educational landscape. In line with this, the intention is to widely disseminate the Europe-wide framework for the QA of e-assessment described in this chapter among the higher education community (HEIs, QA agencies, students, policy-makers) as a medium of support for the QA of e-assessment. Together with the framework, further guidance (in the form of handbooks and/or training) could also be provided on how to best implement procedures that are fit-for-purpose in different contexts and how they can be integrated into regular assessment procedures. Further discussions will be carried out with all stakeholders in higher education in order to closely follow further developments in the field.

Acknowledgements Special thanks go to head panel members António Moreira Teixeira (Universidade Aberta, Portugal), Stephen Jackson (Assessment, Research & Evaluation Associates Ltd., United Kingdom), Esther Andrés (ISDEFE, Spain) and Inguna Blese (University of Latvia), as well as to all the peer-review experts in the project.
The glossary items concerning this chapter have been inspired by the glossary of the Quality Assurance Agency (QAA, UK), 2018. Available from: https://www.qaa.ac.uk/glossary.

References

1. European Association for Quality Assurance in Higher Education: Standards and Guidelines for Quality Assurance in Higher Education (n.d.). Available from: https://enqa.eu/index.php/home/esg/. 11 Feb 2019
2. European Higher Education Area and the Bologna Process (n.d.). Available from: http://www.ehea.info/. 11 Feb 2019
3. Gaebel M, Zhang T (2018) Trends 2018. Learning and teaching in the European higher education area. European University Association aisbl, Brussels
4. Gaebel M, Kupriyanova V, Morais R, Colucci, E (2014) E-learning in European Higher Education Institutions. Results of a mapping survey conducted in October–December 2013. European University Association aisbl, Brussels

5. Huertas E, Biscan I, Ejsing C, Kerber L, Kozlowska L, Marcos S, Lauri L, Risse M, Schörg K, Seppmann G (2018) Considerations for quality assurance of e-learning provision. European Association for Quality Assurance in Higher Education aisbl, Brussels
6. Huertas E, Kelo M (2019) Quality assurance of e-learning in line with the ESG—what key considerations for student-centred learning? Paper presented at *INQAAHE Conference 2019*, Colombo, Sri Lanka
7. Ossiannilsson E, Williams K, Camilleri A, Brown M (2015) Quality models in online and open education around the globe: state of the art and recommendations. International Council for Open and Distance Education—ICDE, Oslo
8. Standards and Guidelines for Quality Assurance in the European Higher Education Area (ESG) (2015) EURASHE, Brussels
9. TeSLA. (n.d.). An adaptive trust-based e-assessment system for learning. Available from: http://tesla-project.eu/. 31 Jan 2019

Glossary

Academic integrity is compliance with the ethical standards of the university in education, research and scholarship. The most relevant aspects of academic integrity for this book relate to the avoidance of cheating and plagiarism.

Accessibility means that people with disabilities have access, on an equal basis with others, to the physical environment, transportation, information and communications technologies and systems (ICT) and other facilities and services.

Acknowledged content copy Citing a text fragment from other authors, indicating the source.

Anonymity Privacy filter to assure that an identity is not identifiable within a given set.

Anti-plagiarism tool Computer program aimed to detect plagiarism between documents.

Assistive technology comprises assistive, adaptive and rehabilitative devices or software for people with disabilities or elderly population.

Audit data are the set of evidences oriented to assist users in decision-making processes. In the context of TeSLA project, they are the set of evidences provided to assist teachers in case of students' dishonest academic behavior is detected.

Authentication Verifying the validity of at least one form of identification.

Authorship Proving the identity of the creator of a piece of work.

Biometrics data Personal data resulting from specific technical processing relating to the physical, physiological or behavioural characteristics of a natural person, which allow or confirm the unique identification of that natural person (Article 4.14 of the GDPR).

Bitcoin was the first implemented blockchain platformand the first cryptocurrency.

Blended learning is the use of several approaches to teaching and learning within a course, it most commonly refers to the combination of e-learning with face-to-face teaching.

D. Baneres et al. (eds.), *Engineering Data-Driven Adaptive Trust-based e-Assessment Systems*, Lecture Notes on Data Engineering and Communications Technologies 34,
https://doi.org/10.1007/978-3-030-29326-0

Bloom's taxonomy is a set of hierarchical classifications of learning objectives in the cognitive, affective and sensory domains. The most used classification is that for the cognitive domain, and the widely used revised version of this classification describes six categories (or levels): remember, understand, apply, analyze, evaluate and create.

Bologna Process is an initiative between countries for the creation of the European Higher Education Area (EHEA) that would include the recognition of qualifications between countries, enable mobility of learners and staff across systems, and that higher education systems would be transparent and comparable.

Certificate Authority is an entity that issues digital certificates.

Cheating sample A sample data (image, audio, document or keystroke pattern) collected during an assessment activity which learner identity has been modified in order to simulate a cheating attempt.

Collaborative learning involves learning activities in which the learner needs to interact with other students in order to achieve the learning outcomes.

Competence is defined in the context of the European Qualifications Framework as the proven ability to use knowledge, skills and personal, social and/or methodological abilities, in work or study situations and in professional and personal development.

Confidentiality Adding restrictions on certain types of data and services, to prevent information disclosure to unauthorized parties.

Consent Any freely given, specific, informed and unambiguous indication of the data subject's wishes by which he or she, by a statement or by a clear affirmative action, signifies agreement to the processing of personal data relating to him or her (Article 4.11 of the GDPR).

Constructive alignment is the principle that the intended learning outcomes, the teaching methods and the assessment methods should be aligned. One way of achieving this is to first define the learning objectives, then the assessment regime that assesses the achievement of those learning objectives, and then, finally, the teaching and learning activities.

Contract cheating is form of cheating in which a student pays a third party to produce work for them for an assessment. A number of online agencies now exist to facilitate this process.

Cross-border provision is such education provision that is provided outside the territory of the home institution, for instance learning provided in a branch campus located in a different country but managed/coordinated by the home institution.

Dashboard is the key visualization component of analytical tools. A dashboard shows the analysis results by means of friendly, clear, concise and intuitive visual interface. Its objective is to make the information more understandable, and to facilitate decision-making processes.

Data controller The natural or legal person, public authority, agency or other body which, alone or jointly with others, determines the purposes and means of the processing of personal data (Article 4.7 of the GDPR).

Data Privacy Leader The Data Privacy Leader has the role of liaising with the Pilot Leaders and advising on any data privacy issues arising during the pilots.

Data processor The natural or legal person, public authority, agency or other body which processes personal data on behalf of the controller (Article 4.8 of the GDPR).

Data subject The identified or identifiable natural person.

Decision-making is the process of reaching a judgment, conclusion or a choice. Nowadays, thanks to analytical tools and methods and techniques proposed in disciplines as business intelligence, data science or learning analytics, amongst others, decision-making is mainly based on data analysis and evidences, instead of beliefs and intuitions.

Diagnostic Assessment is assessment, usually done at the start of a course, which is used to identify a student's knowledge and skills so that an appropriate learning programme can be provided for them.

Distance learning is a course of study that does not involve face-to-face contact between teachers and learners as it uses technology for learning 'e'. See also 'blended learning'.

E-assessment is the use of ICT for the presentation of assessment activities and the recording of responses.

Educational Manager (EM) The EM has an advisory role to the Pilot Leaders in order to establish a common language/understanding through the Educational Framework.

Ethereum is a blockchain platform providing Turing-complete smart contracts.

European Higher Education Area (EHEA) is international collaboration on higher education of 48 countries implementing a common set of commitments: structural reforms and shared tool with the goal of increasing staff and student mobility and the facilitation of employability.

External quality assurance is the systematic monitoring and evaluation of the internal quality assurance system at a higher education institution, and the processes supporting that. Normally conducted by an authorised external organisation as an external review. It can include approaches such as accreditation, audit, evaluation.

External review is a thorough evaluation conducted at a higher education institution coordinated by a quality assurance agency of higher education and carried out by a qualified team of people that are not employed at the higher education institution under review. Also audit, accreditation, institutional evaluation.

Face recognition Biometric tool that validates the identity of a learner using images containing the face of the learner.

Forensic analysis Biometric tool that validates the identity of a learner using documents written by this learner.

Formative assessment is the collection of information about student learning and its evaluation in relation to prior achievement and attainment of learning outcomes so as to allow the teacher or the student to adjust the learning trajectory.

Fuzzy set Uncertain sets in fuzzy logic, whose elements have degrees of membership.

General Data Protection Regulation A regulation in European Union (EU) law on data protection and privacy for all individuals within EU.

Genuine sample A sample data (image, audio, document or keystroke pattern) collected during an assessment activity which learner identity is known and valid.

Higher Education Institution is a university, college or other organisation delivering higher education.

Human-Centred Design (HCD) is recognised as a significant approach for technology development and system evaluation through continual interaction with end-users to ensure that the innovation will address their needs and expectations.

Integrity Preventing fraudulent data alteration.

Key Performance Indicators (KPI) are a set of metrics that result from collecting, analyzing and reporting data that obey a specific goal or analytical question, thus facilitating decision-making processes. KPI usually are visualized by means of dashboards.

Keystroke dynamics recognition Biometric tool that validates the identity of a learner using the timing information captured while the learner is typing on the keyboard.

Learning analytics is a discipline characterized by data-driven decision-making in the learning domain. Classically, it is defined as "the measurement, collection, analysis and reporting of data about students and their contexts, for purposes of understanding and optimizing learning and the environments in which it occurs".

Learning outcomes are skills or knowledge a learner is expected to know, understand or be able to demonstrate after completing a course/programme/degree.

Learning Tools Interoperability (LTI) is a standard that links content and resources to learning platforms.

National Institute of Standards and Technology is an agency of the United States to promote standards and innovation.

N-gram A contiguous sequence of n characters from a given text.

On-line proctoring refers to the digital supervision of an assessment; the student carries out an e-assessment activity and is monitored online during the assessment using one or more of: a webcam, microphone or access to the student's screen.

Paraphrasing Expressing the same meaning with different words.

Peer-review expert is an individual contracted by the coordinating body to the team that reviews the higher education institution/programme under review. He/she should have some experience in the higher education sector/quality assurance.

Peer-to-Peer is a type of computer network, where nodes act both as a server and as a client.

Personal data Any information relating to an identified or identifiable natural person ('data subject'); an identifiable natural person is one who can be identified, directly or indirectly, in particular by reference to an identifier such as a name, an identification number, location data, an online identifier or to one or more factors specific to the physical, physiological, genetic, mental, economic, cultural or social identity of that natural person (Article 4.1 of the GDPR).

Pilot Evaluation Leader the Pilot Evaluation Leader has the role of advising Pilot Leaders on issues arising from the evaluation activities.

Pilot Leader (PL) the PL is in charge of monitoring and controlling all the activities related to the pilots at their own institution.

Pilot Manager (PM) the PM is responsible for the overall planning and coordination of the pilot. The PM chairs the Project Demonstration Board (PDB), which is formed by institutional Pilot Leaders, and which is the principal decision making body for the pilots.

Plagiarism is the presentation of the work of others as if it were one's own original work, without proper acknowledgment.

Plain text file A digital file that only contains ASCII characters.

Processing Any operation or set of operations which is performed on personal data or on sets of personal data, whether or not by automated means, such as collection, recording, organisation, structuring, storage, adaptation or alteration, retrieval, consultation, use, disclosure by transmission, dissemination or otherwise making available, alignment or combination, restriction, erasure or destruction (Article 4.2 of the GDPR).

Profiling Any form of automated processing of personal data consisting of the use of personal data to evaluate certain personal aspects relating to a natural person, in particular to analyse or predict aspects concerning that natural person's performance at work, economic situation, health, personal preferences, interests, reliability, behaviour, location or movements (Article 4.3 of the GDPR).

Pseudonymisation The processing of personal data in such a manner that the personal data can no longer be attributed to a specific data subject without the use of additional information, provided that such additional information is kept separately and is subject to technical and organisational measures to ensure that the personal data are not attributed to an identified or identifiable natural person (Article 4.5 of the GDPR).

Public Key Infrastructure Infrastructure, in which specific trusted entities, called Certification Authorities (CA), are in charge of delivering certificates (i.e., the association between a public key and an identity, within the context of asymmetric cryptography).

Quality Assurance Leader the Quality Assurance Leader has the role of liaising with the Pilot Leaders in reviewing the quality assurance of the courses and assessments in the pilots.

Responsible Research and Innovation (RRI) is a transparent, interactive process by which societal actors and innovators become mutually responsive to each other with a view to the (ethical) acceptability, sustainability and societal desirability of the innovation process and its marketable products (in order to allow a proper embedding of scientific and technological advances in our society). Von Schonberg (2011)

Review Panel is a group of individuals contracted by a coordinating body (usually a quality assurance agency) to carry out an external review. In the context of the TeSLA project, the review panels were contracted by the project consortium.

RSA public-key cryptosystem Created by Rivest, Shamir and Adleman, and based on the mathematical difficulty of factoring very large numbers.

Screen reader is a form of assistive technology. It is a software application that produces text to speech. Screen readers are useful especially for people who are blind or visually impaired.

Standards and Guidelines for Quality Assurance in the European Higher Education Area (ESG) are the expectations for internal and external quality assurance systems in higher education institutions and quality assurance agencies of higher education, adopted by the Ministers of Higher Education in the European Higher Education Area.

Student-centred learning encompasses methods of teaching that shift the focus of instruction from the teacher to the learner, and emphasises an active role of the learner in all aspects of a learning process.

Stylometry The statistical analysis of the writing style.

Summative assessment is assessment, generally undertaken at the end of a learning activity, to make a judgment about a student's achievement; it is often expressed in the form of a course grade. In the case of continuous assessment, a number of smaller assessment tasks may be carried out during the course and the grade on the individual tasks aggregated into a final grade.

Technical Coordinator (TC) the TC is responsible for bringing together information from the various TeSLA instrument developers in relation to the pilots, and for liaising with the Technical Leaders.

Technical Leader (TL) the TL has the role of overseeing the technical implementation of the TeSLA instruments in the institution: liaising directly with TeSLA Technical staff and with the Pilot Leader; dealing with technical queries from teachers/tutors; and devising procedure for dealing with technical queries from students.

Thematic analysis is a form of analysis of qualitative data for identifying, analyzing, and reporting patterns (themes) within data.

Timestamping Authority is an entity that provides a secure timestamp as a service.

Transport Layer Security protocol Protocol, which evolved from Netscape's Secure Sockets Layer (SSL) protocol, and nowadays regulated by IETF's RFC5246 (Dierks & Rescorla 2008).

Trusted Third Party An intermediary entity which facilitates interactions between two parties who both trust the third party.

Universal design for learning provides the framework for creating more robust learning opportunities for all students. It is an inclusive approach to course development and instruction that underlines the access and participation of all students.

Usability is the extent to which a system, product or service can be used by specified users to achieve specified goals with effectiveness, efficiency and satisfaction in a specified context of use.

User experience means the perceptions and responses that result from the use and/or anticipated use of a product, service or system. The concept refers to experience in a broad sense, including all the persons' emotions, beliefs, perceptions, preferences, physical and psychological responses, behaviours and accomplishments that occur before, during and after use.

Verification of authorship Consists of determining if the person who claims to be the author of a document or text is really his or her creator.

Virtual learning environment (VLE) is an interactive site giving access to learning opportunities online. Such environment is normally restricted to registered users such as the learners of an institution. Similar to a Learning Management (LMS) environment.

Voice Recognition Biometric tool that validates the identity of a learner using audio fragments with the voice of the learner.

Web Accessibility Initiative (WAI) is an initiative developed to help make the internet more accessible to people with disabilities.

Web content accessibility guidelines (WCAG) are widely used as design principles for making web content more accessible.

World Wide Web Consortium is a standards organizations for the Internet Web.

X.509 A standard defining the format of public key certificates.

Index

D. Baneres et al. (eds.), *Engineering Data-Driven Adaptive Trust-based e-Assessment Systems*, Lecture Notes on Data Engineering and Communications Technologies 34,
https://doi.org/10.1007/978-3-030-29326-0

Printed in the United States
By Bookmasters